Birkhäuser

Mathematics and Biosciences in Interaction

The Mathematics of Darwin's Legacy

Fabio A.C.C. Chalub
José Francisco Rodrigues
Editors

Birkhäuser

Editors
Fabio A.C.C. Chalub
Departamento de Matemática
Universidade Nova de Lisboa
Quinta da Torre
2829-516 Caparica
Portugal
chalub@fct.unl.pt

José Francisco Rodrigues
Centro de Matemática e Aplicações Fundamentais
Universidade de Lisboa
Av Prof. Gama Pinto 2
1649-003 Lisbon
Portugal
rodrigue@ptmat.fc.ul.pt

ISBN 978-3-0348-0121-8 e-ISBN 978-3-0348-0122-5
DOI 10.1007/978-3-0348-0122-5

Library of Congress Control Number: 2011931958

2000 Mathematics Subject Classification: 92Bxx, 92B05, 92Dxx, 92D15, 92D25

Cover design: deblik, Berlin

Printed on acid-free paper

Springer Basel is part of Springer Science+Business Media

www.birkhauser-science.com

To Alice, who was born together with the idea of this book;
and Renata, without whom neither would be possible (FACCC).

To Mafalda and Francisco, who are beginning to read
while this book appears (JFR).

Contents

Preface

Fabio A.C.C. Chalub and José Francisco Rodrigues

The year 2009 had two important scientific celebrations: "The International Year of Astronomy" and the "Darwin Year". In Astronomy, four hundred years had passed since the first use of the telescope by Galileo Galilei and publication of the first two planetary laws by Johannes Kepler in the book Astronomia nova, published in Prague in 1609. In Biology, the bicentennial of Darwin's birthday and the sesquicentennial of the publication of his book *The Origin of Species*, published in London in 1859, are two important ephemerides of what is now commonly known as the theory of evolution [1]. However, 1809 was also the year of publication in Paris of the book *Philosophie zoologique*, by Jean-Baptiste Lamarck [2], containing an outline of the theory of evolution, although without the key concept of natural selection that was proposed later by Charles Darwin and, independently, by Alfred Russell Wallace.

Darwin's classical book had the great merit of showing that the organization and functionality of living beings comprise a natural process that Science can explain, but which in no sense had a single mathematical model. Nothing vaguely similar to an equation appears on any page. But Darwin respected mathematicians and even once said "I have deeply regretted that I did not proceed far enough at least to understand something of the great leading principles of mathematics; for men thus endowed seem to have an extra sense" (quoted in [3]).

Also around one and a half centuries ago, Gregor Mendel, an Austrian monk and scientist, was studying the reproduction of peas in Brno, a city that is now in the Czech Republic. His work, in which statistics played a central role in predicting how traits were inherited from one generation to the next, led to the formulation of what later became known as Mendel's Laws of Inheritance, which were published in 1866 [4] but were rediscovered only at the beginning of the 20th century. What Mendel devised was the "mechanism of heredity" that was lacking in Darwin's theory. Until then, it was assumed that offspring were a blending of their progenitors. This would make evolution impossible, as variation would very quickly disappear from any population. This was a fundamental objection to Darwin's theory and, as it was only lately recognized, Mendel's laws formed not only the foundation of the modern science of genetics but also found the missing link that made the theory of evolution a mature one, since it is a key ingredient for differential reproduction and, therefore, selection and evolution.

The so-called "Modern Evolutionary Synthesis", made possible only by the active intervention of a generation of great biologists with fundamental training in mathematics and physics, like Ronald A. Fisher, John Haldane, and Sewall Wright, among others, succeeded in merging Darwinian evolution and Mendelian genetics. In fact, the need to make the theory of evolution by natural selection explicitly quantitative was advocated by British biometricians, with the development of statistics as an area of mathematical enquiry, that led to creation of the journal *Biometrika*, by Pearson, Weldon and Galton (cousin of Darwin) in 1901. But that synthesis, in which the fundamental concepts of evolution, selection and mutation were formulated in terms of a mathematical model, took place only in the 1920s and 1930s. An important development in biological modelling with a strong mathematical background that is also worth mentioning was the formulation of a neutral theory of evolution, by Motoo Kimura [5] in the 1960s, in which the vast majority of evolutionary change at the molecular level is caused by random drift of selectively neutral mutants. A second important development was the introduction of evolutionary game theory in Biology, by John Maynard Smith [6] in the mid-1970s, in which the replicator and the replicator-mutator equations play a fundamental role, in particular, giving origin to "Darwinian Dynamics" or "Evolutionary Dynamics" as a mathematical description of the dynamical process of variability, heritability and the struggle to survive and reproduce that underlies natural selection [3, 7].

This briefly sketched story, is, in a certain sense, the starting point of this book; however, this was not the starting point of the relationship between mathematics and biology nor does it cover the whole field of Mathematical Biology, which includes many topics such as population dynamics, theoretical ecology, epidemiology, population genetics, theoretical immunology, neural networks, pattern formation, and genomic or proteomic analysis. That story is in fact much older. However, it is difficult to establish the beginning of this interaction. One of the first references is from the 13th century, when Fibonacci's rabbit problem was formulated in 1202: "Suppose a newly-born pair of rabbits, one male, one female, are put in a field. Suppose that our rabbits never die and that after the first month, females always produce one new pair (one male, one female) every month from the second month on. How many pairs of rabbits will there be after a certain number of months?" The assumptions are so unrealistic that this problem hardly can be considered a problem in biomathematics; actually, it appears as an interesting example of certain mathematical recursion [8]; however, the Fibonacci sequence plays an increasing role in the description of nature.

Despite the fact that Darwin was influenced by Thomas Malthus' "An Essay on the Principle of Population", first published in London in 1798, the model of population growth following a geometric progression was already well known by the mathematician Leonhard Euler. Already in the 18th century he discussed several examples of dynamics of human population and he understood that they correspond to a model of exponential growth [9, 10]. Working with this model, he was able to observe, fifty years before Malthus, that a single couple, living only several hundreds of years ago, was able not only to generate all the human population at the 18th century but also, continuing with the same growth, even to attain so large a total population that the whole Earth could not be fed. He also contributed a chapter to a second edition of a first treatise on demography published in Berlin in 1761.

In an important memoir presented to the Academy of Science of Paris in 1760, Daniel Bernoulli made what is possibly the first use of modern mathematical techniques to solve a biologically relevant problem: the dynamics of smallpox. Bernoulli was ahead of modern epidemiology and divided the population into two categories: the susceptible and the immune (the survivors gain life-time immunity); these groups were modelled using differential equations. In fact, in his model he obtains and solves what we nowadays call a "logistic equation", which is a particular case of Bernoulli's differential equation, named after his uncle Jakob, who discussed it in 1695. Looking at the stationary states of these equations, he was able to project the loss in life-expectancy due to the disease. This had impact in the insurance market, and was also a central question in the introduction of inoculation in France [11].

Population dynamics is one of the most important fields of biomathematics; almost all books on the subject start with a chapter on that topic. We still call "Leslie matrices" the one introduced in the study of structured populations, despite the fact that they have no special attributes from the mathematical point of view [12]. The same thing happens with the (sometimes called) Verhust equation [13], which is just the logistic differential equation already considered and solved by the Bernoullis and is one of the simplest examples of a dynamical system. Perhaps the same cannot be said about the Lotka-Volterra equations, introduced almost simultaneously in 1925 and 1926, respectively, by the American statistician A.J. Lotka and the Italian mathematician Vito Volterra, that describe the interaction between different species and gave rise to a turning-point in mathematical biology in the 20th century, [14]. Other interesting facts with historical references to the interactions between mathematics and biology can be found in [15].

The first mathematical result of interest in evolution and genetics appeared only decades after Darwin. In the first decade of the 20th century, independently, the British mathematician G.H. Hardy [16] and the German doctor W. Weinberg [17] explained why recessive genotypes do not disappear. More precisely, they gave sufficient conditions to make gene frequencies static from one generation to the next. In their ideal model, an equilibrium is attained in a single generation. The knowledge of equilibrium is the baseline against which we can measure change, and evolution is ultimately a theory of (gene frequency) change. Their conditions were no-mutation, no-selection, no-migration, random mating, infinite population. The violation of any of these conditions could, on its own, be responsible for evolution.

Later on, R. Fisher [18] went further and quantified the change, pronouncing what is currently known as "the fundamental theorem of evolution": the rate of change of the mean fitness of a population is equal to the fitness variance at each point in time. This will be discussed in detail in the chapters of this book written by W. Ewens, P. Schuster, and R. Burger. These first three papers will provide the reader with a broad and deep view of models in population genetics.

The book continues with a chapter by P. Jagers studying models for extinction. The starting point will be the Galton-Watson process, initially introduced in the study of extinction of family names. This shows (if someone is not yet convinced) the unifying nature of mathematical knowledge. P. Taylor presents the relations between group theory and homogeneous populations. This provides a consistent framework for generalisation to an entire population of results obtained by studying only one or a few (focal) individuals. Taylor finishes his chapter with a model of altruistic behaviour. This is the starting point of the following chapter, by J. Pacheco, where evolutionary game theory is intensively used to model collective action (in particular, cooperation). Solutions of social dilemmas are probably one of the most important problems we have to face in our daily life.

Yet, there are many different ways to study the evolution of cooperation; two important ones are kin selection and group selection. These models are reviewed and used in V. Jansen's chapter to provide a full understanding of the social behaviour of mice living in haystacks. When different individuals in the same population find different solutions to the same dilemma, we are possibly facing one of the most important problems in evolution: the concept of speciation. So important that the title of Darwin's masterpiece refers directly to it. This is the subject of the chapter by S. Mirrahimi, B. Perthame, E. Bouin and P. Millien and also of S. Méléard's chapter. Models for evolutionary branching, a more general concept, are studied from many different points of view: differential equations, integro-differential equations, stochastic modelling, individual-based models, asymptotic limits... all approaches unified by the concept of "adaptive evolution".

"Adaptive evolution" is also the topic of the last two chapters, respectively, by H. Metz and by M. Gyllenberg, H. Metz and R. Service. These chapters are primarily devoted to meso-evolution, where the focus is the change of traits of individuals in a population. A natural sequel to Metz's chapter, where some elements of an adaptive evolution theory are developed, this final chapter investigates how optimisation approaches fit that point of view.

The book ends with a large and extensive but not exhaustive, bibliography, a merging of all citations that appear in the book. At a first glance, this allows us a rather reasonable overview of the biomathematical literature in the last 150 years. We intend this book to be also a good starting point for anyone interested in working in biomathematics, especially in evolution.

We wish to thank the Calouste Gulbenkian Foundation and the Fundação para a Ciência e Tecnologia (Portugal) for their sponsorship of this work. Their financial support allowed the Portuguese Centro Internacional de Matemática, in collaboration with the European Society for Mathematical and Theoretical Biology, the Centro de Matemática e Aplicações da Universidade Nova de Lisboa and the Centro de Matemática e Aplicações Fundamentais da Universidade de Lisboa, where the conference took place in November 2009, to reunite the scientific and logistic conditions that made this book possible.

Finally we conclude this introduction by adopting Metz's closing sentence and inviting you, interested reader, to join this hard and challenging task of bringing Biology and Mathematics closer and to contribute to the fruitful development of biomathematics!

References

[1] C. Darwin, *On the Origin of Species by Means of Natural Selection or the Preservation of Favoured Races in the Struggle for Life.* John Murray, London, 1859.

[2] J.B. Lamarck, *Philosophie zoologique ou exposition des considérations relatives à l'histoire naturelle des animaux.* Dentu, Paris, 1809.

[3] M.A. Nowak, *Evolutionary Dynamics: Exploring the Equations of Life.* The Belknap Press of Harvard University Press, Cambridge, MA, 2006.

[4] G. Mendel, *Versuche über Pflanzenhybriden.* (*Experiments relating to plant hybridization*). Verh. Naturforsch. Ver. Brünn **4** (1866), 3–17.

[5] M. Kimura, *The Neutral Theory of Molecular Evolution.* Cambridge: University Press, UK, 1983.

[6] J. Maynard Smith, *Evolution and the Theory of Games.* Cambridge University Press, Cambridge, UK, 1982.

[7] T.L. Vincent and J.S. Brown, *Evolutionary Game Theory, Natural Selection, and Darwinian Dynamics.* Cambridge University Press, UK, 2005.

[8] L.E. Sigler, *Fibonacci's Liber Abaci: A Translation into Modern English of Leonardo Pisano's Book of Calculation.* Springer-Verlag, New York, 2002, a translation of Leonardo Pisano's book of 1202 from Latin into modern English.

[9] L. Euler, *Introductio in Analysin Infinitorum,* 1748. *English Translation: John Blanton, Introduction to Analysis of the Infinite*, volume I and II. Springer-Verlag, Berlin, 1988.

[10] L. Euler, *Recherches générales sur la mortalité et la multiplication du genre humain.* Mem. Acad. R. Sci. Belles Lett. (Berlin) **16** (1767), 144–164.

[11] D. Bernoulli, *Essai d'une nouvelle analyse de la mortalité causée par la petite vérole.* Mem. Math. Phy. Acad. Roy. Sci. Paris, année 1760 (1766).

[12] P.H. Leslie, *On the use of matrices in certain population mathematics.* Biometrika **33** (1945), 183–212.

[13] P.F. Verhulst, *Notice sur la loi que la population pursuit dans son accroisement.* Corresp. Math. Phys. **10** (1838), 113–121.

[14] G. Israel, *The emergence of biomathematics and the case of population dynamics, a revival of mechanical reductionism and darwinism.* Science in Context **6** (1993), 469–509.

[15] N. Bacaër, *Histoires de Mathématiques et de Populations.* Editions Cassini, Paris, 2009.

[16] G.H. Hardy, *Mendelian proportions in a mixed population.* Science **28** (1908), 49–50.

[17] W. Weinberg, *Über den Nachweis der Vererbung beim Menschen.* (*On the detection of heredity in man*). Jahreshelfts. Ver. Vaterl. Naturf. Württemb. **64** (1908), 368–382.

[18] R.A. Fisher, *The Genetical Theory of Natural Selection.* Clarendon Press, Oxford, 1930.

Fabio A.C.C. Chalub
Departamento de Matemática

and

Centro de Matemática e Aplicações
Faculdade de Ciências e Tecnologia
Unviersidade Nova de Lisboa
Quinta da Torre
P-2829-516 Caparica, Portugal
e-mail: `chalub@fct.unl.pt`

José Francisco Rodrigues
Universidade de Lisboa/CMAF
Av. Prof. Gama Pinto, 2
P-1649-003 Lisboa, Portugal
e-mail: `rodrigue@fc.ul.pt`

What Changes Has Mathematics Made to the Darwinian Theory?

Warren J. Ewens

Abstract. Mathematics has played a key role in validating the Darwinian theory of evolution by natural selection. Perhaps most importantly it shows that the variation needed for evolution by natural selection is conserved under the Mendelian evolutionary system. It then quantifies the rate at which favorable new genetic types are incorporated into a population by natural selection. Analyses at the whole genome level (the current active area of genetical research) are possible only by the use of mathematics, particularly the use of matrix theory. Finally, it is only by a mathematical analysis, using stochastic process theory, that the effects of random changes in gene frequencies, unavoidable because of the finite size of any population, can be assessed.

Mathematics Subject Classification (2000). Primary 92D25; Secondary 60J70.

Keywords. Selection, variation, evolution, Mendelism, genetics, correlation.

1. Introduction

This chapter has three aims. The first is to give a brief introduction to the history of the Darwinian theory preceding the re-discovery of Mendelian genetics in 1900, with an emphasis on the problems that the theory encountered. It will be shown that these problems are resolved by mathematical methods based on the recognition of the Mendelian hereditary system. This leads to the second aim, which is to give a brief review of the fundamentals of genetics and a description of the Darwinian theory in Mendelian terms. The third aim, by far the most important one, is to give examples of cases where a mathematical approach was central to a formulation the Darwinian theory in genetical terms, resolves problems with that theory which were recognized from the earliest times, as well as fleshes out the theory in a way that would not be possible without mathematics. These aims also form the background to further mathematical analysis of the evolutionary process and current research activities to be found in other chapters of this book.

2. Pre-Mendelian evolutionary theory

Various evolutionary principles and theories were advanced before Darwin's time, but here we consider only his theory of evolution by the action of natural selection, the only evolutionary principle to have survived critical examination. Darwin's theory was put forward in his monumental book [1], now generally called *On the Origin of Species.* The great paradox concerning his theory was that when the book appeared, in 1859, the nature of the hereditary mechanism was unknown, so that Darwin advanced the theory without any knowledge of this essential core element to any evolutionary theory. Worse than this, the prevailing theory of heredity when his book was published, and indeed for more than forty years afterward, was that any particular character in a child, for example the child's blood pressure, is in some sense more or less the average of the blood pressures of the two parents of that child. It is clear under this theory, the so-called "blending" theory of inheritance, that under random mating in the population the variance in blood pressure between individuals would halve in every generation, so that in a comparatively short number of generations all individuals in the population of interest would have essentially the same blood pressure. There would then be no variation for natural selection to act on. Of course we do not observe such uniformity in the present human population, so further argument is needed. Since variation in blood pressure to the degree that is actually observed would only arise from further factors of strong effect which cause the blood pressure of a child to deviate from the average of the blood pressures of the child's parents, the principle that selectively favored parents produce offspring who closely resemble them and thus are themselves selectively favored cannot be sustained. Darwin recognized that the blending theory of inheritance was a major problem for his theory, indeed *the* major problem, and because of this he unfortunately altered subsequent editions of his book in such a way as to substantially alter his theory.

This problem was not resolved until the rediscovery of the Mendelian hereditary mechanism [2] in 1900. A mathematical analysis based on this mechanism leads to the so-called Hardy-Weinberg law [3, 4] described later, and this law shows that there is no intrinsic tendency under Mendelian inheritance for variation (in this case genetic variation) to be lost: once established it stays unchanged (unless forces like selection act, a matter discussed further below). It is the quantal nature of the gene, passed on as a discrete entity from parent to offspring, that resolves Darwin's problem, so that there is no blending involved. It is a pity that Mendel's 1866 paper, although evidently sent to Darwin, was not recognized by him or by anyone else for the revolutionary document that it indeed was.

3. The basics of Mendelism

Genetics is an extremely complex subject, and any attempt to describe it briefly must involve severe simplification, sometimes to the extent of introducing minor distortions from reality. It is sufficient for our immediate purposes to say that genes lie on chromosomes, which are thread-like objects in the cells of any organism. Interest lies mainly in *diploid* organisms, such as humans, who obtain genetic material from two parents, and we consider here only this case. In the case of humans, leaving aside the sex chromosomes (which are XX for females, XY for males), each individual carries 22 pairs of chromosomes, one member of each pair coming from that individual's mother and the other from the father. (In other diploid species numbers different from 22 arise.) We may regard the genes as being like beads on the chromosome "threads". These genes lie at particular positions, or loci, so that at any locus any individual has two genes, one on each of the two chromosomes.

We consider for the moment some specific locus. Genes at this locus are of one or other *allelic type*. For example, at the well-known ABO blood group locus, there are three possible allelic types, or *alleles*, A, B and O, and since any individual carries two genes at this locus, one maternally and one paternally derived, each individual must be either AA, AB, AO, BB, BO or OO. We say that these are the six possible *genotypes* at this locus.

It is important to distinguish between the genotype and the *phenotype* of any individual. In the case of the ABO system, both A and B are *dominant* to O, which implies that the outward appearances of AA and AO individuals are the same, as are the outward appearances of BB and BO individuals. Thus there are only four possible phenotypes at this locus, called A (for AA and AO genotypes), B (for BB and BO genotypes), AB (for AB genotypes) and O (for OO genotypes).

The ABO notation is specific to the ABO gene locus, and to consider the general case it is necessary to introduce a more flexible notation. If there are k possible alleles at some locus A, they are generically denoted here by $A_1, A_2, \ldots, A_k$. These k alleles define $k(k+1)/2$ possible genotypes, $A_1A_1, A_1A_2, \ldots, A_1A_k, A_2A_2, \ldots,$ $A_{k-1}A_k$, A_kA_k. The theory outlined below uses this generic notation.

With this background in place we turn now to evolutionary questions. The genotype frequencies in any daughter generation depend on the mating scheme adopted by the parental generation. We assume initially that random mating applies. Suppose also for the moment that there is no selection, so that the fitness of any individual is independent of his genotype. We also assume no mutation or any other disturbing force. Suppose then that in the parental generation the frequency of the genotype is P_{ii} and of the genotype A_iA_j (for $i \neq j$) is $2P_{ij}$. (It is convenient to describe P_{ij} as the frequency of the *ordered* genotype A_iA_j.) The frequency p_i of the allele A_i is, clearly,

$$p_i = \sum_{j=1}^{k} P_{ij}. \tag{3.1}$$

The Hardy-Weinberg law follows immediately from this. It states that in the daughter generation at the time of its conception the frequency of the genotype A_iA_i is p_i^2 and that of the genotype A_iA_j $(i \neq j)$ is $2p_ip_j$. (If mating is not at random, these values no longer apply. The non-random-mating case is considered below.) Elementary calculations show that the frequency of A_i is p_i, the same value as that applying in the parental generation, and also that the daughter generation genotype frequencies and thus the frequency of the every allele remains unchanged in all future generations. This observation validates the "preservation of variation" comment made in Section 2. Of course, mutation and selection change allelic frequencies from one generation to the next, and random changes will also arise by random sampling, since all populations are finite. Nevertheless all these changes are generally small, and the central importance of the preservation of variation concept remains unaltered.

The Hardy-Weinberg law does not apply under non-random mating, for example under assortative mating (the tendency of like to mate with like). Nevertheless, one important mathematically-derived conclusion applies whatever the form of mating, namely that, again assuming no selection, mutation, or any other disturbing force, allelic frequencies remain unchanged from one generation to the next. This is not true of genotype frequencies, which often change from one generation to the next under non-random mating. While this observation does have some important consequences, the essential feature of the preservation of allelic frequencies, and in this sense of genetic variation, remains. Mendelism is an intrinsically variation-preserving hereditary mechanism.

We now consider the evolutionary process further, introducing complications such as selection and mutation. Suppose first that the fitness of any individual depends only on the genes that he carries at some locus A, at which only two alleles can occur, A_1 and A_2. (By fitness here we mean viability fitness, that is the capacity to survive from conception to reproduction. Fitnesses involving mating success and fertility lead to complicated algebra that we do not go into here.) Denote the fitnesses of the three genotypes A_1A_1, A_1A_2 and A_2A_2 by w_{11}, w_{12} and w_{22} respectively. We assume random mating, so that the frequencies of these three genotypes at the time of conception of any generation are $p^2, 2p(1-p)$ and $(1-p)^2$ respectively, where the frequency of A_1 is p at this time. The so-called *mean fitness* $\overline{w}$ of the population at this time, calculated in the standard statistical fashion for a mean, is given by

$$\overline{w} = p^2 w_{11} + 2p(1-p)w_{12} + (1-p)^2 w_{22}. \tag{3.2}$$

If p' is the frequency of A_1 at the time of conception of the next generation, elementary calculations show that

$$p' = p + \frac{p(1-p)}{\overline{w}} \{w_{11}p + w_{12}(1-2p) - w_{22}(1-p)\}. \tag{3.3}$$

This equation can be used to describe the fundamental micro-evolutionary process, which is the replacement of a "less fit" allele in a population by a "more fit" allele. It can also be used to explain the often observed "standing genetic variation", as is shown later.

Equation (3.3) shows that only the relative (rather than the absolute) values of the w_{ij} are necessary to describe this micro-evolutionary process, and thus we are free to choose one of the three fitnesses to take the value 1. In the case where

$w_{11} > w_{12} > w_{22}$,, so that A_1 is the "more fit" allele, it is convenient to write $w_{11} = 1 + s, w_{12} = 1 + sh, w_{22} = 1$, where $s > sh > 0$. We generally think of the case where s is small, perhaps of order 1%. In this case equation (3.3) shows that, to a close approximation,

$$p' - p = sp(1-p)\{p + h(1-2p)\}. \tag{3.4}$$

If unit time corresponds to one generation, this in turn can be approximated by

$$\frac{\mathrm{d}p}{\mathrm{d}t} = sp(1-p)\{p + h(1-2p\}. \tag{3.5}$$

This equation is easily solved, and the solution provides the trajectory of the increase in the frequency of A_1 over time. It is perhaps more useful to calculate the time $t(p_1, p_2)$ required for this frequency to increase from some value p_1 to some larger value p_2. Clearly

$$t(p_1, p_2) = \int_{p_1}^{p_2} [sp(1-p)\{p + h(1-2p\}]^{-1}\mathrm{d}p. \tag{3.6}$$

Many conclusions can be found from these simple formulae, especially the slow rate of change in the frequency of A_1 when this frequency is either large or small. A collection of results arising from (3.6) and from similar but more complex equations was found by Haldane in the 1920's, and summarized in [5]. These equations bear some similarity to corresponding equations in physics, in that they allow one to predict the future evolution of a system, given the appropriate parameter values and the current state of that system.

An empirical confirmation of equation (3.5) arises in describing the evolution of the melanic form of the peppered moth *Biston betularia* during the 19th century in England. Originally the pale form of this moth was prevalent, but with the rise of industrial pollution and the consequent darkening of the bark of the trees on which these moths settled, the melanic form became selectively favored since it became increasingly difficult for predators to observe the dark form on these trees. Empirical estimates of the selective values s and h were made and it was found that the trajectory of the frequency of the melanic form closely followed that predicted by equation (3.5).

In the case where the heterozygote is the most fit genotype, so that $w_{11} < w_{12} > w_{22}$, there is a point of stable equilibrium where the frequency p^* of A_1 is

$$p^* = \frac{w_{12} - w_{22}}{2w_{12} - w_{11} - w_{22}}. \tag{3.7}$$

This is the case of "heterozygote selective advantage", observed often in reality, and thus this fitness configuration is sufficient to explain the observation of standing genetic variation. (When $w_{11} > w_{12} < w_{22}$ there is an equilibrium frequency again at the same value p^*, but this equilibrium is unstable and thus of little interest.) Thus the Mendelian system can explain not only evolution (in the sense of changes in allelic frequencies) but also the existence of standing genetic variation.

So far we have not considered the possibility of mutation. Genes mutate (usually at a very low rate, of order 10^{-5} or 10^{-6}), so that for some purposes mutation can be ignored. On the other hand mutation eventually is the source of all genetic variation, so that despite these low rates a complete analysis of the evolutionary process must allow for mutational events. Here we assume that an A_1 gene mutates to an A_2 gene with probability u, while an A_2 gene mutates to an A_1 gene with probability v. In the case where there is no selection there is a stable equilibrium of allelic frequencies where the frequency of A_1 is $v/(u+v)$. The case where selection and mutation both arise is of course also of interest. In the case of heterozygote selective advantage there is a stable equilibrium close to that value given in equation (3.7), assuming that selective differences are substantially higher than the mutation rates. The case where $w_{11} > w_{12} > w_{22}$ is perhaps of more interest. Here there is a stable equilibrium frequency of A_1 just less than 1. This situation is relevant when A_2 is a disease allele that is maintained at a low frequency in a population because of recurrent mutation from A_1 to A_2, and is much studied in disease genetics applications.

The above calculations assume only two possible alleles at the locus of interest, and also assume random mating. It is possible to generalize these calculations to allow any number of alleles at the locus and any form of mating, although the analysis becomes more complex. Suppose then that the (viability) fitness of any individual depends only on his genotype at some gene locus A, at which alleles $A_1, A_2, \ldots, A_k$ can arise, and denote the fitness of an individual of genotype A_iA_j by w_{ij}. Consider some parental population at its time of conception, with its genotype frequencies at this time being given as above equation (3.1). These genotype frequencies are not necessarily assumed to be in Hardy-Weinberg form, so that it is not necessarily assumed that $P_{ii} = p_i^2$ and that $P_{ij} = p_ip_j$, since random mating in the preceding generation is not necessarily assumed. The mean population fitness $\overline{w}$ of the population in this generation at the time of its conception is

$$\overline{w} = \sum_i \sum_j P_{ij} w_{ij}. \tag{3.8}$$

Straightforward calculations show that the frequency p_i' of A_i at the time of reproduction of the individuals in the parental generation is

$$p_i' = \frac{1}{\overline{w}} \sum_j P_{ij} w_{ij}. \tag{3.9}$$

Under any form of mating (for example random mating, selfing, partial selfing, assortative mating), p_i' is also the frequency of A_i in the daughter generation at its time of conception. In other words, equation (3.9) can be taken as providing the frequency of A_i in the daughter generation at the time of conception, and this is the interpretation that is normally placed on this equation. It is one component of the full evolutionary description of the changes over one generation of the frequencies of the various genotypes at this locus.

It is not possible to calculate the daughter generation genotype frequencies without knowledge of the mating scheme. This in turn implies that gene frequencies beyond the daughter generation cannot be calculated from genotype frequencies in the parental generation without knowledge of the mating scheme, and thus we are unable to track gene frequency evolution over more than one generation without this knowledge. On the other hand, if random mating can be assumed over all successive generations, this tracking can be carried out.

It was noted above that, in the case where two alleles only are possible at the locus, there will be a point of stable equilibrium with both alleles present at positive frequency if the heterozygote has a higher fitness than does either homozygote. The condition for such an equilibrium when there are k possible alleles at the locus, even assuming random mating, is far more complex. We define an internal equilibrium to arise if all alleles have a positive frequency at that equilibrium. Then the necessary and sufficient condition that there exist an admissible stable equilibrium is that the matrix W, whose typical element is w_{ij}, has exactly one positive eigenvalue and at least one negative eigenvalue, as shown by Kingman [6]. In this case the population evolves from any initial set of allelic frequencies at which all alleles have positive frequency to this equilibrium.

A simple example of this case arises when all heterozygotes $A_iA_j, (i \neq j)$ have fitness 1 and all homozygotes $A_iA_i, (i = 1, 2, \dots, k)$ have fitness $1 - s$, where $1 > s > 0$. In this case it is easily shown that the eigenvalues of the matrix W are $k - s$ (with multiplicity 1) and $-s$ (with multiplicity $k - 1$). These eigenvalues satisfy the condition of the previous paragraph, and (as expected from symmetry arguments) the population evolves to a stable equilibrium where all alleles have frequency k^{-1}.

If there is no admissible stable equilibrium the evolutionary behavior is much more complicated. One useful result of Kingman [6] is that if W has j positive eigenvalues, at most $k-j+1$ alleles can exist at an admissible stable equilibrium. If all allelic frequencies are initially positive, the frequencies of the remaining alleles will approach zero over time under the action of natural selection. Clearly these conclusions can be reached only by a mathematical analysis.

4. Evolutionary principles deriving from a mathematical approach

4.1. The problem

There is a serious problem in evolutionary theory arising from the fact that the fitness of any individual depends on all the genes in his genome, whereas each parent passes on only half of his genes to a child. (Complications due to the sex chromosome are ignored here.) The Darwinian theory is based on the idea that a more "fit" individual leaves, on average, more offspring than a less fit individual, and that the offspring of the more fit individuals in a population inherit this increased fitness from their parents, leading to an increased population frequency of the more fit types. But if a parent only passes on half of his/her genes to an offspring the offspring only partially resembles the parent, and there has to be some modification to the Darwinian theory.

This problem has to be resolved by mathematical methods, and the initial step in making this modification is to introduce the concept of the "fitness" of any allele. This has to be a theoretical, or mathematical, construct, since an allele does not have, in reality, a fitness. Despite this fact, this construct is of the utmost importance, and leads to significant insights into the properties of the evolutionary process under Mendelian inheritance. We now introduce this allelic "fitness" concept in the case of a general number k of alleles at the gene locus of interest.

The concept of "allelic fitnesses" of the alleles $A_i, (i = 1, 2, \ldots, k)$ derives from the concept their average effects. These are defined, respectively for these alleles, as the values of $\beta_1, \beta_2, \ldots, \beta_k$ which minimize the quantity

$$\sum_i \sum_j P_{ij}(w_{ij} - \beta_i - \beta_j)^2. \tag{4.1}$$

This minimization procedure is in effect an attempt to fit the various genotype fitnesses by the sum of two values, each of the two values corresponding to the two alleles in the corresponding genotype. The motivation for this is that, as discussed above, a parent passes on one of the two genes that he has at any gene locus to an offspring, and the minimization procedure leads to a "fitness" $\beta_i + \beta_j$ for an individual of genotype A_iA_j. The values of $\beta_1, \beta_2, \ldots, \beta_k$ found from this weighted least-squares procedure are the solutions of the equations

$$p_i\beta_i + \sum_j P_{ij}\beta_j = \sum_j P_{ij}w_{ij}, \quad (i = 1, 2, \ldots, k). \tag{4.2}$$

Equation (3.9) shows that this equation can be written equivalently as

$$p_i\beta_i + \sum_j P_{ij}\beta_j = \overline{w}p_i', \quad (i = 1, 2, \ldots, k). \tag{4.3}$$

In general the solution (for $\beta_1, \beta_2, \ldots, \beta_k$) of these equations cannot be written down explicitly. Summation over all alleles in (4.3) leads to the equation

$$2\sum_i p_i\beta_i = \overline{w}. \tag{4.4}$$

This equation confirms that we may regard the "fitness" of the genotype A_iA_i as being $2\beta_i$ and the "fitness contribution" of the allele A_i as being β_i. It also shows that the population mean fitness can be calculated as a weighted sum these allelic "fitness contributions", the weights being the frequencies of the respective alleles.

The sum of squares removed by fitting the values of $\beta_1, \beta_2, \ldots, \beta_k$ in (4.1) is the so-called (single locus) additive genetic variance in fitness, denoted by σ_{A}^2. It is of the utmost importance, and is that component of the total variance $\sigma^2 = \sum_i \sum_j P_{ij}(w_{ij} - \overline{w})^2$ in fitness that is explained by genes within genotypes, and is thus often called in the modern literature the "genic" variance. (In view of the fact that an additivity assumption is made in the least-squares procedure, with the "fitness" of the genotype A_iA_j being thought of as $\beta_i + \beta_j$, it might best be called the "additive genic" or "additive allelic" variance.) If values w_i and w_j exist such that, for all (i, j), the fitness w_{ij} can be computed as $w_i + w_j$, then $\beta_i = w_i$ and σ_A^2 is equal to the total variance in fitness. If no such values exist there exists "dominance", or non-additivity, among the fitness values, and this leads to the concept of a non additive or dominance variance σ_D^2, defined simply as $\sigma^2 - \sigma_A^2$.

If the change $p_i' - p_i$ in the frequency of A_i given in (3.3) is denoted δp_i, least-squares theory shows that

$$\sigma_A^2 = 2\overline{w} \sum_i (\delta p_i)\beta_i. \tag{4.5}$$

Since in general explicit formulae for the various β values are not available it is not possible in general to write down an explicit formula for σ_A^2. Fortunately this does not matter for many of the conclusions drawn below. The "genic" nature of σ_A^2 can be seen from the following observation. In the case where only two alleles are possible at the locus, when the frequency of A_1 is at the stable equilibrium point given in (3.7) it is found that $\beta_1 = \beta_2$, so that the two alleles A_1 and A_2 are equally "fit". Further, at this equilibrium, $\sigma_A^2 = 0$. More generally, for an arbitrary number of possible alleles at the locus, evolution in the sense of allelic frequency changes occurs if and only if the additive genetic variance defined implicitly in equation (4.5) is positive. The fact that a parent passes on a gene, and not his entire genotype, to a child, and that at an equilibrium point all alleles are equally "fit", is the key to this observation.

This provides a central mathematically-derived insight into the evolutionary process. The basic Darwinian principle is that variation is necessary for evolution by natural selection: if there is no variation, no-one is more fit than anyone else and evolution by natural selection cannot occur. Thus variation is necessary for Darwinian evolution. However it is not sufficient: what is needed for evolution is *additive genetic* variation. This is a fundamental evolutionary principle. In the following sub-sections we show how the mathematical concepts of average effects and the additive genetic variance further enrich the theory of evolution.

4.2. The correlation between relatives

The analysis of biological data by mathematical and statistical methods began in earnest towards the end of the nineteenth century, and many important statistical concepts, for example correlation and regression, were developed to assist in this analysis. One matter that was extensively studied was the correlation between relatives for various metrical characters, for example height. It is clear to everyone that children to some extent resemble their parents in many such characters, and it became one of the main activities of a group of scientists, soon to become known as biometricians, to quantify and study this resemblance through the statistical concept of correlation. It was found that for almost all characters considered by the biometricians the sib/sib correlation was somewhat above the parent/offspring correlation. This raises the obvious question: "Can this and other correlation patterns, for example uncle/nephew, be explained by Mendelian genetics?" This question can only be addressed by mathematical methods.

This question was taken up by Pearson and various co-workers soon after the rediscovery of Mendelism: see for example [7]. Pearson and Lee made the assumption, common at the time, that complete dominance was the rule at all loci controlling the characters measured (so that the effect on height of carrying two copies of the dominant allele is the same as that when carrying only one copy of the dominant allele). Their calculations led to theoretical correlation values which did not agree with the observed values, and this arose because, unknown to Pearson and Lee, dominance is not a universal phenomenon.

In the treatment above the variances σ_{A}^2 and σ^2 concern the additive and the total variance in fitness. By replacing the fitness values w_{ij} by the corresponding values for any character, we can calculate the additive and the total variance for that character. In this section these variances are assumed to have this more general interpretation.

The first comprehensive treatment of this correlation between relatives problem was that of Fisher [8], who realized that the universal assumption of dominance was not appropriate for many characteristics. (This is a good place to introduce Fisher, whose work in this and other areas is often discussed below. Fisher was the leading theoretical population geneticist of the 20th century, whose work transformed the subject and who introduced many of its key concepts, including that of the additive genetic variance referred to above.)

Fisher showed, in the simple case where the character in question is determined by the genes at a single locus, mating is at random and there is no environmental component to variation, that the following formulae hold:

$$\text{correlation (parent/offspring)} = \frac{1}{2}\left\{\frac{\sigma_{\mathrm{A}}^2}{\sigma^2}\right\}, \tag{4.6}$$

$$\text{correlation (sib/sib)} = \frac{1}{2}\left\{\frac{\sigma_{\mathrm{A}}^2}{\sigma^2}\right\} + \frac{1}{4}\left\{\frac{\sigma_{\mathrm{D}}^2}{\sigma^2}\right\}, \tag{4.7}$$

$$\text{correlation (uncle/nephew)} = \frac{1}{4}\left\{\frac{\sigma_{\mathrm{A}}^2}{\sigma^2}\right\}, \tag{4.8}$$

$$\text{correlation (double first cousins)} = \frac{1}{4}\left\{\frac{\sigma_{\mathrm{A}}^2}{\sigma^2}\right\} + \frac{1}{16}\left\{\frac{\sigma_{\mathrm{D}}^2}{\sigma^2}\right\}, \tag{4.9}$$

together with various similar formulae. A comparison of equations (4.6) and (4.7) shows that these two mathematically-derived formulae agree with the empirical correlations observed by the biometricians described above.

There is an elegant simple way, devised by Malécot [9], to arrive at these and other correlations. We consider two individuals, X and Y, and define x_{f} as the gene that X received from his father and x_m as the gene that he received from his mother, with, for individual Y, y_{f} and y_{m} being defined similarly. We use the symbol "≡" to denote "identical by descent", and define

$$P_{\mathrm{ff}} = \mathrm{Prob}(x_{\mathrm{f}} \equiv y_{\mathrm{f}}), \qquad P_{\mathrm{fm}} = \mathrm{Prob}(x_{\mathrm{f}} \equiv y_{\mathrm{m}}),$$
$$P_{\mathrm{mf}} = \mathrm{Prob}(x_{\mathrm{m}} \equiv y_{\mathrm{f}}), \qquad P_{\mathrm{mm}} = \mathrm{Prob}(x_{\mathrm{m}} \equiv y_{\mathrm{m}}). \tag{4.10}$$

Malécot showed that when the two parents of any individual are unrelated,

$$\mathrm{correlation}(X,Y) = \frac{1}{2}\left(P_{\mathrm{ff}} + P_{\mathrm{fm}} + P_{\mathrm{mf}} + P_{\mathrm{mm}}\right)\frac{\sigma_{\mathrm{A}}^2}{\sigma^2} + \left(P_{\mathrm{ff}}P_{\mathrm{mm}} + P_{\mathrm{fm}}P_{\mathrm{mf}}\right)\frac{\sigma_{\mathrm{D}}^2}{\sigma^2}. \tag{4.11}$$

This elegant formula provides a simple method for deriving correlations for any two related individuals, and we now use it to re-derive (4.6) and (4.7).

Consider first the parent-offspring correlation, with X being the parent and Y the offspring. Since the mother and father are assumed to be unrelated, $P_{\mathrm{mm}} = P_{\mathrm{fm}} = 0$. Also $P_{\mathrm{ff}} = P_{\mathrm{mf}} = \frac{1}{2}$, and insertion of these values into (4.11) yields (4.6). If X and Y are full sibs, $P_{\mathrm{ff}} = P_{\mathrm{mm}} = \frac{1}{2}$, $P_{\mathrm{fm}} = P_{\mathrm{mf}} = 0$, and insertion of these values in (4.11) gives (4.7). Equations (4.8) and (4.9) can be found equally easily. Many other interesting conclusions can be drawn from equation (4.11). One of these is that ancestral line correlations do not contain the term involving σ_{D}^2. Thus for example the great-grandfather/great-grandson correlation is $(1/8)\sigma_{\mathrm{A}}^2/\sigma^2$, and more generally each ancestral line correlation decreases by a factor of $\frac{1}{2}$ with each additional generation separating the two individuals of interest.

In the parent-offspring correlation (4.6), the factor $\frac{1}{2}$ arises because the offspring receives only half his genes from the parent, and the factor $\sigma_{\mathrm{A}}^2/\sigma^2$ arises because the parent can only pass on an "allelic value" contribution for the character in question.

Of course essentially all measured characteristics such as height and weight are controlled by the genes at many loci, not just one locus. Also, in respect of various measurements in man (for example height) mating is not at random: tall people tend to marry tall people, and so on. Further, variation caused by the environment has to be considered. The theory has been generalized to cover these cases, but is too complex to give here. Nevertheless, observed correlations among humans for many characters followed the general pattern provided by equations (4.6)–(4.9), and thus the mathematical development provides further evidence for the importance of the Mendelian hereditary scheme.

The relevance of these calculations extends beyond evolutionary considerations. In plant and animal breeding programs the ratio $\sigma_{\mathrm{A}}^2/\sigma^2$ is called the *heritability* of a trait. It has been shown above that the additive genetic variance σ_{A}^2 has an evolutionary significance arising from the passage of genes from parent to offspring. The value of the heritability for any trait indicates, to a plant or animal breeder, the extent to which his breeding program can be expected to improve the trait of interest.

4.3. The Fundamental Theorem of Natural Selection

Ever since it was first put forward by in Fisher [10], the "Fundamental Theorem of Natural Selection" (henceforth referred to as the FTNS) has provoked as much controversy, and caused as much misunderstanding, as perhaps any other result in evolutionary population genetics. There are two aspects to the controversy surrounding the FTNS. The first is essentially mathematical: what does the theorem actually state? The second is biological: what is its biological relevance? Here we focus on the first question; for an extensive discussion of the second question see [11]. Before doing this, it is useful to make two background comments. First, Fisher saw himself as casting the main principles of Darwinian evolution in Mendelian and mathematical terms, and the FTNS, stated by him as holding the supreme position in the biological sciences, was a key component of this effort. Second, Fisher had an essentially "gene's-eye" view of evolution, and as shown below the FTNS has a gene's-eye flavor to it. This viewpoint bears on the questions of the correct level at which to describe evolution and of the appropriate unit of selection, matters which are addressed extensively in the biological literature, but which also require a mathematical treatment for their full consideration.

Fisher's various presentations of the FTNS were not consistent with each other, and the following distillation of these presentations, that "The rate of increase in mean fitness of any population at any time is equal to its additive genetic variance in fitness at that time", is generally accepted as the statement of the theorem. However, even this distilled version can be interpreted in several ways. The two main interpretations, the "classical" and the "modern", are described below. The classical interpretation is perhaps the more interesting biologically, and thus is of main interest to biologists. The modern interpretation is mathematically far deeper and is thus primarily of interest to mathematicians.

We consider first the classical interpretation. In the early years of population genetics theory various simplifying assumptions were, of necessity, made. One of these was that the individuals in a population mate at random. A second assumption often made was that, in studying the evolution of allelic frequencies at any locus through the effects of mutation and selection, all other loci in the genome can be ignored and the locus of interest treated in isolation. A third assumption, often made in connection with the second, is that the fitness of any individual depends only on the allelic types of the two genes that he carries at a single gene locus and is independent of the allelic types of the genes carried in the remainder of the genome. Some of the theory given above, and also the classical interpretation of the FTNS, reflect these simplifying (and unrealistic) assumptions. They lead to the following (classical) interpretation of the FTNS. If an arbitrary number of different allelic types is allowed at some single gene locus, if the fitness of any individual depends only on his genotype defined by these alleles, and if these genotype fitnesses are fixed constants, then assuming mating is random, the population mean fitness will increase from one generation to the next, or at least remain constant. The most straightforward proof of this classical version of the theorem, under these assumptions, was given by Kingman [12]. Kingman further showed that when $\overline{w} = 1, \Delta\overline{w} \approx \sigma_A^2$, where $\Delta\overline{w}$ is the change in the mean fitness between parental and offspring generations and σ_A^2 is the parental generation additive genetic variance in fitness. The level of approximation involved in this statement can be seen from the fact that if single locus genotype fitnesses differ from each other by a small term of order δ and if $\overline{w} = 1$, the value of $\Delta\overline{w}$ differs from σ_A^2 by an extremely small term (of order δ^3).

It is easy to find examples for which mean fitness decreases between parental and offspring generations if random mating is not the case. Thus the random mating requirement is essential for the classical version of the theorem. It is also a standard result of population genetics theory that population mean fitness can decrease from one generation to the next, even under random mating, if (as is the case in practice) the fitness of any individual depends on the allelic types of the genes that he carries at more than one gene locus. Thus the assumption that fitness depends on the genes at one gene locus is also essential for the classical version of the theorem. The classical interpretation of the FTNS is attractive in that it appears to quantify in Mendelian terms the two prime themes of the Darwinian theory, namely that variation is needed for evolution by natural selection and that evolution by natural selection is a process of steady improvement in the population. Also, cases where mean fitness decreases from one generation to the next are either comparatively rare, and when these decreases arise they are often small. When fitness differentials are small, the population mean fitness "usually" increases under random mating when fitness depends on the genes at many loci, and when $\overline{w} = 1$ the change in mean fitness is "usually" approximately equal to σ_A^2, thus generalizing Kingman's result given above [13, 14].

We now turn to the modern interpretation of the FTNS. The assumptions made in, and the conclusion of, the classical version of the FTNS contradict various claims that Fisher made about the theorem. First, the fact that $\Delta\overline{w}$ is not exactly equal to σ_A^2 contradicts Fisher's claim that the FTNS is an exact result and not an approximation. Second, the fact that the population mean fitness can decrease under non-random mating contradicts his claim that the FTNS is true under any form of mating. Finally, the fact that even under random mating the population mean fitness can decrease when the fitness of any individual depends on the genes that he carries at more than one locus contradicts the claim by Fisher that the FTNS holds when fitness depends on the allelic types of the genes carried by any individual at all loci in the genome. There are thus severe difficulties in reconciling the classical version of the FTNS with Fisher's explicitly stated views. The modern interpretation of the theorem resolves all these difficulties since it is an exact result, holds under non-random mating, and applies when the fitness of any individual depends on the allelic types of all the genes in his genome.

The modern interpretation of the FTNS was first proposed by Price [15]. Price claimed that Fisher was not interested in the total change of mean fitness ($\Delta\overline{w}$ above), but rather only in that part of the total change due to natural selection, or (more or less equivalently) due to changes in allelic frequencies. Here we refer to this as the "partial change" in mean fitness. To define this change we consider first the simple case where fitness values depend on the genotype of an individual at one gene locus only. Random mating is *not* assumed.

Suppose then that the fitness of any individual depends entirely on his genotype at a single locus at which may occur alleles $A_1, A_2, \ldots, A_k$. As above we denote the frequency of the genotype A_iA_j at the time of conception of the parental generation by P_{ii} (when $i = j$) and $2P_{ij}$ (when $i \neq j$). These frequencies are not necessarily in Hardy-Weinberg form, since random mating is not assumed. The frequency p_i of the allele A_i at this time is $\sum_j P_{ij}$. The fitness of an individual of individuals of the genotype A_iA_j is w_{ij}, and the mean population fitness is $\sum_i \sum_j P_{ij}w_{ij}$. As noted above, Fisher's main evolutionary focus was on the genes in any individual at the locus of interest, not the genotypes, since it is a gene and not the genotype that is passed on from parent to child at that locus. He therefore thought of the mean population fitness as being given not by the above expression but by an expression involving the average effects of the various alleles as defined implicitly in (4.2), namely as

$$\sum_i \sum_j P_{ij}(\beta_i + \beta_j). \tag{4.12}$$

This change of viewpoint is however a purely conceptual, since the two expressions $\sum_i \sum_j P_{ij}w_{ij}$ and $\sum_i \sum_j P_{ij}(\beta_i + \beta_j)$ can be shown to be numerically identical. Despite this identity, the new conceptualization (4.12) is central to Fisher's view of the between-generation partial change in mean fitness. This was conceived as the between-generation change in the expression in (4.12) brought about by changes in the genotype frequencies P_{ij}, with the changes in the average effects β_i and β_j (which do in fact occur) being ignored. This generation-to-generation partial change in mean fitness, denoted by $\Delta_{\rm P}(\overline{w})$, (the suffix "P" denoting "partial") is, clearly,

$$\Delta_{\rm P}(\overline{w}) = \sum_i \sum_j (P'_{ij} - P_{ij})(\beta_i + \beta_j) = \sum_i \sum_j (\delta P_{ij})(\beta_i + \beta_j), \tag{4.13}$$

where P'_{ij} is the daughter generation frequency of the genotype A_iA_j, defined as for the parental generation value at its time of conception, and δP_{ij} is the between generation change in the ordered frequency of this genotype. The extreme right-hand term in (4.13) is easily shown to be $2\sum_i (\delta p_i)\beta_i$, and equation (4.5) then shows that the partial change in mean fitness is exactly $\sigma^2_{\rm A}/\overline{w}$, whether or not random mating occurs. This result, involving no approximations, is the modern interpretation of the single-locus FTNS.

The parallel whole-genome statement of the theorem, applying when the fitness of any individual depends in an arbitrary way on all the genes in the genome, can be found as follows. Assume that the various (gigantically large number of) possible whole-genome genotypes are listed in some agreed order as genotypes $1, 2, \ldots, s, \ldots, S$. The "time of conception" frequency of the typical genotype s in the parental generation is denoted by g_s and the fitness of this genotype by w_s. Thus the parental generation population mean fitness, denoted (for the whole-genome as for the one-locus case) by $\sum_s g_s w_s$.

As in the one locus case, the average effects of the various alleles at the various loci in the genome are defined by a least squares procedure. These average effects of all the alleles in the genome are determined by minimizing the sum of squares

$$\sum_g g_s \left\{ w_s - \sum c_i^{\mathrm{A}} \beta_i^{\mathrm{A}} \right\}^2 . \tag{4.14}$$

In the expression (4.14), β_i^{A} is the average effect of A_i at the typical gene locus A, the outer sum is taken over all whole-genome genotypes and the inner sum is taken, for each whole-genome genotype, over all alleles at all loci in the genome contained within that genotype, with $c_i^{\mathrm{A}} = 1, 2$ or 0 depending on whether A_i arises once, twice or not at all within the genotype g_s, at the locus A.

It is not necessary to give explicit formulae for the various β values defined by this least-squares procedure: indeed they can only be expressed implicitly as the (unique) solution of a gigantic set of simultaneous equations. As in the one-locus case, the Fisher's "gene's-eye" view of the fitness of the typical whole-genome genotype s is not its actual fitness, but instead is the linear combination $\sum c_{ai}\beta_{ai}$, defined as above. In parallel with the one-locus case analysis, the mean fitness of the population is now thought of as being

$$\sum_g g_s \left\{ \sum c_i^{\mathrm{A}} \beta_i^{\mathrm{A}} \right\}, \tag{4.15}$$

which (as in the corresponding one-locus case) is numerically identical to that given by the standard definition of mean fitness, here $\sum_s g_s w_s$ given above.

Again in parallel with the one-locus case, the partial change $\Delta_{\mathrm{P}}\overline{w}$ in mean fitness is defined as the change in the expression (23) derived solely from the changes Δg_s in the various whole-genome genotype frequencies and ignoring the changes in the β values, namely

$$\sum_g \Delta g_s \left\{ \sum c_i^{\mathrm{A}} \beta_i^{\mathrm{A}} \right\}. \tag{4.16}$$

The resulting expression can be shown to be equal to $\sigma_{\mathrm{A}}^2/\overline{w}$, where σ_{A}^2 now denotes the whole-genome additive genetic variance, defined in a manner extending that for the one-locus case. This simple and exact result is the modern interpretation of the whole-genome FTNS [16, 17]. It is true whatever the mating scheme. It is inconceivable that this "gene's-eye" view of evolution could have been obtained by anything other than a mathematical treatment. Further, the mathematical treatment provides an insight into evolutionary principles not obtainable in any other way.

5. Optimality principles

It is natural in evolutionary genetics to attempt to follow a long-established practice in physics, associated there often with least-action principles, and to ask: "What is optimized under the allelic frequency changes brought about by natural selection?"

In considering this question we start with a case where a frequently encountered "natural" optimality principle is not true. This incorrect principle is most easily stated in the continuous-time analogue of the analysis in the previous subsection, using differential equations generalizing that in (3.5). We consider here only the simple case of a population that mates at random and for which the fitness of any individual depends on the genes at a single locus which admits k alleles $A_1, A_2, \ldots, A_k$. It was stated above that in this case the population mean fitness increases (or at worst remains constant) from one generation to the next. This incorrect principle is, in effect, that natural selection acts in such a way as to maximize the rate of increase of the population mean fitness. This, however, is not the trajectory taken by natural selection, even in this simple case, as shown by Svirezhev and Passekov [18, pp. 105-6]. Equivalently, and employing the frequently-used concept that the population mean fitness represents a dimension in a k-dimensional Euclidean space for which the other $k-1$ dimensions represent the frequencies of the alleles $A_1, A_2, \ldots, A_{k-1}$, (the frequency of A_k being implied by the frequencies of the remaining alleles), natural selection does not act in such a way that allelic frequencies follow a path of steepest ascent up the mean fitness surface in this k-dimensional space. On the other hand, if one employs a non-Euclidean geometry whose distance measure is discussed below, then in this non-Euclidean space the path of steepest ascent *is* taken.

Is there an optimality principle where random mating is not assumed? We discuss this question first in the one-locus case assumed in the previous paragraph. Consider a parental and a daughter generation at their respective times of conception, and form the set of inter-generational changes $\delta_j (j = 1, 2, \ldots, k)$ in the frequencies of the various alleles into a column vector δ. Write the parent generation average effects $\beta_1, \beta_2, \ldots, \beta_k$ conformally as a column vector β and similarly write the parent generation gene frequencies $p_1, p_2, \ldots, p_k$ as a column vector p. Define D as diagonal matrix with j^{th} diagonal element p_j and P as a matrix whose (i, j) element is P_{ij}, the parent generation frequency of the ordered genotype A_{ij}. Then equation (4.3) may be written in matrix terms as

$$(D + P)\beta = \overline{w}(p + \delta). \tag{5.1}$$

From this, immediately,

$$\delta = (D + P)\beta(\overline{w})^{-1} - p. \tag{5.2}$$

Now consider any arbitrary vector d of gene frequency changes whose j^{th} element is d_j. (This vector is of course subject to the requirement that all allelic frequencies must be non-negative and that all frequencies must add to 1.) The biologically natural (squared) distance measure between parental and daughter generation frequencies referred to above is $d'(D+P)^{-1}d$. It follows from (5.2) and some algebra that in the case of the natural selection changes δ in gene frequency, this distance measure takes the value $\sigma_{\mathrm{A}}^2/(2\overline{w}^2)$.

The discussion following equation (4.4) shows that the partial change in mean fitness following the arbitrary changes $d_j, (j = 1, 2, \ldots, k)$ is $2\sum_j d_j\beta_j$. It can then be shown that subject to the condition that the changes d in gene frequency between parental and daughter generations leads to the same distance $\sigma_{\mathrm{A}}^2/(2\overline{w}^2)$ as that achieved by the natural selection, the changes which maximize the partial increase in mean fitness are the natural selection changes given in the vector δ. This is the desired optimizing principle of natural selection.

A parallel result continues to apply at the whole-genome level when the matrix $D+P$ is changed to one appropriate to that level. Here the required matrix is of the form $D+P+Q$, where D is a gigantic diagonal matrix displaying along its diagonal the various allelic frequencies at all the various loci in the genome, P is a conformal gigantic block diagonal matrix, with the various blocks containing single locus genotype frequencies at the various gene loci in the genome, and Q is an equally gigantic matrix, best described by Castilloux and Lessard [16], involving various two-locus frequencies. The vector δ of natural selection allelic frequency changes is now defined as the set of changes for all alleles in the entire genome and the arbitrary set of changes for all alleles in the entire genome is correspondingly denoted by d. The biologically natural (squared) distance measure between parental and daughter generation frequencies for this arbitrary set of changes in this whole-genome case is $d'(D+P+Q)^{-1}d$, and for the natural selection set of changes δ this distance takes the $\sigma_{\mathrm{A}}^2/(2\overline{w}^2)$, where σ_{A}^2 is now the whole genome additive genetic variance. The conclusion of the previous paragraph, starting "It can then be shown..." then follows word for word with now d, δ and σ_{A}^2 having their whole-genome interpretations.

It is impossible to imagine this whole-genome result being reached without using a mathematical approach. Although the result is not involved with validating the Darwinian theory (the main theme of this chapter), it does flesh out our understanding of evolution under natural selection, that is under the Darwinian paradigm.

6. The stochastic theory

In the calculations given above it is implicitly assumed that the population of interest is infinite in size. However, all populations are of course finite in size, and this implies that some modifications have to be made to these calculations. The main modification is that changes in allelic frequencies must be viewed as being part of a stochastic, rather than a deterministic, process. It is necessary, in order to arrive at an evaluation of the importance of the stochastic factor, to set up a stochastic model which reasonably describes the behavior of a population in the stochastic case. Perhaps more than in any other part of the theory the choice of a model here is somewhat arbitrary, and we do not pretend that Nature necessarily follows at all closely the model given below (or indeed any other model available in the literature).

We consider, as the simplest possible case, a diploid population of fixed size N. Suppose first that no selective difference exist between the two alleles A_1 and A_2 possible at a certain locus A and that there is no mutation. There are $2N$ genes in the population in any generation, and it is sufficient to center our attention on the number X of A_1 genes. Clearly in any generation X takes one or other of the values $0, 1, \ldots, 2N$, and we denote the value assumed by X in generation t by $X(t)$.

We must now assume some specific model which describes the way in which the genes in generation $t+1$ are derived from the genes in generation t. The model which we consider assumes that the genes in generation $t+1$ are derived by sampling with replacement from the genes of generation t. This means that the number $X(t+1)$ is a binomial random variable with index $2N$ and parameter $X(t)/2N$. More explicitly, given that $X(t) = i$, the probability p_{ij} that $X(t+1) = j$ is assumed to be given by

$$p_{ij} = \binom{2N}{j} (i/2N)^j \{1 - (i/2N)\}^{2N-j}, \quad i, j = 0, 1, 2, \ldots, 2N. \tag{6.1}$$

This is the so-called Wright-Fisher model, first investigated by Fisher [19] and Wright [20]. It is clear that $X(\cdot)$ is a Markovian random variable with transition matrix $P = \{p_{ij}\}$, so that Markov chain theory can be used to discuss some of its properties. Perhaps the most important of these is that whatever the value $X(0)$, eventually $X(\cdot)$ will take either the value 0 or $2N$, and once this happens there will be no further change in the value of $X(\cdot)$, genetic variation is permanently lost, and no further evolution is possible at this locus. In view of the importance of the preservation of genetic variance referred to above, it is therefore natural to attempt to find the mean number of generations pass before fixation of one or other allele occurs. This mean depends on the value of $X(0)$, the initial number of A_1 genes, and as it happens, no simple explicit formula for this mean exists. On the other hand, it is known that the formula

$$-4N \{q \log q + (1-q) \log(1-q)\}, \tag{6.2}$$

where $q = X(0)/(2N)$, provides an extremely accurate approximation to the required mean number of generations.

Except in cases where the initial number of A_1 genes is either very small or very large, this mean is clearly of order N generations and is thus large in a large population. In practice the assumptions of the model (6.1), in particular that no mutation occurs during from the initial time until the time of loss of variation, cannot be expected to hold for this long period of time, and thus this result may be thought of as a stochastic analogue of the "variation-preserving" property of infinite genetic populations shown by the Hardy–Weinberg law. Darwin's theory is thus not compromised by the finite size of the population.

A second question arises when selection exists. Suppose that, as in the discussion leading to equation (3.3), the fitness of the genotype A_1A_1 is w_{11}, of A_1A_2 is w_{12} and of A_2A_2 is w_{22}. Suppose that in some generation there are i A_1 genes, and put $p = i/(2N)$. The number j of A_1 genes in the following generation is a random variable, and from (3.2) and (3.3) it is reasonable to replace the probability p_{ij} in (6.1) by

$$p_{ij} = \binom{2N}{j} (p')^j \{1 - p'\}^{2N-j}, \quad i, j = 0, 1, 2, \ldots, 2N, \tag{6.3}$$

with p'_i defined as in (3.3). It is clear from standard Markov chain theory that $X(\cdot) = 0$ and $X(\cdot) = 2N$ are absorbing states, and attention now focuses on the probability that (for example) the second of these states is eventually entered when $w_{11} > w_{12} > w_{22}$, that is when A_1 is selectively favored over A_2.

No exact form for this probability is known. However, an extremely accurate approximation is available. The simplest case arises when fitnesses assume the "linear" form where $w_{11} = 1 + s, w_{12} = \frac{1}{2}s, w_{22} = 1$, where $s > 0$. This is that the required probability is approximately

$$\frac{1 - e^{-2Nsq}}{1 - e^{-2Ns}}. \tag{6.4}$$

It is of some interest to use this approximate formula to get some idea of the effect of the selective differences on the probability of fixation of A_1. Suppose for example that $N = 10^5$, $s = 10^{-4}$, and $q = 0.5$. Then from (6.4) the required probability is about 0.999955. By contrast, for $s = 0$ this probability is 0.5. Evidently the rather small selective advantage 0.0001 is nevertheless large enough in evolutionary terms to have a significant effect on the fixation probability. Clearly this occurs because, while selection might have only a minor effect in any generation, the number of generations until fixation occurs is so very large that the cumulative effect of selection is considerable. Here is a case where mathematics is needed to validate a key aspect of the Darwinian theory, since a selective difference of 10^{-4} is, except in rare cases, too small to be observed in laboratory experiments.

7. Other matters

The focus in this chapter, as its title suggests, is on the use of mathematics to validate and then extend and flesh out the Darwinian theory of evolution by natural selection. However, the use of mathematics in population genetics now far extends beyond this validation program. Perhaps the most active research area in population genetics does not concern processes going forward in time (the direction, of course, of actual evolution and also of the topics discussed in this chapter), but focuses on retrospective questions, looking backward in time. Here the great volumes of data now becoming available enable us to answer questions like: "When did the lines leading to humans and chimpanzees diverge? How can we use current genetic information to detect signatures of selection in past times?" Mathematics is now playing as essential a role in answering these and other data-driven questions as it played in the past in validating and extending the Darwinian theory.

References

[1] C. Darwin, *On the Origin of Species by Means of Natural Selection or the Preservation of Favoured Races in the Struggle for Life.* John Murray, London, 1859.

[2] G. Mendel, *Versuche über Pflanzenhybriden.* (*Experiments relating to plant hybridization*). Verh. Naturforsch. Ver. Brünn **4** (1866), 3–17.

[3] G.H. Hardy, *Mendelian proportions in a mixed population.* Science **28** (1908), 49–50.

[4] W. Weinberg, *Über den Nachweis der Vererbung beim Menschen.* (*On the detection of heredity in man*). Jahreshelfts. Ver. Vaterl. Naturf. Württemb. **64** (1908), 368–382.

[5] J.B.S. Haldane, *The Causes of Evolution.* Longmans Green, London, 1932.

[6] J.F.C. Kingman, *A mathematical problem in population genetics.* Proc. Camb. Phil. Soc. **57** (1961), 574–582.

[7] K. Pearson and A. Lee, *On the laws of inheritance in man. I. Inheritance of physical characters.* Biometrika **2** (1903), 357–462.

[8] R.A. Fisher, *The correlation between relatives on the supposition of Mendelian inheritance.* Trans. of the Roy. Soc. of Edinburgh **52** (1918), 399–433.

[9] G. Malécot, *Les Mathématiques de l'Hérédité.* Masson, Paris, 1948.

[10] R.A. Fisher, *The Genetical Theory of Natural Selection.* Clarendon Press, Oxford, 1930.

[11] W.J. Ewens, *What is the gene trying to do?* Brit. J. Philos. Sci. **62** (2011), 155–176.

[12] J.F.C. Kingman, *A matrix inequality.* Quart. J. Math. **12** (1961), 78–80.

[13] T. Nagylaki, *Selection in One- and Two-Locus Systems.* Springer, Berlin, 1977.

[14] T. Nagylaki, *Theoretical Population Genetics.* Springer-Verlag, Berlin, 1992.

[15] G. Price, *Fisher's "Fundamental Theorem" made clear.* Ann. Hum. Genet. **36** (1972), 129–140.

[16] W.J. Ewens, *An optimizing principle of natural selection in evolutionary population genetics.* Theoret. Pop. Biol. **42** (1992), 333–346.

[17] A.M. Castilloux and S. Lessard, *The fundamental theorem of natural selection in Ewens' sense (case of many loci).* Theoret. Pop. Biol. **48** (1995), 733–742.

[18] Y.M. Svirezhev and V.P. Passekov, *Fundamentals of Mathematical Evolutionary Genetics.* Kluwer, Dordrecht, 1990.

[19] R.A. Fisher, *On the dominance ratio.* Proc. Roy. Soc. Edin. **42** (1922), 321–341.

[20] S. Wright, *Evolution in mendelian populations.* Genetics **16** (1931), 97–159.

Warren J. Ewens
Department of Biology
The University of Pennsylvania
Philadelphia, PA 19104, USA
e-mail: `wewens@sas.upenn.edu`

The Mathematics of Darwin's Theory of Evolution: 1859 and 150 Years Later

Peter Schuster

Abstract. A mathematical formulation of Darwin's theory of evolutionary optimization through variation and selection is derived in terms of conventional ODEs that can be interpreted as chemical kinetics of evolution. Variation in form of mutation and recombination operates on genotypes being DNA or RNA sequences, whereas phenotypes, which are represented by organisms or molecular structures, are the target of selection. The impact of recombination on optimization is briefly sketched. Differential equations modelling selection in populations with correct replication and mutation are derived from the molecular mechanisms of polynucleotide replication. The analysis of these ODEs reveals restrictions of the optimization principle caused by mutation. Error propagation over generations sets a limit to mutation rates in evolution, which manifests itself in the form of a phase transition-like phenomenon characterized as error threshold. Conditions on fitness landscapes for the occurrence of error thresholds derived from numerical investigations are presented: Smooth fitness landscapes show no error thresholds but gradual transitions, sufficiently steep landscapes and rugged landscapes sustain error thresholds. Sharp transitions are also found with realistic landscapes combining ruggedness and neutrality. Lethal mutants may lead to extinction of populations and set another upper limit to mutation rates in form of an extinction threshold through lethal mutagenesis.

Mathematics Subject Classification (2000). Primary 92D25; Secondary 80A30.

Keywords. Error threshold, genotype-phenotype mapping, lethal mutagenesis, population dynamics, neutrality, quasispecies, sequence space.

1. Preamble

This paper has been presented at a meeting that celebrated the 150 years anniversary of Charles Darwin's famous book on the *Origin of Species* [1]. Darwin was presumably the greatest naturalist that has ever lived, a highly talented observer,

This contribution reports work that was performed at Vienna University and at the Santa Fe Institute with financial support from the *Austrian Fonds zur Förderung der wissenschaftlichen Forschung (FWF)*.

and a genius in intuition. He was certainly not a fan of mathematics and his centennial treatise of evolution does not contain a single formula. Here we shall make a *gedankenexperiment*: How might Charles Darwin have formulated his theory if he were a mathematician? The knowledge on mathematics related to evolution was sparse or non existing at Darwin's time. The mechanism of inheritance and the origin of variation were completely unknown to Charles Darwin and his contemporary biologists. Darwin's speculations on blending of the parental properties in the offspring were completely off the point. Gregor Mendel did the first careful experiments on plants [2] and drew the right conclusions concerning inheritance: The genomes of both parents are split into pieces and recombined in the offspring thereby conserving the order of genes but choosing more or less randomly from father or mother. Mendel had an education in mathematics and physics and with this background he was in the position to discover a statistical law – a regularity that becomes evident only when sufficiently many experiments are superimposed and correctly evaluated. Recombination is one common source of phenotypic variation in sexually reproducing species. Mutation, the second source of variation that is occurring in both, asexual and sexual species, and was identified as a change in the nucleotide sequence of the genetic message. Although Gregor Mendel gave the correct interpretation of his experiments and discovered the idealized principle of recombination, a full understanding of sexual reproduction without the insights from cellular and molecular biology is not possible. Mutation cannot be understood at all without molecular knowledge.

2. Selection and optimization

Charles Darwin's principle of natural selection is a powerful abstraction from observations, which provides insight into the origin of changing species. Species or populations don't multiply but individuals do, either directly in asexual species, like viruses, bacteria or protists, or in sexual species through pairings of individuals with opposite sex. Variability of individuals in populations is an empirical fact that can be seen easily in everyday life. Within populations the variants are subjected to natural selection and those having more progeny prevail in future generations. The power of Darwin's abstraction lies in the fact that neither the shape and the structure of individuals nor the mechanism of inheritance are relevant for selection unless they have an impact on the number of offspring. Otherwise Darwin's approach had been doomed to fail since his imagination of inheritance was incorrect. Indeed Darwin's principle holds simultaneously for highly developed organisms, for primitive unicellular species like bacteria, for viruses and even for reproducing molecules in cell-free assays.

Molecular biology provided a powerful possibility to study evolution in its simplest form outside biology: Replicating ribonucleic acid molecules (RNA) in cell-free assays [3] play natural selection in its purest form: In the test tube, evolution, selection, and optimization are liberated from all unnecessary complex features, from obscuring details, and from unimportant accessories. Hence, *in vitro* evolution can be studied by the methods of chemical kinetics. The parameters determining the "fitness of molecules" are replication rate parameters, binding constants, and other measurable quantities, which can be determined independently of *in vitro* evolution experiments, and constitute an alternative access to the determination of the outcome of selection. Thereby "survival of the fittest" is unambiguously freed from the reproach of being the mere tautology of "survival of the survivor". In addition, *in vitro* selection turned out to be extremely useful for the synthesis of molecules that are tailored for predefined purposes. A new area of applications called evolutionary biotechnology branched off evolution in the test tube. Examples for evolutionary design of molecules are [4, 5] for nucleic acids, [6, 7] for proteins, and [8] for small organic molecules.

The section starts by mentioning a few examples of biological applications of mathematics before Darwin (Subsection 2.1), we derive and analyze an ODE describing simple selection with asexual species (Subsection 2.2), and consider the effects of variable population size (Subsection 2.3). The next Subsection 2.4 analyzes optimization in the Darwinian sense, Subsection 2.5 presents a brief account of Fisher's selection equation and his fundamental theorem of natural selection and eventually we consider generic properties of typical growth functions (Subsection 2.6).

2.1. Counting and modelling before Darwin

The first mathematical model that seems to be relevant for evolution was conceived by the medieval mathematician Leonardo Pisano also known as Fibonacci. His famous book *Liber abaci* has been finished and published in the year 1202 and was translated into modern English eight years ago [9]. Among several other important contributions to mathematics in Europe Fibonacci discusses a model of rabbit multiplication in *Liber abaci*. Couples of rabbits reproduce and produce young couples of rabbits according to the following rules:

(i) Every adult couple has a progeny of one young couple per month,
(ii) a young couple grows to adulthood within the first month and accordingly begins producing offspring in the second months,
(iii) rabbits live forever, and
(iv) the number of rabbit couples is updated every month.

The model starts with one young couple (**1**), nothing happens during maturation of couple **1** in the first month and we have still one couple in the second month. In the third month, eventually, a young couple (**2**) is born and the number of couples increases to two. In the fourth month couple **1** produces a new couple (**3**) whereas couple **2** is growing to adulthood, and we have three couples now. Further rabbit counting yields the Fibonacci sequence:[1]

month	0	1	2	3	4	5	6	7	8	9	...
#couples	0	1	1	2	3	5	8	13	21	34	...

It is straightforward to derive a recursion for the rabbit count. The number of couples in month $(n+1)$, f_{n+1}, is the sum of two terms: The number of couples in month n, because rabbits don't die, plus the number of young couples that is identical to the number of couples in month $(n-1)$:

$$f_{n+1} = f_{n-1} + f_n \quad \text{with} \quad f_0 = 0 \ \text{ and } \ f_1 = 1\,. \tag{2.1}$$

With increasing n the ratio of two subsequent Fibonacci numbers converges to the golden ratio, $f_{k+1}/f_k = (1+\sqrt{5})/2$ (For a comprehensive discussion of the Fibonacci sequence and its properties see [12, pp. 290–301] or, e.g., [13]).

In order to proof this convergence we make use of a matrix representation of the Fibonacci model:

$$\mathrm{F}^n \begin{pmatrix} f_0 \\ f_1 \end{pmatrix} = \begin{pmatrix} f_n \\ f_{n+1} \end{pmatrix} \quad \text{with } \mathrm{F} = \begin{pmatrix} f_0 & f_1 \\ f_1 & f_2 \end{pmatrix} \text{ and } \mathrm{F}^n = \begin{pmatrix} f_{n-1} & f_n \\ f_n & f_{n+1} \end{pmatrix}.$$

The matrix representation transforms the recursion into an expression that allows for direct computation of the elements of the Fibonacci sequence.

$$f_n = (1 \quad 0)\ \mathrm{F}^n \begin{pmatrix} f_0 \\ f_1 \end{pmatrix} = (1 \quad 0) \begin{pmatrix} f_{n-1} & f_n \\ f_n & f_{n+1} \end{pmatrix} \begin{pmatrix} f_0 \\ f_1 \end{pmatrix}. \tag{2.2}$$

Theorem 2.1 (Fibonacci convergence). *With increasing n the Fibonacci sequence converges to a geometric progression with the golden ratio as factor,* $q = (1+\sqrt{5})/2$.

Proof. The matrix F is diagonalized by the transformation $T^{-1} \cdot \mathrm{F} \cdot \mathrm{T} = \mathrm{D}$ with $\mathrm{D} = \begin{pmatrix} \lambda_1 & 0 \\ 0 & \lambda_2 \end{pmatrix}$. The two eigenvalues of F are: $\lambda_{1,2} = (1 \pm \sqrt{5})/2$. Since F is a symmetric matrix the $\mathbb{L}^2$-normalized eigenvectors of F, $(\mathbf{e}_1, \mathbf{e}_2) = \mathrm{T}$, form an orthonormal set,

$$T = \begin{pmatrix} \frac{1}{\sqrt{1+\lambda_1^2}} & \frac{1}{\sqrt{1+\lambda_2^2}} \\ \frac{\lambda_1}{\sqrt{1+\lambda_1^2}} & \frac{\lambda_2}{\sqrt{1+\lambda_2^2}} \end{pmatrix} \quad \text{and} \quad T \cdot T' = \begin{pmatrix} 1 & 0 \\ 0 & 1 \end{pmatrix}$$

with T′ being the transposed matrix, and $\mathrm{T}^{-1} = \mathrm{T}'$. Computation of the nth power of matrix F yields

$$\mathrm{F}^n = T \cdot \mathrm{D}^n \cdot \mathrm{T}' = \mathrm{T} \cdot \begin{pmatrix} \lambda_1^n & 0 \\ 0 & \lambda_2^n \end{pmatrix} \cdot \mathrm{T}' = \frac{1}{\sqrt{5}} \begin{pmatrix} \lambda_1^{n-1} - \lambda_2^{n-1} & \lambda_1^n - \lambda_2^n \\ \lambda_1^n - \lambda_2^n & \lambda_1^{n+1} - \lambda_2^{n+1} \end{pmatrix},$$

from which the expression for f_n is obtained by comparison with (2.2)

$$f_n = \frac{1}{\sqrt{5}}(\lambda_1^n - \lambda_2^n)\,. \tag{2.3}$$

[1]According to Parmanand Singh [10] the Fibonacci numbers were invented earlier in India and used for the solution of various problems (see also Donald Knuth [11]).

Because $\lambda_1 > \lambda_2$ the ratio converges to zero: $\lim_{n\to\infty} \lambda_2^n/\lambda_1^n = 0$, and the Fibonacci sequence is approximated well by $f_n \approx \frac{1}{\sqrt{5}}\, q^n$ with $q = (1+\sqrt{5})/2$. □

Since λ_2 is negative the Fibonacci sequence alternates around the geometric progression. Expression (2.3) is commonly attributed to the French mathematician Jacques Binet [14] and named after him. As outlined in ref. [12, p. 299] the formula has been derived already hundred years before by the great Swiss mathematician Leonhard Euler [15] but was forgotten and rediscovered.

Thomas Robert Malthus was the first who articulated the ecological and economic problem of population growth following a geometric progression [16]: Animal or human populations like every system capable of reproduction grow like a geometric progression provided unlimited resources are available. The resources, however, are either constant or grow – as Malthus assumes – according to an arithmetic progression if human endeavor is involved. The production of nutrition, says Malthus, is proportional to the land that is exploitable for agriculture and the gain in the area of fields will be a constant in time – the increase will be the same every year. An inevitable result of Malthus' vision of the world is the pessimistic view that populations will grow until the majority of individuals will die premature of malnutrition and hunger. Malthus could not foresee the *green revolutions* but he was also unaware that population growth can be faster than exponential – sufficient nutrition for the entire human population is still a problem. Charles Darwin and his younger contemporary Alfred Russel Wallace were strongly influenced by Robert Malthus and took form population theory that in the wild, where birth control does not exist and individuals fight for food, the major fraction of progeny will die before they reach the age of reproduction and only the strongest will have a chance to multiply.

Leonhard Euler introduced the notions of the exponential function in the middle of the eighteenth century [17] and set the stage for modelling populations by means of differential equations. Simple reproduction results in exponential growth of a population with $N(t)$ individuals:

$$\frac{\mathrm{d}N}{\mathrm{d}t} = rN \quad \text{and} \quad N(t) = N_0 \exp(rt) \quad \text{with} \quad N_0 = N(0)\,, \tag{2.4}$$

where the parameter r is commonly called Malthus or growth parameter.

Presumably not known to Darwin, the mathematician Pierre François Verhulst complemented the concept of exponential growth by the introduction of finite resources [18, 19, 20]. The Verhulst equation is of the form[2]

$$\frac{\mathrm{d}N}{\mathrm{d}t} = rN\left(1 - \frac{N}{K}\right)\,, \tag{2.5}$$

where $N(t)$ again denotes the number of individuals of a species $\mathcal{X}$, and K is the *carrying capacity* of the ecological niche or the ecosystem. Equation (2.5) can be

[2]The Verhulst equation is also called logistic equation and its discrete analogue is the logistic map, a standard model to demonstrate the occurrence of deterministic chaos in a simple system. The name *logistic equation* was coined by Verhulst himself in 1845.

integrated by means of partial fractions and reads

$$N(t) = N_0 \, \frac{K}{N_0 + \big(K - N_0\big)\exp(-rt)} \ . \tag{2.6}$$

Apart from the initial condition N_0, the number of individuals $\mathcal{X}$ at time $t = 0$, the logistic equation has two parameters: (i) the Malthusian parameter or the growth rate r and (ii) the carrying capacity K of the ecological niche or the ecosystem. A population of size N_0 grows exponentially at short times: $N(t) \approx N_0 \exp(rt)$ for $K \gg N_0$ and t sufficiently small. For long times the population size approaches the carrying capacity asymptotically: $\lim_{t\to\infty} N(t) = K$.

The two parameters r and K are taken as criteria to distinguish different evolutionary strategies: Species that are r-selected exploit ecological niches with low density, produce a large number of offspring each of which has a low probability to survive, whereas K-selected species are strongly competing in crowded niches and invest heavily in few offspring that have a high probability of survival to adulthood. The two cases, r- and K-selection, are the extreme situations of a continuum of mixed selection strategies. In the real world the r-selection strategy is an appropriate adaptation to fast changing environments, whereas K-selection pays in slowly varying or constant environments.

2.2. The selection equation

The logistic equation can be interpreted in a different way that is useful for the forthcoming analysis: In the second term – $-(N/K)\,rN$ – the expression rN/K is identified with a constraint for limiting growth: $rN/K \equiv \phi(t)$,

$$\frac{\mathrm{d}N}{\mathrm{d}t} = N\Big(r - \phi(t)\Big) \ , \tag{2.5$'$}$$

The introduction of $\phi(t)$ gives room for other interpretations of constraints than carrying capacities of ecosystems. For example, $\phi(t)$ may be a dilution flux in laboratory experiments on evolution in flow reactors [21, pp. 21–27].

Equation (2.5′) can be used now for the derivation of a selection equation in the spirit of Darwin's theory. The single species $\mathcal{X}$ is replaced by several variants forming a population, $\Pi = \{\mathcal{X}_1, \mathcal{X}_2, \ldots, \mathcal{X}_n\}$; in the language of chemical kinetics competition and selection are readily cast into a reaction mechanism consisting of n independent, simple replication reactions:

$$(\mathsf{M}) + \mathcal{X}_j \;\xrightarrow{\;\;f_j\;\;}\; 2\,\mathcal{X}_j \ , \quad j = 1, 2, \ldots, n \ . \tag{2.7}$$

The symbol M denotes the material from which $\mathcal{X}_j$ is synthesized (It is put in parentheses, because we assume that it is present in access and its concentration is constant therefore). The numbers of individuals of the variants are denoted by $N_j(t)$, or in vector notation $\mathbf{N}(t) = \big(N_1(t), N_2(t), \ldots, N_n(t)\big)$ with $\sum_{i=1}^n N_i(t) = C(t)$. The same common carrying capacity is defined for all n variants:

$$\lim_{t\to\infty} \sum_{i=1}^{n} N_i(t) = \lim_{t\to\infty} C(t) = K \ .$$

The Malthus parameters are given here by the fitness values $f_1, f_2, \ldots, f_n$, respectively. For individual species the differential equations take on the form

$$\frac{\mathrm{d}N_j}{\mathrm{d}t} = N_j\left(f_j - \frac{C}{K}\phi(t)\right)\,;\; j = 1, 2, \ldots, n \quad \text{with}$$
$$\phi(t) = \frac{1}{C}\sum_{i=1}^{n} f_i N_i(t) \tag{2.8}$$

being the mean fitness of the population. Summation over all species yields a differential equation for the total population size

$$\frac{\mathrm{d}C}{\mathrm{d}t} = C\left(1 - \frac{C}{K}\right)\phi(t)\,. \tag{2.9}$$

Stability analysis is straightforward: From $\mathrm{d}C/\mathrm{d}t = 0$ follow two stationary states of Equation (2.9): (i) $\bar{C} = 0$ and (ii) $\bar{C} = K$.[3] For conventional stability analysis we calculate the (1×1) Jacobian and obtain for the eigenvalue

$$\lambda = \frac{\partial(\mathrm{d}C/\mathrm{d}t)}{\partial C} = \phi(t) - \frac{C}{K}\left(2\phi(t) - K\frac{\partial\phi}{\partial C}\right) - \frac{C^2}{K}\frac{\partial\phi}{\partial C}\,.$$

Insertion of the stationary values yields $\lambda_{(i)} = \phi > 0$ and $\lambda_{(ii)} = -\phi < 0$, state (i) is unstable and state (ii) is asymptotically stable. The total population size converges to the value of the carrying capacity, $\lim_{t\to\infty} C(t) = \bar{C} = K$.

Equation (2.9) can be solved exactly yielding thereby an expression that contains integration of the constraint $\phi(t)$:

$$C(t) = C(0)\frac{K}{C(0) + \big(K - C(0)\big)\exp(-\Phi)} \quad \text{with} \quad \Phi = \int_0^t \phi(\tau)\mathrm{d}\tau\,,$$

where $C(0)$ is the population size at time $t = 0$. The function $\Phi(t)$ depends on the distribution of fitness values within the population and its time course. For $f_1 = f_2 = \cdots = f_n = r$ the integral yields $\Phi = rt$ and we retain Equation (2.6). In the long time limit Φ grows to infinity and $C(t)$ converges to the carrying capacity K.

At constant population size $C = \bar{C} = K$ Equation (2.8) becomes simpler

$$\frac{\mathrm{d}N_j}{\mathrm{d}t} = N_j\Big(f_j - \phi(t)\Big)\,;\quad j = 1, 2, \ldots, n\,. \tag{2.8$'$}$$

and can be solved exactly by means of the integrating factor transformation [22, p. 322ff.]:

$$Z_j(t) = N_j(t)\exp\left(\int_0^t \phi(\tau)\,\mathrm{d}\tau\right) \quad \text{or} \quad \mathbf{Z}(t) = \mathbf{N}(t)\exp\left(\int_0^t \phi(\tau)\,\mathrm{d}\tau\right),$$

[3] There is also a third stationary state defined by $\phi = 0$. For strictly positive fitness values, $f_i > 0\,\forall\, i = 1, 2, \ldots, n$, this condition can only be fulfilled by $N_i = 0\,\forall\, i = 1, 2, \ldots, n$, which is identical to state (i). If some f_i values are zero – corresponding to lethal variants – the respective variables vanish in the infinite time limit because of $\mathrm{d}N_i/\mathrm{d}t = -\phi(t)\,N_i$ with $\phi(t) > 0$.

solving the linear differential equation $\mathrm{d}\mathbf{Z}/\mathrm{d}t = \mathrm{F}\cdot\mathbf{Z}$, where F is a diagonal matrix containing the fitness values f_j $(j = 1, 2, \ldots, n)$ as elements, and using the trivial equality $\mathbf{Z}_0 = \mathbf{N}_0$ we obtain for the components of $\mathbf{N}$:

$$N_j(t) = N_j(0)\exp(f_j t)\,\frac{C}{\sum_{i=1}^{n} N_i(0)\exp(f_i t)};\ j = 1, 2, \ldots, n\,. \tag{2.10}$$

Equation (2.10) encapsulates Darwinian selection and will be discussed in detail in Section 2.4.

2.3. Variable population size

Now we shall show that the solution of Equation (2.8) describes internal equilibration for constant **and** variable population sizes as long as the population does neither explode nor die out [23]. The validity of theorem 2.2 that will be proven below is not restricted to constant fitness values f_j and hence we can replace them by general growth functions $G_j(N_1, \ldots, N_n) = G_j(\mathbf{N})$ or fitness functions $F_j(\mathbf{N})$ with $G_j(\mathbf{N}) = F_j(\mathbf{N})N_j$ in the special case of replicator equations [24]: $\mathrm{d}N_j/\mathrm{d}t = N_j\big(F_j(\mathbf{N}) - \Psi(t)\big)$ where $\Psi(t)$ comprises both, variable total concentration and constraint.

Time-dependence of the conditions in the ecosystem can be introduced in two ways: (i) variable carrying capacity, $K(t) = \bar{C}(t)$, and (ii) a constraint or flux[4] $\varphi(t)$, where flux refers to some specific physical device, for example to a flow reactor. Constraints and fluxes may correspond to unspecific or specific migration.[5] Considering time-dependent carrying capacity and variable constraints simultaneously, we obtain

$$\frac{\mathrm{d}N_j}{\mathrm{d}t} = G_j(\mathbf{N}) - \frac{N_j}{K(t)}\varphi(t);\ j = 1, 2, \ldots, n\,. \tag{2.11}$$

Summation over all variants $\mathcal{X}_j$ and restricting the analysis to an equilibrated total concentration $C \approx \bar{C} = K$ yields a relation between the time-dependencies of flux and total concentration:

$$\begin{aligned} \varphi(t) &= \sum_{i=1}^{n} G_j(\mathbf{N}) - \frac{\mathrm{d}C}{\mathrm{d}t} \quad \text{or} \\ C(t) &= C(0) + \int_0^t \left(\sum_{i=1}^{n} G_j(\mathbf{N}) - \varphi(\tau)\right)\mathrm{d}\tau\,. \end{aligned} \tag{2.12}$$

Theorem 2.2 (Equilibration in populations of variable size). *Evolution in populations of changing size approaches the same internal equilibrium as evolution in*

---

[4]There is a slight difference in the definitions of the fluxes ϕ and φ: $\phi(t) = \varphi(t)/C(t)$.

[5] *Unspecific migration* means that the numbers N_j of individuals for each variant $\mathcal{X}_j$ decrease (or increase) proportional to the numbers of individuals currently present in the population, $\mathrm{d}N_j = kN_j\mathrm{d}t$. *Specific migration* is anything else. In a flow reactor, for example, we have a dilution flux corresponding to unspecific emigration and an influx of one or a few molecular species corresponding to specific immigration into the reactor.

populations of constant size provided the growth functions are homogeneous functions of degree γ in the variables N_j. Up to a transformation of the time axis, stationary and variable populations have identical trajectories provided the population size stays finite and does not vanish.

Proof. Normalized variables, $x_i = N_i/C$ with $\sum_{i=1}^{n} x_i = 1$, are introduced in order to separate of population growth, $C(t)$, and population internal changes in the distribution of variants $\mathcal{X}_i$. From Equations (2.11) and (2.12) with $C = \bar{C} = K$ and $N_j = Cx_j$ follows:

$$\frac{\mathrm{d}x_j}{\mathrm{d}t} = \frac{1}{C}\left(G_j(C\mathbf{x}) - x_j \sum_{i=1}^{n} G_i(C\mathbf{x})\right) \; ; \; j = 1, 2, \ldots, n \; . \tag{2.13}$$

The growth functions are assumed to be homogeneous of degree γ in the variables[6] N_j: $G_j(\mathbf{N}) = G_j(C\mathbf{x}) = C^{\gamma}\, G_j(\mathbf{x})$. and we find

$$\frac{1}{C^{\gamma-1}} \frac{\mathrm{d}x_j}{\mathrm{d}t} = G_j(\mathbf{x}) \; - \; x_j \sum_{i=1}^{n} G_i(\mathbf{x}); \; j = 1, 2, \ldots, n \; ,$$

which is identical to the selection equation in normalized variables for $C = 1$. For $\gamma = 1$ the concentration term vanishes and populations of constant and variable size have identical trajectories and equilibrium points. In case $\gamma \neq 1$ the two systems are the same up to a transformation of the time axis:

$$\mathrm{d}\tilde{t} = C^{\gamma-1}\, \mathrm{d}t \quad \text{and} \quad \tilde{t} = \tilde{t}_0 + \int_{\tilde{t}_0}^{\tilde{t}} C^{\gamma-1}(t)\, \mathrm{d}t \; ,$$

where $\tilde{t}_0$ is the time corresponding to $t = 0$ (commonly $\tilde{t}_0 = 0$). From Equation (2.13) we expect instabilities at $C = 0$ and $C = \infty$. □

The instability at vanishing population size, $\lim C \to 0$, is of practical importance for modelling drug action on viral replication. In the case of lethal mutagenesis [26, 27] medication aims at eradication of the virus population, $C \to 0$, in order to terminate the infection of the host. At the instant of virus extinction Equation (2.8) is no longer applicable. More details will be discussed in Section 3.7.

2.4. Optimization

Since systems with growing and stationary population size are identical for homogeneous growth function of degree $\gamma = 1$ by Theorem 2.2, we shall use normalized or internal coordinates except in Subsection 3.7. The ODE is of the form

$$\begin{aligned} &\frac{\mathrm{d}x_j}{\mathrm{d}t} = f_j x_j - x_j \phi(t) = x_j\Big(f_j - \phi(t)\Big); \; j = 1, 2, \ldots, n \quad \text{with} \\ &\phi(t) = \sum\nolimits_{i=1}^{n} f_i x_i \; , \end{aligned} \tag{2.14}$$

[6]The degree γ is determined by the mechanism of reproduction. For sexual reproduction according to Ronald Fisher's selection equation (2.19) [25] we have $\gamma = 2$. Asexual reproduction discussed here fulfils $\gamma = 1$.

the solution is derived in the same way as in case of Equation (2.8):

$$x_j(t) = \frac{x_j(0)\exp(f_j t)}{\sum_{i=1}^{n} x_i(0)\exp(f_i t)}\,;\; j = 1, 2, \ldots, n\,. \tag{2.15}$$

The use of normalized variables, $\sum_{i=1}^{n} x_i = 1$, implies that the unit simplex, $\mathbb{S}_n^{(1)} = \{0 \le x_i \le 1\, \forall\, i = 1, \ldots, n \wedge \sum_{i=1}^{n} x_i = 1\}$, is the physically accessible domain. All boundaries of the simplex – corners, edges, faces, etc. – are invariant sets, since $x_j = 0 \Rightarrow \mathrm{d}x_j/\mathrm{d}t = 0$ by Equation (2.14).

For a discussion of selection and optimization we shall assume here that all fitness values f_j are different and that without loosing generality we rank them:

$$f_1 > f_2 > \cdots > f_{n-1} > f_n\,. \tag{2.16}$$

The variables $x_j(t)$ fulfil two time limits:

$$\lim_{t\to 0} x_j(t) = x_j(0)\ \forall\ j = 1, 2, \ldots, n \ \text{ by definition, and}$$

$$\lim_{t\to\infty} x_j(t) = \begin{cases} 1 \text{ iff } j = 1 \\ 0\ \forall\ j = 2, \ldots, n\,. \end{cases}$$

In the long time limit the population becomes homogeneous and contains only the fittest genotype $\mathcal{X}_1$. The process of selection is illustrated best by differential fitness, $f_j - \phi(t)$, the second factor in the ODE (2.14): The constraint $\phi(t) = \sum_{i=1}^{n} f_i x_i = \bar{f}$ represents the mean fitness of the population. The population variables x_l of all variants with a fitness below average, $f_l < \phi(t)$, decrease whereas the variables x_h with $f_h > \phi(t)$ increase. As a consequence the average fitness $\phi(t)$ in increasing too and more sequences fall below the threshold for survival. The process continues until the fittest variant is selected.

Optimization of mean fitness can be proved also without referring to differential fitness:

Theorem 2.3 (Optimization of mean fitness). *The mean fitness*

$$\phi(t) = \bar{f} = \sum\nolimits_{i=1}^{n} f_i x_i \quad \textit{with} \quad \sum\nolimits_{i=1}^{n} x_i = 1$$

in a population as described by Equation (2.14) *is non-decreasing.*

Proof. The time-dependence of the mean fitness or flux ϕ is given by

$$\begin{aligned} \frac{\mathrm{d}\phi}{\mathrm{d}t} &= \sum_{i=1}^{n} f_i \dot{x}_i = \sum_{i=1}^{n} f_i \left(f_i x_i - x_i \sum_{j=1}^{n} f_j x_j \right) = \\ &= \sum_{i=1}^{n} f_i^2 x_i - \sum_{i=1}^{n} f_i x_i \sum_{j=1}^{n} f_j x_j \ = \\ &= \overline{f^2} - \left(\bar{f}\right)^2 = \mathrm{var}\{f\} \ge 0\,. \end{aligned} \tag{2.17}$$

Since a variance is always nonnegative, Equation (2.17) implies that $\phi(t)$ is a non-decreasing function of time. □

The condition $\mathrm{var}\{f\} = 0$ is met only by homogeneous populations. The one containing only the fittest variant $\mathcal{X}_1$ has the largest possible mean fitness: $\bar{f} = \phi_{\max} = f_1 = \max\{f_j; j = 1, 2, \ldots, n\}$. ϕ cannot increase any further and hence, it was been optimized by the selection process. The state of maximal fitness of population $\Pi = \{\mathcal{X}_1, \ldots, \mathcal{X}_n\}$, $\mathbf{x}|_{\max\{\phi(\Pi)\}} = \{x_1 = 1, x_i = 0 \,\forall\, i = 2, \ldots, n\} = \mathbf{P}_1$, is the unique stable stationary state, and all trajectories starting from initial conditions with nonzero amounts of $\mathcal{X}_1$, $x_1 > 0$, have $\mathbf{P}_1$ as ω-limit. An illustration of the selection process with three variants is shown in Figure 2.1: The trajectories are plotted on the unit simplex $\mathbb{S}_3^{(1)}$.

Gradient systems [28, p. 199] facilitate the analysis of the dynamics, they obey the equation

$$\frac{\mathrm{d}\mathbf{x}}{\mathrm{d}t} = -\mathrm{grad}\{V(\mathbf{x})\} = -\nabla V(\mathbf{x}) \tag{2.18}$$

and fulfil criteria that are relevant for optimization:

(i) The eigenvalues of the linearization of (2.18) evaluated at the equilibrium point are real.
(ii) If $\bar{\mathbf{x}}_0$ is an isolated minimum of V then $\bar{\mathbf{x}}_0$ is an asymptotically stable solution of (2.18).
(iii) In $\mathbf{x}(t)$ is a solution of (2.18) that is not an equilibrium point, then $V(\mathbf{x}(t))$ is a strictly decreasing function and the trajectories are perpendicular to the constant level sets of V.
(iv) Neither periodic nor chaotic solutions of (2.18) do exist.

As easily seen from Figure 2.1 the trajectories of (2.14) are not perpendicular to the constant level sets of $\phi(\mathbf{x})$ and hence, Equation (2.14) is not a gradient system in the strict sense. With the definition of a generalized inner product corresponding to a Riemannian metric [29], however, the selection equation can be visualized as a generalized gradient and oscillations or deterministic chaos can be excluded [30].

2.5. Fisher's selection equation

Sexual reproduction introduces obligatory recombination of genotypes into the selection equation. A sexually reproducing organism carries two copies of every gene. Contrasting asexual reproduction where the whole genome is replicated and transferred to progeny in one piece, sexual reproduction is accompanied by partitioning the two parental genomes into pieces and by organized recombination of genes into the genomes of the progeny.[7] On the population level genes are chosen from a set of n variants called the gene pool: $\mathcal{A} = \{A_1, \ldots, A_n\}$. Every specific (single) locus on the genome is occupied by two alleles. Ronald Fisher, the great scholar

[7] *Organized* means here that each offspring carries two *alleles* for every gene. An *allele* A_j is a variant of a specific gene. One allele of the offspring is one of the two maternal alleles, the second one comes from the father who also carries two alleles of the gene. The position of a gene on the genome or chromosome is called the *locus*. The genome consists of several chromosomes. There are two classes of chromosomes: autosomes and sex chromosomes. In humans the males carry two different sex chromosomes, X and Y, with different genes and this constitutes an exception of the two alleles per gene rule.

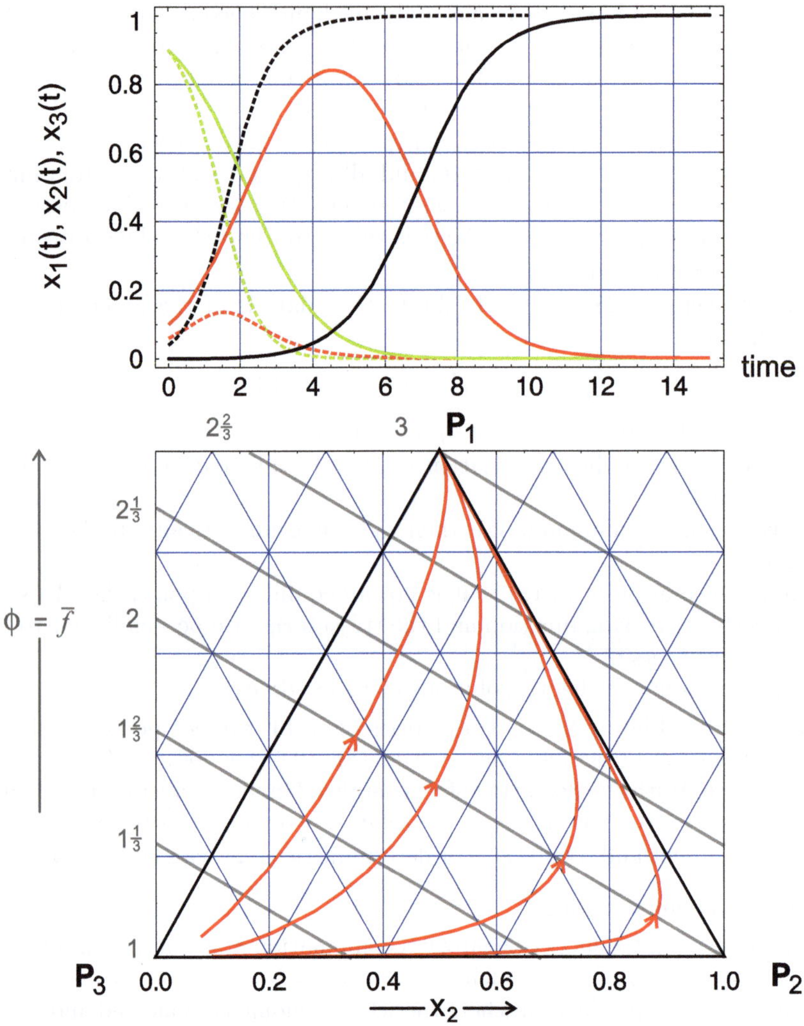

FIGURE 2.1. **Selection on the unit simplex.** In the upper part of the figure we show solution curves $\mathbf{x}(t)$ of Equation (2.15) with $n = 3$. The parameter values are: $f_1 = 3\,[\mathrm{t}^{-1}]$, $f_2 = 2\,[\mathrm{t}^{-1}]$, and $f_3 = 1\,[\mathrm{t}^{-1}]$, where $[\mathrm{t}^{-1}]$ is an arbitrary reciprocal time unit. The two sets of curves differ with respect to the initial conditions:

(i) $\mathbf{x}(0) = (0.02, 0.08, 0.90)$, dotted curves, and

(ii) $\mathbf{x}(0) = (0.0001, 0.0999, 0.9000)$, full curves.

Color code: $x_1(t)$ black, $x_2(t)$ red, and $x_3(t)$ green. The lower part of the figure shows parametric plots $\mathbf{x}(t)$ on the unit simplex $\mathbb{S}_3^{(1)}$. Constant level sets of $\phi(\mathbf{x}) = \bar{f}$ are shown in grey.

of population genetics, presented the first mathematical unification of Darwin's theory of natural selection and Mendel's laws of inheritance [25]. Since population genetics will be treated extensively in other contributions, only a very brief account on Fisher's the selection equation and his fundamental theorem of natural selection will be given here. The selection equation describes the evolution of the allele distribution at a single locus:

$$\frac{\mathrm{d}x_j}{\mathrm{d}t} = x_j \left(\sum_{i=1}^{n} a_{ji}\, x_i - \phi(t) \right) \; ; \; j = 1, \ldots, n \quad \text{with}$$
$$\phi(t) = \sum_{j=1}^{n} \sum_{i=1}^{n} a_{ji}\, x_i\, x_j \quad \text{and} \quad \sum_{i=1}^{n} x_i = 1 \; . \tag{2.19}$$

The (normalized) variables x_j represent the allele frequencies, the parameters a_{ji} represent the fitness values for the allele combination $[A_j \cdot A_i]$, and the constraint $\phi(t)$ conserves the normalization condition. Fisher's selection equation (2.19) is a replicator equation [24] with $F_j(\mathbf{x}) = \sum_{i=1}^{n} a_{ji} x_i$ being a linear function and $G_j(\mathbf{x})$ is homogeneous with $\gamma = 2$. Since the fitness of an allele combination is assumed to be independent of the descendence of the allele on an autosome – it does not matter whether a particular allele stems from the paternal or the maternal chromosome, the allele combinations $[A_j \cdot A_i]$ and $[A_i \cdot A_j]$ have identical fitness, and the matrix $\mathrm{A} = \{a_{ji}; i, j = 1, 2, \ldots, n\}$ is symmetric.

The introduction of mean rate parameters $\bar{a}_i = \sum_{j=1}^{n} a_{ij} x_j$ facilitates the forthcoming calculation. The time-dependence of ϕ is now given by

$$\begin{aligned}
\frac{\mathrm{d}\phi}{\mathrm{d}t} &= \sum_{i=1}^{n} \sum_{j=1}^{n} a_{ij} \left(\frac{\mathrm{d}x_i}{\mathrm{d}t} \cdot x_j + x_i \cdot \frac{\mathrm{d}x_j}{\mathrm{d}t} \right) = 2 \sum_{i=1}^{n} \sum_{j=1}^{n} a_{ji} \cdot x_i \cdot \frac{\mathrm{d}x_j}{\mathrm{d}t} \\
&= 2 \sum_{i=1}^{n} \sum_{j=1}^{n} a_{ji} \cdot x_i \left(\sum_{k=1}^{n} a_{jk} x_j x_k - x_j \sum_{k=1}^{n} \sum_{\ell=1}^{n} a_{k\ell} x_k x_\ell \right) \\
&= 2 \sum_{j=1}^{n} x_j \sum_{i=1}^{n} a_{ji} x_i \sum_{k=1}^{n} a_{jk} x_k - 2 \sum_{j=1}^{n} x_j \sum_{i=1}^{n} a_{ji} x_i \sum_{k=1}^{n} x_k \sum_{\ell=1}^{n} a_{k\ell} x_\ell \\
&= 2 \left(\langle \bar{a}^2 \rangle - \langle \bar{a} \rangle^2 \right) = 2 \; \mathrm{var}\{\bar{a}\} \geq 0 \; .
\end{aligned} \tag{2.20}$$

Again we see that the constraint $\phi(t)$ is a non-deceasing function of time, and it approaches an optimal value on the simplex. This result is often called Fisher's fundamental theorem of evolution (see, for example, [31]).

The physically relevant part of $\mathbb{R}^n$ with non-negative values for all variables is the unit simplex $\mathbb{S}_n^{(1)}$ and – as in case of the selection equation (2.14) for asexual reproduction – all boundary sets of the simplex are invariant for Fisher's selection equation (2.19). Both equations, (2.14) and (2.19), have also in common that the selection constraint $\phi(t)$ is a non-decreasing function of time t, and accordingly $\phi(t)$ is optimized. There is, however, also an important difference between the two selection equations: The selected state is unique in the asexual case, whereas

Fisher's selection equation may converge to different states for different initial conditions. A straightforward example is sufficiently large fitness of the homozygotes $[A_i \cdot A_i]$ and $[A_j \cdot A_j]$ compared to the heterozygote $[A_j \cdot A_i]$, $(a_{ii}, a_{jj}) > a_{ij}$. Then, either of the two homozygotes – corresponding to the corners $\mathbf{P}_i$ and $\mathbf{P}_j$ of the simplex – may be selected and the outcome of the selection process depends on the initial state $\mathbf{x}(0) = \mathbf{x}_0$. In case the heterozygote has higher fitness than the homozygotes,[8] $a_{ij} > (a_{ii}, a_{jj})$, Fisher's selection equation (2.19) predicts that the selected stationary allele distribution becomes 0.5 : 0.5. Heterozygote selection in real populations is more complicated, since a homogeneous population of heterozygotes is not compatible with random mating. In this case the resulting diploid combinations, $[A_i \cdot A_i]$, $[A_j \cdot A_i]$ and $[A_j \cdot A_j]$, are formed with a ratio of 0.25:0.5:0.25. Homozygotes will not vanish completely unless they are lethal. A real world example of overdominance with one practically lethal homozygote is sickle-cell anemia (for an overview of the highly complex phenotypes of this disease see [32, 33]).

In the simple version presented here, Fisher's fundamental theorem of natural selection is identified with optimization of $\phi(t)$ in the selection equation (2.19). Unfortunately – but fortunately for population geneticists and theoretical biologists, because a whole plethora of interesting mathematical problems emerged and still emerges from the mathematics of recombination – Fisher's selection equation is a single locus model and holds on the genome level only for the unrealistic assumption of independent genes. Two and more locus models with gene interaction turned out to be much more complicated and no generally valid optimization principle was reported so far: Natural selection in the sense of Charles Darwin is an extremely powerful optimization heuristic but no theorem (see also Subsection 3.3). Nevertheless, Fisher's fundamental theorem is much deeper than the toy version that has been presented here. The interested reader is referred to a few, more or less arbitrarily chosen references from the enormous literature on this issue [34, 35, 36].

2.6. Growth functions and selection

It is worth considering different classes of growth functions $z(t)$ and the behavior of long time solutions of the corresponding ODEs. An intimately related problem concerns population dynamics: What is the long time or equilibrium distribution of genotypes in a normalized population, $\lim_{t\to\infty} \mathbf{x}(t)$ provided the initial distribution has been $\mathbf{x}_0$? Is there a universal long time behavior, for example selection, coexistence or cooperation, that is characteristic for certain classes of growth functions?

The differential equation describing unlimited growth,

$$\frac{\mathrm{d}z}{\mathrm{d}t} = f \cdot z^n \tag{2.21}$$

[8]The two situations are called *underdominance* or homozygote advantage and *overdominance* or heterozygote advantage, respectively.

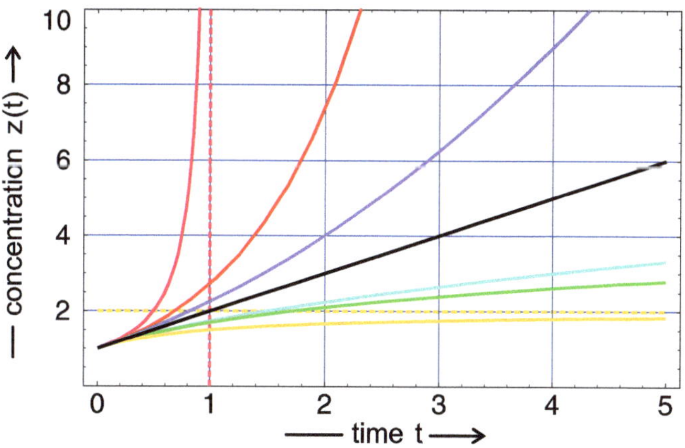

FIGURE 2.2. **Typical functions describing unlimited growth**. All functions are normalized in order to fulfil the conditions $z_0 = 1$ and $\mathrm{d}z/\mathrm{d}t|_{t=0} = 1$. The individual curves show hyperbolic growth ($z(t) = 1/(1-t)$; magenta; the dotted line indicates the position of the instability), exponential growth ($z(t) = \exp(t)$; red), parabolic growth ($z(t) = (1+t/2)^2$; blue), linear growth ($z(t) = 1+t$; black), sublinear growth ($z(t) = \sqrt{1+2t}$; turquoise), logarithmic growth ($z(t) = 1 + \log(1+t)$; green), and sublogarithmic growth ($z(t) = 1 + t/(1+t)$; yellow; the dotted line indicates the maximum value $z_{\max}$: $\lim_{t\to\infty} z(t) = z_{\max}$).

yields two types of general solutions for the initial value $z(0) = z_0$

$$z(t) = \left(z_0^{1-n} + (1-n)ft\right)^{1/(1-n)} \qquad \text{for } n \neq 1 \text{ and} \tag{2.21a}$$

$$z(t) = z_0 \cdot e^{ft} \qquad \text{for } n = 1\,. \tag{2.21b}$$

In order to make the functions comparable we normalize them in order to fulfil $z_0 = 1$ and $\mathrm{d}z/\mathrm{d}t|_{t=0} = 1$. According to Equations (2.21) this yields $z_0 = 1$ and $f = 1$. The different classes of growth functions indicated by different colors in Figure 2.2 are characterized by the following behavior:

(i) Hyperbolic growth requires $n > 1$; for $n = 2$ it yields the solution curve of the $z(t) = 1/(1-t)$. Characteristic is the existence of an instability in the sense that $z(t)$ approaches infinity at some critical time, $\lim_{t\to t_{\mathrm{cr}}} z(t) = \infty$ with $t_{\mathrm{cr}} = 1$. The selection behavior of hyperbolic growth is illustrated by the Schlögl model:[9] $\mathrm{d}z_j/\mathrm{d}t = f_j z_j^2$; $j = 1,2,\ldots,n$. Depending on the initial conditions each of the replicators $\mathcal{X}_j$ can be selected. $\mathcal{X}_m$ the species with the highest replication parameter, $f_m = \max\{f_i; i = 1,2,\ldots,n\}$ has the largest basin of attraction and the highest probability to be selected. After selection has occurred a new species $\mathcal{X}_k$ is extremely unlikely to replace the current

[9] The Schlögl model is tantamount to Fisher's selection equation with diagonal terms only: $f_j = a_{jj}$; $j = 1,2,\ldots,n$ [37].

species $\mathcal{X}_m$ even if its replication parameter is substantially higher, $f_k \gg f_m$. This phenomenon is called *once-for-ever selection.*

(ii) Exponential growth is observed for $n = 1$ and described by the solution $z(t) = e^t$. It represents the most common growth function in biology. The species X_m having the highest replication parameter, $f_m = \max\{f_i; i = 1, 2, \ldots, N\}$, is always selected, $\lim_{t\to\infty} z_m = 1$. Injection of a new species $\mathcal{X}_k$ with a still higher replication parameter, $f_k > f_m$, leads to selection of the fitter variant $\mathcal{X}_k$.

(iii) Parabolic growth occurs for $0 < n < 1$ and for $n = 1/2$ has the solution curve $z(t) = (1 - t/2)^2$. It is observed, for example, in enzyme free replication of oligonucleotides that form a stable duplex, i.e., a complex of one plus and one minus strand [38]. Depending on parameters and concentrations coexistence or selection may occur [39].

(iv) Linear growth follows from $n = 0$ and takes on the form $z(t) = 1 + t$. Linear growth is observed, for example, in replicase catalyzed replication of RNA at enzyme saturation [40].

(v) Sublinear growth occurs for $n < 0$. In particular, for $n = -1$ gives rise to the solution $y(t) = (1 + 2t)^{1/2} = \sqrt{1 + 2t}$.

In addition we mention also two additional forms of weak growth that do not follow from Equation (2.21):

(vi) Logarithmic growth that can be expressed by the function $z(t) = z_0 + \ln(1 + ft)$ or $z(1) = 1 + \ln(1 + t)$ after normalization, and

(vii) sublogarithmic growth modeled by the function $z(t) = z_0 + ft/(1 + ft)$ or $z(t) = 1 + t/(1 + t)$ in normalized form.

Hyperbolic growth, parabolic growth, and sublinear growth constitute families of solution curves that are defined by a certain parameter range (see figure 2.2), for example a range of exponents, $n_{\text{low}} < n < n_{\text{high}}$, whereas exponential growth, linear growth and logarithmic growth are critical curves separating zones of characteristic growth behavior: Logarithmic growth separates growth functions approaching infinity in the limit $t \to \infty$, $\lim_{t\to\infty} z(t) = \infty$ from those that remain finite, $\lim_{t\to\infty} z(t) = z_\infty < \infty$, linear growth separates concave from convex growth functions, and exponential growth eventually separates growth functions that reach infinity at finite times from those that don't.

3. Mutation, selection, and optimization

Molecular biology was initiated when Watson and Crick published their centennial paper on the structure of deoxyribonucleic acid (DNA) [41]. Further development provided information on the chemistry of life at a breathtaking pace [42]. A closer look on the structure of DNA revealed the discrete nature of base pairing – two nucleotides make a base pair that fits into the double helix or they don't. With this restriction the natural nucleobases allow for only four pairings: AT, TA, GC, and CG. This fact is already sufficient for an rough understanding of the molecular basis of genetics: Genetic information is of digital nature and multiplication of information is tantamount to copying. Mutation, the process that leads to innovation in evolution, was identified with imperfect reproduction or erroneous copying. Molecular insights into recombination occurring during meiotic cell division[10] provided straightforward explanations for the deviations from Mendel's idealized ratios of offspring with different appearance.

3.1. Complementary replication of nucleic acids

The device for cellular DNA replication is highly involved as the replication complex consists of more than twenty protein enzymes performing a concerted reaction that makes two double stranded molecules from one double stranded molecule. Simpler DNA replication and common RNA replication use the principle of complementary strand template completion: A single strand is completed to a double helix nucleotide after nucleotide (Figure 3.1) whereby the complementary strand is synthesized leading to the mechanism for complementary synthesis (in the unlimited growth case):

$$(M) + \mathcal{X}^- \xrightarrow{f^-} \mathcal{X}^+ + \mathcal{X}^- \quad \text{and}$$

$$(M) + \mathcal{X}^+ \xrightarrow{f^+} \mathcal{X}^- + \mathcal{X}^+ \tag{3.1}$$

$$\text{lead to} \quad \frac{dz^+}{dt} = f^- z^- \quad \text{and} \quad \frac{dz}{dt} = f^+ z^+ \, .$$

Transformation of variables,

$$\zeta = \frac{z^+}{\sqrt{f^+}} + \frac{z^-}{\sqrt{f^-}} \quad \text{and} \quad \eta = \frac{z^+}{\sqrt{f^+}} - \frac{z^-}{\sqrt{f^-}} \, ,$$

yields the solutions

$$\eta(t) = \eta(0) \exp(-ft) \quad \text{and} \quad \zeta(t) = \zeta(0) \exp(+ft) \tag{3.1'}$$

with $f = \sqrt{f^+ f^-}$. The combined variable $\eta(t)$ describes the internal equilibration of plus- and minus-strand and $\zeta(t)$ is the growth function of the plus-minus ensemble. It is worth noticing that the fitness value of the ensemble is the geometric mean of the individual fitness values of the two strands, f^+ and f^-.

Complementary replication of DNA is the basis of the polymerase chain reaction (PCR) [43], which is a standard technique in molecular genetics. RNA replication in cells infected by several classes of RNA viruses and replication in cell free media follow also a complementary replication mechanism and usually involve only one or very few enzymes. The kinetic reaction mechanism of RNA

[10]Two different classes of cell divisions are distinguished: (i) *Mitosis* leading from one *diploid* cell to two diploid cells, where diploid defines the fact that the cells carries two sets of chromosomes, and (ii) *meiosis* leading from one diploid cell to four *haploid* cells constituting the germ line. Haploid expresses the fact that the cell carries only one set of chromosomes. Diploidy of the cells is restored when one paternal and one maternal haploid germ line cell fuse to yield a diploid zygote.

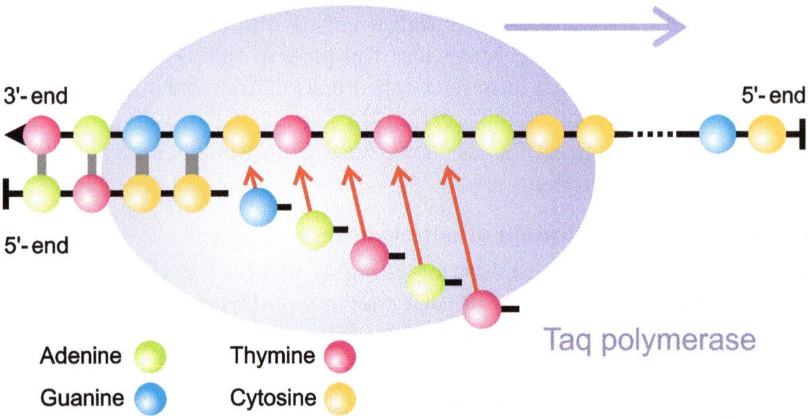

FIGURE 3.1. **Sketch of template induced DNA replication.** Template induced polymerization of nucleic acids (DNA or RNA) follows the same logical principle: The polymerase binds to the $3'$-end of the template molecule and synthesizes the complementary strand in the direction from the $5'$-end to the $3'$-end by adding nucleotide after nucleotide. Considering the incorporation of individual nucleotides as independent reaction steps, the accuracy of correct replication of template $\mathcal{X}_j$ is simply: $Q_{jj} = q_1 \cdot q_2 \cdot \cdots \cdot q_\ell$, where ℓ denotes the chain length of the polynucleotide and q_j is the accuracy of correct incorporation at position "j". An important example of a DNA polymerizing single enzyme is the thermostable DNA polymerase isolated from the bacterium *Thermus aquaticus*. It replicates single stranded DNA and is used commonly for DNA amplification in the polymerase chain reaction (PCR) technique [43].

replication *in vitro* has been studied in great detail [40, 44, 45]: Under suitable conditions, excess replicase and nucleotide triphosphates (ATP, UTP, GTP, and CTP), the concentration of the RNA plus-minus ensemble grows exponentially (Figure 3.2). The population maintains exponential growth if consumed materials are replenished either by a suitable flow device or by serial transfer, which consists of repeated transfer of small quantities of the current reaction mixture into fresh reaction medium [3]. The condition of exponential growth is trivially fulfilled for cellular life, because replication and cell divisions are strongly coupled mechanistically through cellular metabolism, and selection relevant multiplication occurs at the level of cells. With virus replication in the host cell the situation is more involved: Inside the cell replication is exhausting the reservoir of building blocks for the virus specific biomolecules and conditions of exponential growth exist only in the early stages of cellular infection. Formation of virus particles, so-called virions, however, introduces exponential growth and selection of fitter variants.

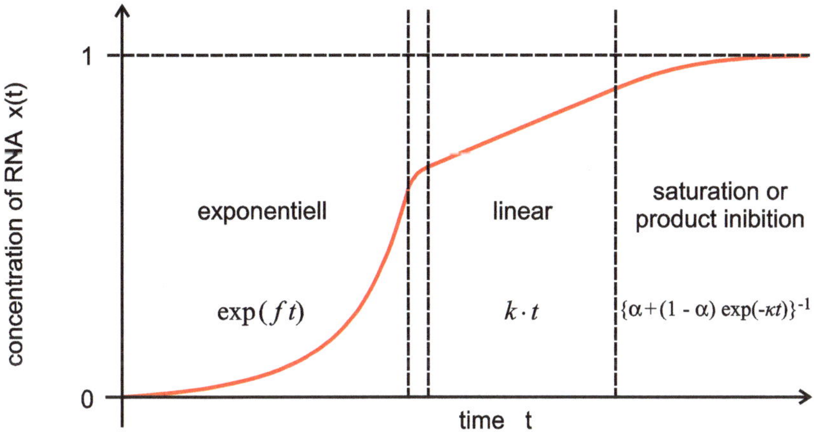

FIGURE 3.2. **Kinetics of RNA replication in a closed system.** The time course of specific RNA replication by $Q\beta$-replicase shows three distinct growth phases: (i) an exponential phase, (ii) a linear phase, and (iii) a phase characterized by saturation through product inhibition [40, 44, 45]. The experiment is initiated by transfer of a very small sample of RNA suitable for replication into a medium containing $Q\beta$-replicase (R) and the activated monomers, ATP, UTP, GTP, and CTP in excess (consumed materials are not replenished in this experiment). In the phase of exponential growth there is shortage of RNA templates, every free RNA molecule is instantaneously bound to an enzyme molecule and replicated, and the corresponding over-all kinetics follows $\mathrm{d}x/\mathrm{d}t = f \cdot x$ resulting in $x(t) = x_0 \cdot \exp(ft)$. In the linear phase the concentration of template is exceeding that of enzyme, every enzyme molecule in engaged in replication, and over-all kinetics is described by $\mathrm{d}x/\mathrm{d}t = k' \cdot e_0^{(\mathrm{R})} = k$, wherein $e_0^{(\mathrm{R})}$ is the total enzyme concentration, and this yields after integration $x(t) = x_0 + kt$. Further increase in RNA concentration slows down the dissociation of product (and template) RNA from the enzyme-RNA complex and leads to a phenomenon known as product inhibition of the reaction. It can be approximated by a Verhulst-type saturation term: $x(t) = \alpha / \big(\alpha + (1-\alpha)\exp(-\kappa t)\big)$. At the end, all enzyme molecules are blocked by RNA in complexes and no more RNA synthesis is possible, $x(t) \to 1$.

The repertoire of naturally occurring changes in genomes is very rich and ranges from point mutations, insertions and deletions to duplications of genes and whole genomes or other large scale genome rearrangements. For the sake of simplicity only single point mutations will be considered here. This, however, is sufficient to be able to reach every sequence from every other sequence of the same length through a finite number of mutation steps. Subsection 3.2 introduces the kinetic

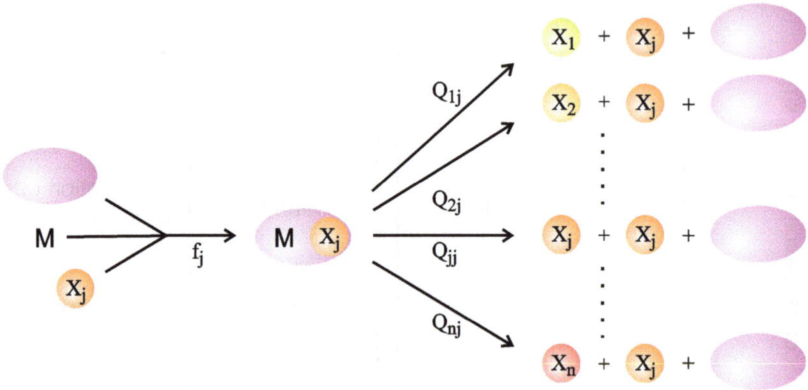

FIGURE 3.3. **A schematic view of replication and mutation.** The replication device (violet) binds the template nucleic acid molecule (DNA or RNA; $\mathcal{X}_j$, orange) and initiates copying of the genetic information with a rate parameter f_j. The reaction has n different channels and yields a correct copy with frequency Q_{jj} or leads through mutation to one of the $(n-1)$ variants $\mathcal{X}_k$ (spectrum of colors) with frequency Q_{kj} whereby $Q_{jj} \gg Q_{kj} \, \forall \, k \neq j$ is required for stable inheritance. Stoichiometry of replication requires $\sum_{i=1}^{n} Q_{ij} = 1$, since the product has to be either correct or incorrect. Replication requires activated monomers M – in nature in form of the deoxynucleoside-triphosphates (dATP, dTTP, dGTP, and dCTP in DNA) or nucleoside-triphosphates (ATP, UTP, GTP, and CTP in RNA) – and suitable reaction conditions.

differential equation for replication and mutation, and derives the solution. Different from selection based on error free replication, optimization of mean fitness $\bar{f} = \phi$ is no longer a global property but restricted to some region on the unit simplex $\mathbb{S}_n^{(1)}$ (Subsection 3.3). Correct reproduction and mutation at the molecular level are seen as parallel chemical reactions (Figure 3.3). In order to guarantee inheritance, correct copying must occur more frequently than mutation. In Subsection 3.4 we shall cast this intuitive statement into a quantitative expression and in Subsection 3.5 the analytical results are supplemented by data from numerical computations. Eventually, we consider neutrality in replication (Subsection 3.6) and lethal mutations (Subsection 3.7).

3.2. The mutation selection equation

Replication leading to correct copies and mutations is properly described by the overall mechanism

$$(\mathsf{M}) + \mathcal{X}_j \xrightarrow{Q_{ij} f_j} \mathcal{X}_i + \mathcal{X}_j \ ; \quad i,j = 1,2,\ldots,n \ , \tag{3.2}$$

that is cast by chemical kinetics into the differential equation:[11]

$$\frac{\mathrm{d}x_j}{\mathrm{d}t} = \sum_{i=1}^{n} Q_{ji} f_i x_i - \phi(t) x_j \, , \; j = 1, 2, \ldots, n \qquad \text{with}$$

$$\phi(t) = \sum_{i=1}^{n} \sum_{j=1}^{n} Q_{ji} f_i x_i = \sum_{i=1}^{n} f_i x_i \, , \tag{3.2$'$}$$

$$\text{or} \quad \frac{\mathrm{d}\mathbf{x}}{\mathrm{d}t} = (\mathrm{Q} \cdot \mathrm{F} - \phi(t))\, \mathbf{x} \quad \text{in vector notation} \, ,$$

where $\mathbf{x}$ is an n-dimensional column vector; Q and F are $n \times n$ matrices. The matrix Q contains the mutation probabilities – Q_{ji} referring to the production of $\mathcal{X}_j$ as an error copy of template $\mathcal{X}_i$ – and F is a diagonal matrix whose elements are the replication rate parameters or fitness values f_j. The product $\mathrm{Q} \cdot \mathrm{F} = \mathrm{W}$ is defined as the *value matrix* W, since it encapsulated the selective values of variants.

The form of the mutation selection equation portrays the schematic mechanism shown in Figures 3.1 and 3.3: Initiation of replication, propagation along the template $\mathcal{X}_j$, and termination are modeled by an overall rate parameter f_j that accounts also for the constant concentrations of building blocks M. After initiation the enzyme progresses stepwise along the polynucleotide chain, each nucleotide incorporation opens κ reaction channels where κ is the number of different nucleotide bases ($\kappa = 2$ holds for binary sequences and $\kappa = 4$ for natural DNA or RNA molecules) and hence, correct replication and mutation are parallel chemical reactions. One reaction channel incorporates the correct nucleotide whereas $\kappa - 1$ channels produce mutants, the matrix elements Q_{ji} with $j = 1, \ldots, n$ determine the probability to obtain $\mathcal{X}_j$ as a (correct or incorrect) copy of $\mathcal{X}_i$ and by conservation of probabilities we have $\sum_{j=1}^{n} Q_{ji} = 1$.

Equation (3.2) can be transformed into a linear ODE by means of integrating factor transformation and is thereby reduced to the following eigenvalue problem [46, 47]:

$$\boldsymbol{z}(t) = \boldsymbol{x}(t) \cdot \exp\left(\int_0^t \phi(\tau) \mathrm{d}\tau \right) \, ,$$

$$\frac{\mathrm{d}\boldsymbol{z}}{\mathrm{d}t} = \mathrm{Q} \cdot \mathrm{F}\, \boldsymbol{z} = \mathrm{W}\, \boldsymbol{z} \, , \quad \text{and}$$

$$\mathrm{W} = \mathrm{B} \cdot \Lambda \cdot \mathrm{B}^{-1} \quad \text{or} \quad \Lambda = \mathrm{B}^{-1} \cdot \mathrm{W} \cdot \mathrm{B} \, ,$$

with Λ being a diagonal matrix, whose elements are the ordered eigenvalues of W, $\lambda_1 \geq \lambda_2 \geq \cdots \geq \lambda_n$. The calculation of the solutions x_j yields by straightforward insertion:

$$x_j(t) = \frac{\sum_{k=1}^{n} b_{jk} \sum_{i=1}^{n} h_{ki} x_i(0) \exp(\lambda_k t)}{\sum_{l=1}^{n} \sum_{k=1}^{n} b_{lk} \sum_{i=1}^{n} h_{ki} x_i(0) \exp(\lambda_k t)} \, , \; j = 1, 2, \ldots, n \, . \tag{3.3}$$

[11] The building blocks M are put in parentheses because they are assumed to be present in excess and therefore do not appear as variables in the ODEs.

The new quantities in this equation are the elements of the two transformation matrices:

$$\mathrm{B} = \{b_{jk}; j,k = 1,\ldots,n\} \quad \text{and}$$
$$\mathrm{B}^{-1} = \{h_{kj}; k,j = 1,2,\ldots,n\}$$

The columns of B and the rows of B^{-1} represent the right-hand and left-hand eigenvectors of the matrix W. For example we have

$$\zeta_1 = \begin{pmatrix} b_{11} \\ b_{21} \\ \vdots \\ b_{n1} \end{pmatrix} .$$

Assuming a unique largest eigenvalue, $\lambda_1 > \lambda_2 \geq \lambda_3 \geq \cdots \geq \lambda_n$ (see Theorem 3.1), the stationary solution contains only the contributions of the largest eigenvector, ζ_1:

$$\lim_{t\to\infty} x_j(t) = \bar{x}_j = \frac{b_{j1}\sum_{i=1}^{n} h_{1i}x_i(0)}{\sum_{k=1}^{n} b_{k1}\sum_{i=1}^{n} h_{1i}x_i(0)} , \ j = 1,\ldots,n . \tag{3.4}$$

In other words, ζ_1 describes the stationary distribution of mutants and represents the genetic reservoir of an asexually reproducing species similarly to the gene pool of a sexual species. For this reason ζ_1 has be called *quasispecies*, its properties will be discussed now in Subsection 3.3.

3.3. Mutation and optimization

Before considering the optimization problem we shall derive the biologically relevant properties of quasispecies.

Theorem 3.1 (Unique and strictly positive quasispecies). *If all single point mutations yield non-lethal variants, the largest eigenvector of matrix* W*, the quasispecies* ζ_1*, will be unique and all elements of* ζ_1 *will be strictly positive.*

Proof. The matrix F is a diagonal matrix with strictly positive elements. The matrix Q contains ℓ strictly positive diagonal elements representing the replication accuracies Q_{jj} ($j = 1,\ldots,\ell$) and strictly positive elements for all single point mutations that can be reached from some specific (initial) sequence $\mathcal{X}_0$ by one replication event. These are all $(\kappa-1)\ell$ sequences $\mathcal{X}_k$ with Hamming distance[12] $d_\mathrm{H}(\mathcal{X}_k,\mathcal{X}_0) = d^\mathrm{H}_{k0} = 1$ and accordingly, Q has $\ell(2\kappa-1)$ strictly positive and $\ell(\ell-2\kappa+1)$ zero entries. Since multiplication with a diagonal matrix leaves zero entries unchanged, $\mathrm{W} = \mathrm{Q}\cdot\mathrm{F}$ has the same zero elements as Q. Cumulative consecutive replication events as described by the mutation matrices $\mathrm{Q}^2, \mathrm{Q}^3, \ldots,$ provide nonzero mutation frequencies for all sequences $\mathcal{X}_k$ of Hamming distances

[12]The Hamming distance counts the number of positions in which two aligned sequences differ[48].

$d^{\mathrm{H}}_{k0} = 2, 3, \ldots$, and eventually after ℓ replications the matrix Q^{ℓ} is strictly positive: Q is a primitive matrix, and so is W.

Accordingly, Perron-Frobenius theorem [49, pp. 3–11] for primitive matrices applies for matrix W and there exists an largest eigenvalue λ_1 such that

(i) λ_1 is real and strictly positive,
(ii) associated with λ_1 are one strictly positive left and one strictly positive right eigenvector, $\tilde{\boldsymbol{\zeta}}_1$ and $\boldsymbol{\zeta}_1$,
(iii) the eigenvectors associated with λ_1 are unique to constant multiples,
(iv) $\lambda_1 > |\lambda_k|$ holds for all $\lambda_k \neq \lambda_1$, and
(v) λ_1 is a simple root of the characteristic equation of W.

Items (ii), (iii), (iv), and (v) meet the conjectures of the theorem. □

In other words, the quasispecies theorem states that the mutation selection problem as modeled by equation (3.2) has a unique long time or equilibrium solution, the quasispecies, and no variant vanishes in the long time limit. Both conditions root the model safely in chemical kinetics and thermodynamics.

The eigenvectors belonging to the eigenvalues λ_k with $k \neq 1$ have no direct physical meaning but they are important for an understanding of optimization in the presence of mutation. For the purpose of illustration we consider $\mathbb{L}^1$ normalized variables on the unit simplex $\mathbb{S}_n^{(1)}$ and assume that the eigenvalues of W are real.[13] A point on $\mathbb{S}_n^{(1)}$ is defined by the vector $\mathbf{x}$, likewise we define a simplex $\boldsymbol{\Sigma}_n^{(1)}$ in the eigenspace of W and its points by the vectors $\boldsymbol{\xi} = (\xi_1, \ldots, \xi_n)$. The unit vectors $\mathbf{P}_j = (x_j = 1, x_i = 0\, \forall\, i \neq j)$ in concentration space are replaced by unit vectors $\Xi_k = (\xi_k = 1, \xi_i = 0\, \forall\, i \neq k)$.

Definition 3.2. The **eigensimplex** $\boldsymbol{\Sigma}_n^{(1)}$ is a unit simplex of unit vectors Ξ_k, $k = 1, \ldots, n$ in $\mathbb{R}^n$ spanned by the eigenvectors $\boldsymbol{\zeta}_k$ of the value matrix W: $\boldsymbol{\Sigma}_n^{(1)} = \{0 \leq \xi_i \leq 1\, \forall\, i = 1, \ldots, n \wedge \sum_{i=1}^{n} \xi_i = 1\}$.

Theorem 3.3 (Restricted optimization of average fitness). *The average fitness* $\phi(t) = \bar{f} = \sum_{i=1}^{n} f_i x_i$ *with* $x_i \geq 0$ *and* $\sum_{i=1}^{n} x_i = 1$ *in a population described by the mutation selection equation* (3.2) *is nondecreasing on the intersection* $\mathbb{S}_n^{(1)} \cup \boldsymbol{\Sigma}_n^{(1)}$, *denoted as* optimization cone.

Proof. Points on the simplex $\mathbb{S}_n^{(1)}$ fulfil the conditions $x_i \geq 0$ and $\sum_{i=1}^{n} x_i = 1$. Equation 3.2 transformed by diagonalization of $\mathrm{W} = \mathrm{B} \cdot \Lambda \cdot \mathrm{B}^{-1}$ becomes

$$\frac{\mathrm{d}\boldsymbol{\zeta}_k}{\mathrm{d}t} = \boldsymbol{\zeta}_k \Big(\lambda_k - \phi(t) \Big) , \quad k = 1, \ldots, n ,$$

which is identical to Equation (2.14) and hence, Theorem 2.3 holds on the simplex $\boldsymbol{\Sigma}_n^{(1)}$. Both conditions are fulfilled on the intersection of the two simplices. □

[13]The matrix W may have complex conjugate pairs of eigenvalues without violating Perron-Frobenius theorem. Special cases of W with complex eigenvalues can be constructed but are not compatible with realistic assumptions on mutation frequencies Q_{ji}.

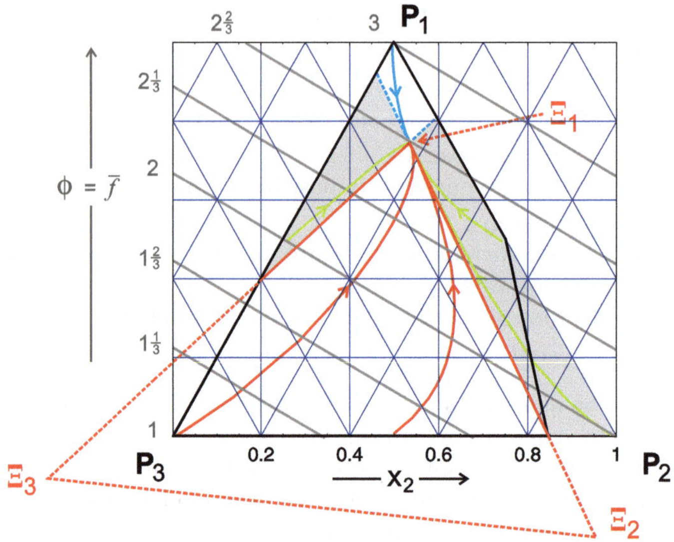

FIGURE 3.4. **The quasispecies on the unit simplex**. Shown is the case of three variables (x_1, x_2, x_3) on $\mathbb{S}_3^{(1)}$. The point representing pure quasispecies $\boldsymbol{\Xi}_1$, is shown together with the points for the other two eigenvectors, $\boldsymbol{\Xi}_2$ and $\boldsymbol{\Xi}_3$. The simplex is partitioned into an *optimization cone* ($\mathbb{S}_n^{(1)} \cup \boldsymbol{\Sigma}_n^{(1)}$; lower white area) where the mean replication rate $\bar{f}(t)$ is nondecreasing, and three other zones where $\bar{f}(t)$ may also decrease. Here, $\mathcal{X}_1$ is chosen to be the master sequence, the sequence with the highest fitness value. Solution curves are presented as parametric plots $\mathbf{x}(t)$. The mean replication rate $\bar{f}(t)$ is monotonously increasing along red trajectories, monotonously decreasing along the blue trajectory (upper white area), and not necessarily monotonous along green trajectories (grey areas). The parameter values are: $f_1 = 2.1\,[\mathrm{t}^{-1}]$, $f_2 = 2.0\,[\mathrm{t}^{-1}]$, and $f_3 = 1.9\,[\mathrm{t}^{-1}]$, the Q-matrix was assumed to be bistochastic with the elements $Q_{ii} = 0.98$ and $Q_{ij} = 0.01$ for $i, j = \{1, 2, 3\}$. The eigenvalues and eigenvectors of W are:

k	λ_k	b_{1k}	b_{2k}	b_{3k}
1	2.065	0.742	0.165	0.093
2	1.958	-0.248	1.078	0.170
3	1.857	-0.103	-0.224	1.327

The optimization principle does not hold outside the optimization cone, $\mathbb{S}_n^{(1)} \cup \boldsymbol{\Sigma}_n^{(1)}$. It can be shown that $\phi(t)$ is nonincreasing in the cone defined by $\boldsymbol{\Theta} = \{\xi_1 \geq 1, \xi_i \leq 0\,\forall j = 2, \ldots, n\}$ [50]. This fact can be easily made plausible by assuming as initial condition a homogenous population of the *master sequence*, the sequence corresponding to the fittest phenotype. All variants have lower fitness and accordingly, mutation leads to a reduction in the mean fitness $\bar{f}$, which is decreasing as the quasispecies with $\bar{f} = \lambda_1 < f_1$ is approached in the long time limit. In the remaining sections of the unit simplex no predictions on $\phi(t)$ can be made, special cases with non-monotonous behavior can be constructed [50]. The somewhat sophisticated conditions of optimization in the replication mutation system are illustrated by means of a numerical example in Figure 3.4.

3.4. Mutation rates and error thresholds

In general, the mutation rates are not tunable but they can be varied within certain limits by applying suitable experimental assays. In order to illustrate the mutation rate dependence of quasispecies and to subject it to mathematical analysis, a simplifying model called *uniform error rate* model is adopted [51]. The error rate per nucleotide and replication, p, is assumed to be independent of the position and the nature of the nucleotide exchange: $\mathsf{A}\rightarrow\mathsf{U}$, $\mathsf{A}\rightarrow\mathsf{G}$ or $\mathsf{A}\rightarrow\mathsf{C}$ occur with the same frequency p and the total error rate at a given position is $3p$. Then the elements of the mutation matrix Q depend only on three quantities: The chain length of the sequence to be replicated, ℓ, the error frequency p and the Hamming distance between the template, $\mathcal{X}_i$, and the newly synthesized sequence, $\mathcal{X}_j$, denoted by d_{ij}^{H},

$$Q_{ji} = \big(1-(\kappa-1)p\big)^{\ell-d_{ij}^{\mathrm{H}}} \cdot p^{d_{ij}^{\mathrm{H}}} = \big(1-(\kappa-1)p\big)^{\ell}\,\varepsilon^{d_{ij}^{\mathrm{H}}} \quad \text{with} \qquad \varepsilon = \frac{p}{1-(\kappa-1)p}\,, \tag{3.5}$$

with κ being the size of the nucleotide alphabet ($\kappa = 4$ for natural polynucleotides corresponding to $\{\mathsf{A},\mathsf{U}(\mathsf{T}),\mathsf{G},\mathsf{C}\}$). The explanation of Equation (3.5) follows from Figure 3.1: The two sequences differ in d_{ij}^{H} positions and hence $\ell - d_{ij}^{\mathrm{H}}$ nucleotides have to be copied correctly, each one contributing a factor $1-(\kappa-1)p$, and d_{ij}^{H} errors with frequency p have to be made at certain positions. Since the Hamming distance is a metric, we have $d_{ij}^{\mathrm{H}} = d_{ji}^{\mathrm{H}}$, and within the approximation of the uniform error rate model the mutation matrix Q is symmetric.

For $p = 0$ we encounter the selection case (2.15): The species of highest fitness, the master sequence $\mathcal{X}_1$, is selected and all other variants disappear in the long time limit. The other extreme is random replication where correct and incorrect incorporations of digits are equally probable and occur with frequency $\tilde{p} = \kappa^{-1}$. Then all elements of matrix Q are equal to $\kappa^{-\ell}$, and the uniform distribution $\boldsymbol{\Upsilon} = \{\bar{x}_j = n^{-1}\,\forall\, j = 1,2,\ldots,n \text{ with } n = \kappa^{\ell}\}$ is the eigenvector corresponding to the largest eigenvalue $\lambda_1 = \kappa^{-\ell}\sum_{i=1}^{n} f_i$ (all other eigenvalues of W vanish). In the whole range $0 \le p \le \kappa^{-1}$ the stationary distribution changes from the homogeneous population, $\Xi_1 = \{\bar{x}_1 = 1, \bar{x}_j = 0\,\forall\, j = 2,\ldots,n\}$ to the uniform distribution $\boldsymbol{\Upsilon}$.

Between the two extremes the function $\bar{x}_1(p)$ was approximated by Manfred Eigen through neglect of back-flow from mutants to the master sequence. He obtained for $\mathrm{d}x_1/\mathrm{d}t = 0$ [51]:

$$\bar{x}_1 = Q_{11} - \frac{\bar{f}_{-1}}{f_1} = Q_{11} - \sigma_1^{-1} \quad \text{with} \quad \bar{f}_{-1} = \frac{\sum_{i=1}^{n} f_i \bar{x}_i}{1 - \bar{x}_1} . \tag{3.6}$$

The quantity $\sigma_1 = f_1/\bar{f}_{-1}$ is denoted as the *superiority* of the master sequence $\mathcal{X}_1$. In this rough, zeroth-order approximation the frequency of the master sequence becomes zero at a critical value of the mutation rate parameter, $p_{\max}$, for constant chain length ℓ or at a maximal chain length $\ell_{\max}$ for constant replication accuracy p,

$$p_{\max} \approx \frac{\ln \sigma_1}{(\kappa - 1)\,\ell} \quad \text{or} \quad \ell_{\max} \approx \frac{\ln \sigma_1}{(\kappa - 1)\,p} ,$$

respectively. The critical replication accuracy has been characterized as the *error threshold* of replication. Numerically computed error thresholds remind of a phase transition in which the quasispecies changes from a mutant distribution centered around a master sequence to the uniform distribution. In other words the solution that becomes exact at $p = \tilde{p}$ is closely approached at $p_{\max}$ already. For the purpose of illustration for a superiority of $\sigma_1 = 1.1$ and a chain length of $\ell = 100$ we obtain $p_{\max} = 0.00032$ compared to $\tilde{p} = 0.5$.

Both relations for the error threshold, the maximum replication accuracy and the maximum chain length, were found to have practical implications: (i) RNA viruses replicate at mutation rates close to the maximal value [52]. A novel concept for the development of antiviral drugs makes use of this fact and aims at driving the virus population to mutation rates above the error threshold [53]. (ii) There is a limit in chain length for faithful replication that depends on the replication machinery: The accuracy limit of enzyme-free replication is around one error in one hundred nucleotides, RNA viruses with a single enzyme and no proof reading can hardly exceed accuracies of one error in 10 000 nucleotides, and DNA replication with repair on the fly reaches one error in 10^8 nucleotides. For prokaryotic DNA replication with post-replication repair the accuracy increases to $10^{-9} - 10^{-10}$, which is roughly one mutation in 300 duplications of bacterial cells, and for eukaryotes similar fractions of mutations were reported despite much longer genomes [54].

3.5. Numerical computations on error thresholds

The approximation of the error threshold through neglect of mutational backflow (3.6) caused the results to be largely independent of the distribution of the fitness values of mutants, since only the mean fitness, $\bar{f}_{-m}$, enters the expressions, in contrast to numerical computations, which suggested that the appearance of an error threshold and its shape are strongly dependent on details of the fitness landscape (see [55] and [21, pp. 51–60]). The influence of the distribution of fitness values will be considered in two steps:

(i) different fitness values are applied for error classes being sets of sequences with the same Hamming distances from the master sequence $\mathcal{X}_1$,

$$\begin{aligned} \mathcal{Y}_k &= \{\mathcal{X}_j \,|\, d^{\mathrm{H}}_{j1} = k\} \;, \\ y_k &= \sum\nolimits_j x_{j\,|\,\mathcal{X}_j \in \mathcal{Y}_k} \quad \text{and} \end{aligned} \tag{3.7}$$

(ii) different fitness values are assigned to individual sequences.

In the first case all sequences $\mathcal{X}_j$ with Hamming distance $d^H_{1,j} = k$ fall into the *error class* 'k'. Although the assumption that all sequences in a given error class have identical fitness is not justified on the basis of molecular data, it is frequently used in population genetics. Calculations applying this assumption turned out to be useful for an understanding of the threshold phenomenon.

Five model landscapes, which are characterized by their fitness matrices $\mathrm{F} = \{F_{ij} = f_i \cdot \delta_{ij}\}$, were applied here:

(i) the single-peak landscape: $f(\mathcal{Y}_0) = f_0$ and $f(\mathcal{Y}_j) = f_n \,\forall\, j \neq 1$,
(ii) the hyperbolic landscape: $f(\mathcal{Y}_j) = f_0 - (f_0 - f_n)(n+1)j/\big(n(j+1)\big) \,\forall\, j$,
(iii) the step-linear landscape: $f(\mathcal{Y}_j) = f_0 - (f_0 - f_n)j/k \,\forall\, j = 0, \ldots, k$ and $f(\mathcal{Y}_j) = f_n \,\forall\, j = k+1, \ldots, n$,
(iv) the multiplicative landscape: $f(\mathcal{Y}_j) = f_0(f_n/f_0)^{j/n} \,\forall\, j$, and
(v) the additive or linear landscape: $f(\mathcal{Y}_j) = f_0 - (f_0 - f_n)j/n \,\forall\, j$.

Examples for the dependence of the quasispecies distribution on the error rate, $\mathcal{Y}(p)$, on different landscapes are shown in Figure 3.5.

The analysis of error thresholds on different landscapes revealed three separable features: (i) a steep decay in the frequency of the master sequence – $x_1(p) \to 0$ in the zeroth-order approximation (3.6), (ii) a phase transition-like sharp change in the mutant distribution, and (iii) the transition leads from the quasispecies to the uniform distribution. All three phenomena coincide on the single-peak landscape. Characteristic for most hyperbolic landscapes is a steep decay of the master sequence (i) and an abrupt transition in the distribution of sequences according to (ii) but – in contrast to the single-peak landscape – the transition does not lead to the uniform distribution but to another distribution that changes gradually into the uniform distribution, which becomes the exact solution at the point $p = \tilde{p}$. The step-linear landscape illustrates the separation of the decay range (i) and the phase transition to the uniform distribution (ii and iii). In particular, the phase transition point $p_{\max}$ shifts towards higher values of p when the position of the step (at error class k) moves towards higher error-classes, whereas the decay of the master sequence ($\mathcal{X}_1$) moves in opposite direction. The additive and the multiplicative landscape, the two landscapes that are often used in population genetics, do not sustain threshold-behavior. On these two landscapes the quasispecies is transformed smoothly with increasing p into the uniform distribution.

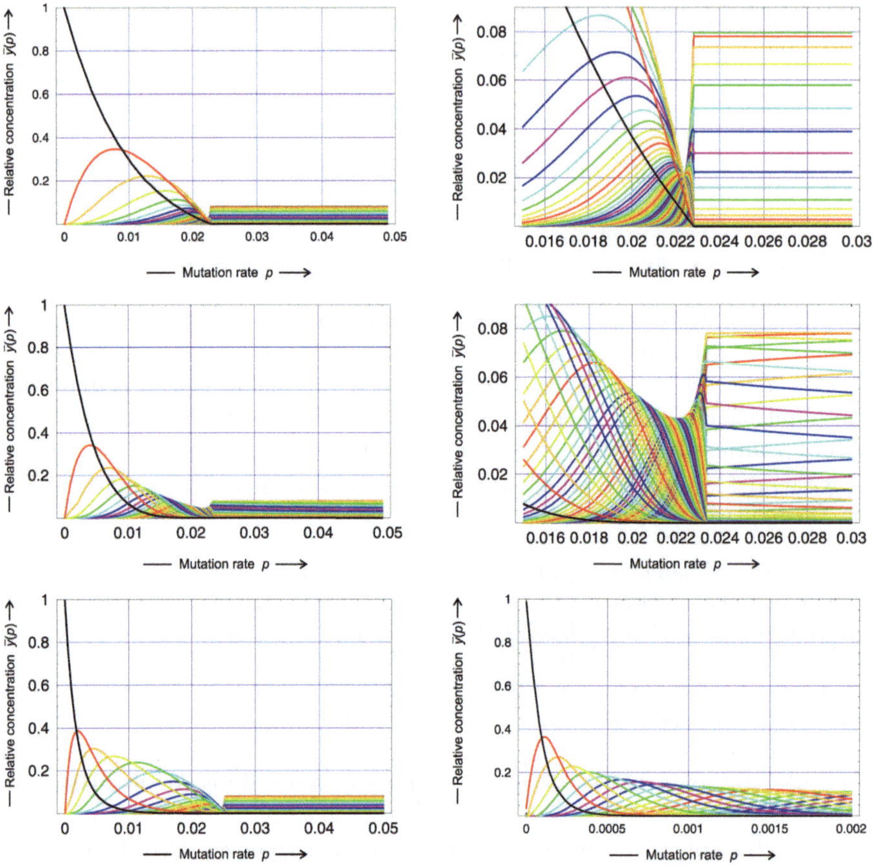

FIGURE 3.5. **Error thresholds on model landscapes**. Relative stationary concentrations of entire error classes $\bar{y}_k(p)$ are plotted as functions of the mutation rate p (different error classes $\mathcal{Y}_k$ are color coded, $k = 0$: black, $k = 1$: red, $k = 2$: yellow, $k = 3$: chartreuse, k=4: green, etc). Top row: single-peak fitness landscape (enlarged on the right-hand side), conditions (i), (ii), and (iii) coincide. Middle row: hyperbolic landscape (enlarged on the right-hand side), the phase transition leads to a distribution that changes gradually into the uniform distribution, (i) has an offset to the left of (ii). Bottom row: step-linear landscape on the l.h.s. meets condition (ii) and (iii) but (i) has a large offset to the left, and additive landscape on the r.h.s. does not sustain an error threshold at all. Parameters used in the calculations: $\ell = 100$, $f_1 = f_0 = 10$, $f_n = 1$ (except the hyperbolic landscape where we used $f_n = 0.9091$ in order to have $\bar{f}_{-m} = 1$ as for the single peak landscape), and $k = 5$ for the step-linear landscape.

Realistic fitness landscapes as derived from the properties of biomolecules are characterized by two features: (i) ruggedness and (ii) neutrality.[14] Error thresholds on realistic fitness landscapes can be modeled by the assumption of a randomly scattered distribution of fitness values within a given band of width d for all sequences except the master sequence:

$$f(\mathcal{X}_j) = \bar{f}_n + 2d(f_0 - f_n)\big(\eta_{\mathrm{rnd}}(j) - 0.5\big) - 1,\; j = 2, \ldots, \kappa^\ell \,. \tag{3.8}$$

In this expression "$\eta_{\mathrm{rnd}}(j)$" is a random number drawn from a random number generator with uniform distribution of numbers in the range $0 \le \eta_{\mathrm{rnd}}(j) \le 1$ with j being the index of the consecutive calls of the random function, and d is the band width of fitness values. The two limiting cases are: the full band width $d = 1$, where individual fitness values may be as large as f_0, the value for the master sequence, and the single peak landscape with $d = 0$. The computational capacities of today allow for studies of error thresholds at the resolution of individual sequences up to chain lengths $n = 10$. Further increase in computational power raises expectation to be able to reach $n = 20$, which in case of binary sequences is tantamount to the diagonalization of $10^6 \times 10^6$ matrices.

Transitions between two quasispecies were found first by computational analysis [50]: The master sequence on quasispecies I with a larger fitness value, $f_1^{(I)} > f_1^{(II)}$, has a smaller mutational backflow than quasispecies II, $\sum_{i=2}^n Q_{1i}^{(I)} x_i^{(I)} < \sum_{k=2}^n Q_{1k}^{(II)} x_k^{(II)}$. Since the mutational backflow increases with increasing error rate p, there may exist a value $p = p_{\mathrm{trans}} < p_{\max} \ll \tilde{p}$ at which the two quasispecies exchange stability. The sharpness of the transition depends primarily on the Hamming distance between the two master sequences, $d^{\mathrm{H}}(\mathcal{X}_1^{(I)}, \mathcal{X}_1^{(II)}) = d^{\mathrm{H}}_{I,II}$: The larger the distance the sharper is the transition – numerical computations with binary sequences of chain lengths $\ell = 10$ have shown that a Hamming distance of $d^{\mathrm{H}} > 6$ is sufficient for hard transitions. Several transitions of this kind may occur before the error threshold. In Figure 3.6 we show an example of a fitness landscape that sustains three transitions between four quasispecies. A sharp transition with Hamming distance $d^{\mathrm{H}}_{I,II} = 8$ between the master sequences is presented on the l.h.s. of the figure. Two more consecutive transitions are shown on the l.h.s: A sharper one with $d^{\mathrm{H}}_{II,III} = 7$ and a smoother one with $d^{\mathrm{H}}_{III,IV} = 6$. A large mutational backflow requires a small difference in fitness between the master sequence and its mutants and hence, the transitions with increasing p between quasispecies lead from steeper to flatter region on the fitness landscape. Examples are common and constructed straightforwardly. Thirteen years later the phenomenon has been rediscovered with digital organisms and named *survival of the flattest* [56].

Two more questions are important in the context of fitness values at a resolution of individual sequences: (i) How does the dispersion of fitness values expressed

---

[14] *Ruggedness* implies that nearby sequences may lead to identical but also to very different structures. By the same token functions like fitness values may be the same or very different for close by lying genotypes. *Neutrality* means that a certain fraction of different genotypes measured by the degree of neutrality, λ, have properties that cannot be distinguished by selection.

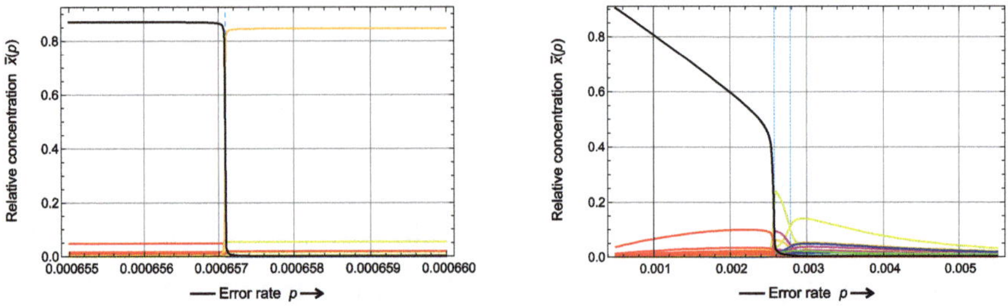

FIGURE 3.6. **Transitions between quasispecies**. Both figures show transitions on a landscape defined by Equation (3.8) for 1024 binary sequences of chain length $\ell = 10$. The transition on the l.h.s. involves the sequences $\mathbf{0} \to \mathbf{1003}$ (decimal equivalent of the binary sequence of chain length $\ell = 10$) with $d^{\mathrm{H}}_{\mathbf{0},\mathbf{1003}} = 8$ and is extremely sharp. The plot on the r.h.s shows to more transitions $\mathbf{0} \to \mathbf{923}$ and $\mathbf{923} \to \mathbf{247}$ ($d^{\mathrm{H}}_{\mathbf{0},\mathbf{923}} = 7$, $d^{\mathrm{H}}_{\mathbf{923},\mathbf{247}} = 6$). Parameters: $f_0 = 1.1$, $f_n = 1.0$, and the seed $s = 229$ for the random number generator *legacy* of *Mathematica*. The band width was chosen $d = 1.0$ (l.h.s.) and $d = 0.995$ (r.h.s.). The fitness values are: $f_{\mathbf{0}} = 1.1$, $f_{\mathbf{1003}} = 1.09999$ and 1.09949 for the two bandwidth, $f_{\mathbf{923}} = 1.09921$, and $f_{\mathbf{247}} = 1.09834$ for $d = 0.995$, respectively.

by the band width d change the characteristics of the error threshold and (ii) what happens if two more sequences have the same maximal fitness value $f_1 = f_0$. The answer to question (i) follows readily from the computed results (Figure 3.7): The position at which the frequency of the master sequence decays to zero, $x_1(p) \to 0$, migrates towards smaller f-values with increasing band width d. This observation agrees well with the expectations, since the fitness value closest to f_1 increases for broader bands (at constant $\bar{f}_{-1}$). In other words, the gap in fitness values between the fittest variant and the fittest but one variant becomes smaller. The scatter of fitness values at the same time broadens the band of curves for the sequences belonging to one error class. Both phenomena are easily recognized in the three plots on the l.h.s of Figure 3.7.

3.6. Neutral variants

Degeneracy of fitness values occurs when two or more genotypes have the same fitness and this is commonly denoted as *neutrality* in biology. An investigation of the role of neutrality requires an extension of Equation (3.8). A certain fraction of sequences, expressed by the degree of neutrality λ, is assumed to have the highest fitness value f_0 and the fitness values of the remaining fraction $1 - \lambda$ are assigned as in the non-neutral case (3.8). This random choice of neutral sequences together with a random dispersion of the other fitness values yields an interesting result: Random selection in the sense of Motoo Kimura's neutral theory of evolution [57] occurs only for sufficiently distant fittest sequences. The case of vanishing mutation

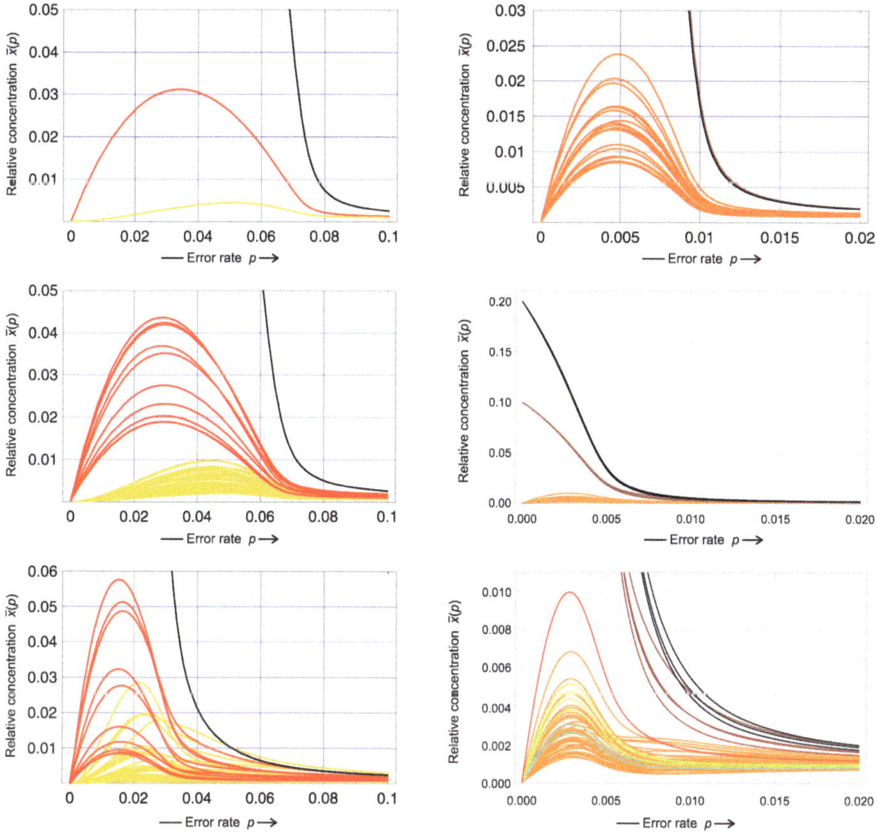

FIGURE 3.7. **Error thresholds on realistic landscapes.** The plots on the l.h.s. show the error threshold for a landscape defined by Equation (3.8) for 1024 binary sequences of chain length $\ell = 10$. With increasing d (top: $d = 0$, single peak landscape, middle: $d = 0.5$, and bottom: $d = 0.925$) the scatter of sequences within the same error class increases and the error threshold defined by sharp decline of the master sequence, $\bar{x}_1(p) \to 0$ migrates towards smaller error rates. The plots on the r.h.s. refers to landscapes with neutrality. The topmost plot shows the frequencies of the master sequences in a neutral network of two nearest neighbors, $\mathcal{X}_\mathbf{0}$ (black) and $\mathcal{X}_{\mathbf{64}}$ (red, almost entirely hidden behind the black curve), with the 18 one-error mutants surrounding the two sequences (orange). The other plots refer to a network of seven nearest neighbors (Figure 3.8; three inner sequences, $\mathcal{X}_{\mathbf{248}}$, $\mathcal{X}_{\mathbf{760}}$ and $\mathcal{X}_{\mathbf{728}}$ (black), and four outer sequences, $\mathcal{X}_{\mathbf{184}}$, $\mathcal{X}_{\mathbf{504}}$, $\mathcal{X}_{\mathbf{600}}$ and $\mathcal{X}_{\mathbf{729}}$ (brown)). Strong coupling pertains until the population reaches the error threshold. The spectrum of one error mutants is shown in the plot at the bottom. Parameters, l.h.s: $f_0 = 2, f_n = 1, s = 491$, and r.h.s: $f_0 = 1.1, f_n = 1.0, d = 0.5$, topmost plot: $\lambda = 0.01, s = 367$, middle and bottom: $\lambda = 0.1, s = 229$, color code see Figure 3.8.

rate, $\lim p \to 0$, has been studied analytically [50] for two neutral genotypes, $\mathcal{X}_j$ and $\mathcal{X}_k$ and different Hamming distance d^{H}_{jk}. The exact results are

(i) $d^{\mathrm{H}}_{jk} = 1$: $\lim_{p\to 0} \frac{\bar{x}_j}{\bar{x}_k} = 1$ or $\lim_{p\to 0} \bar{x}_j = \lim_{p\to 0} = \bar{x}_k = 0.5$,

(ii) $d^{\mathrm{H}}_{jk} = 2$: $\lim_{p\to 0} \frac{\bar{x}_j}{\bar{x}_k} = \alpha$ or $\lim_{p\to 0} \bar{x}_j = \alpha/(1-\alpha)$, $\lim_{p\to 0} \bar{x}_k = 1/(1-\alpha)$, with some value α and

(iii) $d^{\mathrm{H}}_{jk} \geq 3$: $\lim_{p\to 0} \bar{x}_1 = 1$, $\lim_{p\to 0} \bar{x}_2 = 0$ or $\lim_{p\to 0} \bar{x}_1 = 0$, $\lim_{p\to 0} \bar{x}_2 = 1$.

In full agreement with the exact result [50] we find that two fittest sequences of Hamming distance $d^{\mathrm{H}} = 1$ – being two nearest neighbors in sequence space – are selected as a strongly coupled pair with equal frequency of both members. Numerical results demonstrate that this strong coupling occurs not only for small mutation rates but extends over the whole range of p-values from $p = 0$ to the error threshold $p = p_{\max}$ (Figure 3.7). For clusters of more than two nearest neighbor sequences, the frequencies of the individual members of the cluster can be obtained from the largest eigenvector of the adjacency matrix. Pairs of fittest sequences with Hamming distance $d^{\mathrm{H}} = 2$ – being two next nearest neighbors with two sequences in between – are also selected together but the ratio of the two frequencies is different from one. Again strong coupling extends from zero mutation rates up to the error threshold $p = p_{\max}$. For fittest sequences with $d^{\mathrm{H}} \geq 3$ random selection chooses one sequence arbitrarily and eliminates all others as predicted by the Kimura's neutral theory of evolution. An example of a network of seven strongly coupled sequences is shown in Figure 3.8. The stationary frequencies of the seven sequences are readily obtained from the largest eigenvector and have the simple form: $\bar{x}_{\text{inner}} = 2\bar{x}_{\text{outer}}$ where by the three inner sequences (black) and the four outer sequences have identical frequencies. Deviation from the exact result for $p \to 0$ can be seen only in the enlargement (plot at the bottom) and are small indeed. Strong coupling of fittest sequences manifests itself, for example, in virology in form of systematic deviations from consensus sequences of populations as they are indeed found in nature through systematic sequencing of populations.

3.7. Lethal mutants

Lethal mutants are mutants with zero fitness, $f = 0$, which are unable to replicate and which may lead to extinction of the population. The model that has been used to investigate mutation and selection at constant population size is not applicable as concentrations may go to zero and a more detailed physical setup is required for the description. Different setups were discussed and analyzed rigorously before [30] and the existence of an *extinction threshold* for autocatalytic processes like replication reactions was proved (see also [21, pp. 18–27]). We choose here a continuously stirred flowreactor (CSTR) as an appropriate device: An influx of reactants, for example monomeric building blocks M at concentration m_0 and polymerase to compensate for enzyme degradation in case of RNA replication, is installed to replace consumed materials, the reactor is well stirred in order to guarantee instantaneous mixture of the reactor content, the influx is compensated by an unspecific outflux, and the flow rate r is assumed to be tunable. The mechanism

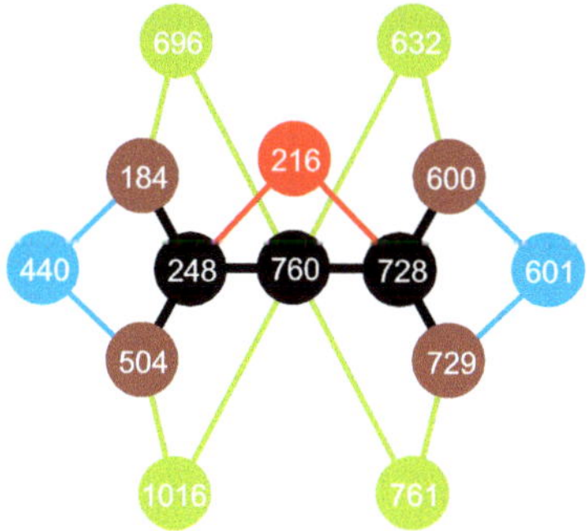

FIGURE 3.8. **A neutral network of replicating genotypes.** Shown is the network of seven neutral sequences being nearest neighbors (black and brown). Seven out of the 51 one error mutants of the neutral sequences occupy special positions that are coupled to two members of the network. As seen in the plot in Figure 3.7 (r.h.s., bottom) only $\mathcal{X}_{\mathbf{216}}$ (red) has substantially higher frequency and the curves of the other six special sequences are embedded in the band of the one error class.

for replication and mutation in the flowreactor with $\mathcal{X}_1$ being the only replicating variant is of the form

$$\begin{aligned} \star \quad &\xrightarrow{\;m_0\, r\;} \quad \mathsf{M}\,, \\ \mathsf{M} + \mathcal{X}_1 \quad &\xrightarrow{\;Q_{j1}k_1\;} \quad \mathcal{X}_j + \mathcal{X}_1\,; \quad j = 1,2,\ldots,n\,, \\ \mathsf{M} \quad &\xrightarrow{\;r\;} \quad \oslash\,, \quad \text{and} \\ \mathcal{X}_j \quad &\xrightarrow{\;r\;} \quad \oslash\,; \quad j = 1,2,\ldots,n\,. \end{aligned} \tag{3.9}$$

where $\star$ and $\oslash$ refer to influx and outflux, respectively. Reaction kinetics leads to the differential equation

$$\begin{aligned} \frac{\mathrm{d}m}{\mathrm{d}t} &= -\left(\sum_{i=1}^{n} Q_{j1}k_1c_1\right) m + r(m_0 - m) \\ \frac{\mathrm{d}c_j}{\mathrm{d}t} &= Q_{j1}k_1 m c_1 - r c_j\,,\; j = 1,2,\ldots,n\,. \end{aligned} \tag{3.10}$$

In order to be able to handle lethal mutants properly we have to go back to absolute concentrations $c_j = x_j c(t)$ with $\sum_{j=1}^{n} c_j = c$, the variable m is the concentration of the building blocks M in the reactor, and the rate parameter k_1 is related to the fitness through $f_1 = k_1 m$.

Calculation of stationary states is straightforward and yields two solutions, (i) the state of extinction with $\bar{m} = m_0$ and $\bar{c}_j = 0\;\forall\, j = 1,2,\ldots,n$, and (ii) a state of quasispecies selection consisting of $\mathcal{X}_1$ and its mutant cloud at the concentrations $\bar{m} = r/(Q_{11}k_1)$, $\bar{c}_1 = Q_{11}m_0 - r/k_1$, and $\bar{c}_j = \bar{c}_1(Q_{j1}/Q_{11})$ for $j = 2,\ldots,n$.

Stability analysis yields a straightforward result: A nonzero population is stable below a critical flow rate $r < r_{\rm cr} = Q_{11} k_1 m_0$. The master sequence $\mathcal{X}_1$ and all its mutants vanish for the condition $r = r_{\rm cr}$ above which we have extinction. As an example for the distribution of genotypes we compute a maximum error rate for the uniform error rate model (3.5):

$$Q_{11} = (1-p)^{\ell} \quad \text{and}$$
$$Q_{j1} = p^{d^{\rm H}_{j1}} \cdot (1-p)^{\ell - d^{\rm H}_{j1}} ,$$

where $d^{\rm H}_{j1}$ is the Hamming distance between the two sequences $\mathcal{X}_j$ and $\mathcal{X}_1$. Instead of the superiority σ of the master sequence, which diverges since $\bar{k}_{-m} = 0$ because of $k_2 = \cdots = k_n = 0$, we use the carrying capacity of the flowreactor, η, which can be obtained straightforwardly as

$$\eta = \frac{k_1 m_0}{r} .$$

The value of p, at which the stationary concentration of the master sequence $\bar{c}_1(p)$ and all other mutants vanishes, represents the analogue to the error threshold (3.6), and for the sake of clearness it is referred to as *extinction threshold.* Using the approximation $\ln(1-p) \approx -p$ we obtain:

$$\hat{p}_{\rm max} \approx \frac{\ln \eta}{\ell} \quad \text{for small } p . \tag{3.11}$$

The major difference between the error threshold (3.6) and the extinction threshold (3.11) concerns the state of the population at values $p > p_{\rm max}$: Replication with non-zero fitness of mutants leads to the uniform distribution whereas the population goes extinct in the lethal mutant case. Accordingly, the transformation to relative concentrations fails and Equation (3.2) is not applicable. In Figure 3.9 we show an example for the extinction threshold with $\ell = 20$ and $\eta = 2$. The extinction threshold is calculated from Equation (3.11) to occur at $\hat{p}_{\rm max} = 0.03466$ compared to an exact value of 0.03406. In the figure we see also a comparison of the curves for the master sequence and the one error class for the single peak landscape and the lethality model. The agreement of the two curves for the master sequences is not surprising, since the models were adjusted to coincide in the values $\bar{c}_1(0) = 1$ and $\hat{p}_{\rm max} = p_{\rm max} = \ln 2/20$. The curves for the one error class show some difference that is caused by the lack of mutational backflow from higher mutants in case of lethal variants.

It is important to note that a quasispecies[15] can exist also in cases where the Perron-Frobenius theorem is not fulfilled as in the current example of lethal mutants: Only genotype $\mathcal{X}_1$ has a positive fitness value, $f_1 > 0$ and $f_2 = \cdots = f_n = 0$, and hence only the entries $W_{k1} = Q_{k1} f_1$ of matrix W are nonzero and

[15] Although this distribution is not derived under constant population size, it will be called *quasispecies* here.

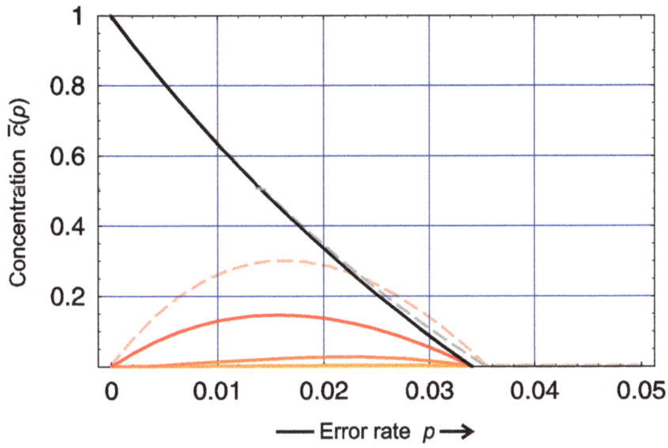

FIGURE 3.9. **Lethal mutants and replication errors.** The model for lethal mutants corresponding to a single peak landscape with $f_1 = 2$ and $f_2 = \cdots = f_n = 0$ is studied in the flowreactor. The concentrations of the master sequence (black) and the mutant classes (red, dark orange, light orange, etc.; full lines) are shown as functions of the error rate p. The parameters were chosen to be $\ell = 20$, $m_0 = 2$, and $\eta = 2$. The plots are compared to the curves for the master sequence and the one error class (grey, light red; broken curves) for a single peak landscape with $f_1 - 2$, $f_2 - \cdots - f_n - 1$, $\ell - 20$, $\sigma - 2$. The single peak landscape has been chosen such that the error threshold coincides with the extinction threshold at $\hat{p}_{\max} = p_{\max} = \ln 2/20$.

hence

$$\mathrm{W} = \begin{pmatrix} W_{11} & 0 & \ldots & 0 \\ W_{21} & 0 & \ldots & 0 \\ \vdots & \vdots & \ddots & \vdots \\ W_{n1} & 0 & \ldots & 0 \end{pmatrix} \quad \text{and} \quad \mathrm{W}^k = W_{11}^k \begin{pmatrix} 1 & 0 & \ldots & 0 \\ \frac{W_{21}}{W_{11}} & 0 & \ldots & 0 \\ \vdots & \vdots & \ddots & \vdots \\ \frac{W_{n1}}{W_{11}} & 0 & \ldots & 0 \end{pmatrix} .$$

Accordingly, W is not primitive in this example, but under suitable conditions $\bar{\mathbf{x}} = (Q_{11}, Q_{21}, \ldots, Q_{n1})$ is a stable stationary mutant distribution and for $Q_{11} > Q_{j1} \, \forall j = 2, \ldots, n$ (correct replication occurs more frequently than a particular mutation) genotype $\mathcal{X}_1$ is the master sequence. On the basis of a rather idiosyncratic mutation model consisting of a one-dimensional chain of mutants [58] the claim has been raised that no quasispecies can be stable in presence of lethal mutants, and hence no error thresholds can occur. A more recent paper [59, 60] used more realistic high-dimensional mutation models and presented numerically computed examples of perfect error thresholds in the presence of lethal mutants. A more detailed discussion of lethal mutagenesis is found in [26, 27, 60] but for a full understanding of the phenomenon investigations of realistic landscapes are indispensable.

4. Perspectives

The Darwinian principle of variation and selection has been attributed with two features that were studied here by means of simple models: (i) optimization of mean fitness in populations and (ii) uniqueness of the result of a selection process. Both features are not necessarily fulfilled neither in model systems nor in nature. In Table 1 a comparison is shown for the three models described in this chapter. Only the simple selection equation (2.14) sustains both features as mean fitness is optimized and in absence of neutrality the outcome of the selection process is unique. When mutation is included in Equation (3.2′) the unique outcome as represented by the quasispecies is guaranteed but the mean fitness may also decrease under certain circumstances. Fisher's selection equation (2.19) describes recombination on a single locus and mean fitness is optimized according to the fundamental theorem but no unique outcome is guaranteed – different initial conditions may lead to different allele combinations. Extrapolation of Fisher's model to multiple loci requires independence of genes and gene functions that is completely unrealistic. Interacting genes, in general, fulfil none of the two criteria. Therefore, the Darwinian principle is no theorem but a very powerful optimization heuristic that finds various applications in many disciplines from engineering to social sciences [61].

The mathematics of the ODE based model systems is fairly simple and straightforward. When applied to real situations the models gain enormous complexity essentially for two reasons: (i) any comprehensive model should cover a sufficiently large section of sequence space and this leads to gigantic dimensionality – the number of possible genotypes is 4^ℓ DNA or RNA sequences of chain length ℓ – that cannot be handled without drastic simplifications, and (ii) suitable models that allow for applications require realistic assumptions on the structure of fitness landscapes. Simple model landscapes assuming, for example, additivity of mutation effects or assigning equal fitness to all genotypes at the same distance from a reference sequence lead to wrong results. Ruggedness and neutrality shape fitness landscapes in reality and this can be taken into account properly only by models making use of random assignments, examples are the approach described here (Subsection 3.5) and the frequently used NK-model [62, pp. 40–60]. Direct calculation of fitness values for individual genotypes is possible at present only

TABLE 1. Optimization behavior in simple genetic systems.

Phenomenon	Optimization	Unique outcome
Selection	**yes**	**yes**
Mutation and selection	**no**	**yes**
Recombination and selection		
independent genes	**yes**	**no**
interacting genes	**no**	**no**

for RNA molecules within certain approximations [63, 64]. *In silico* evolution on these landscapes yields results that are in general agreement with those reported here. The great challenge for the future, however, is the construction of fitness landscapes that are based on solid experimental information. Apart from molecular *in vitro* systems, where such data are hard to get but within reach, systematic studies on virus infection in hosts are very promising for this goal.

References

[1] C. Darwin, *On the Origin of Species by Means of Natural Selection or the Preservation of Favoured Races in the Struggle for Life.* John Murray, London, 1859.

[2] G. Mendel, *Versuche über Pflanzenhybriden.* (*Experiments relating to plant hybridization*). Verh. Naturforsch. Ver. Brünn **4** (1866), 3–17.

[3] S. Spiegelman, *An approach to the experimental analysis of precellular evolution.* Quart. Rev. Biophys. **4** (1971), 213–253.

[4] G.F. Joyce, *Directed evolution of nucleic acid enzymes.* Annu. Rev. Biochem. **73** (2004), 791–836.

[5] S. Klussmann (ed.), *The Aptamer Handbook. Functional Oligonucleotides and Their Applications.* Wiley-VCh Verlag, Weinheim, DE, 2006.

[6] S. Brakmann and K. Johnsson (eds.), *Directed Molecular Evolution of Proteins or How to Improve Enzymes for Biocatalysis.* Wiley-VCH, Weinheim, DE, 2002.

[7] C. Jäckel, P. Kast, and D. Hilvert, *Protein design by directed evolution.* Annu. Rev. Biophys. **37** (2008), 153–173.

[8] S.J. Wrenn and P.B. Harbury, *Chemical evolution as a tool for molecular discovery.* Annu. Rev. Biochem. **76** (2007), 331–349.

[9] L.E. Sigler, *Fibonacci's Liber Abaci: A Translation into Modern English of Leonardo Pisano's Book of Calculation.* Springer-Verlag, New York, 2002, a translation of Leonardo Pisano's book of 1202 from Latin into modern English.

[10] P. Singh, *The so-called Fibonacci numbers in ancient and medieval India.* Historia Mathematica **12** (1985), 229–244.

[11] D.E. Knuth, *The Art of Computer Programming. Vol. I: Fundamental Algorithms.* Third edition, Addison-Wesley Publishing Co., Reading, MA, 1997.

[12] R.L. Graham, D.E. Knuth, and O. Patashnik, *Concrete Mathematics.* Second edition, Addison-Wesley Publishing Co., Reading, MA, 1994.

[13] R.A. Dunlap, *The Golden Ratio and Fibonacci Numbers.* World Sceintific, Singapore, 1997.

[14] J. Binet, *Mémoire sur l'intégration des équations linéaires aux différences finies, d'un ordre quelconque, à coefficients variables.* Comptes Rendue Hebdomadaires des Séances de l'Académie des Sciences (Paris) **17** (1843), 559–567.

[15] L. Euler, *Theoremata arithmetica nova methodo demonstrata.* Novi Comentarii Scientiarum Imperialis Petropolitanæ **8** (1760), 74–104, reprinted in Euler's *Commentationes Arithmeticæ Collectæ*, Vol. 1, 274–286 and in Euler's *Opera Omnia*, Series 1, Vol. 2, 531–555.

[16] T.R. Malthus, *An Essay on the Principle of Population, as it Affects the Future Improvement of Society, with Remarks on the Speculations of Mr Godwin, M. Condorcet and Other Writers.* John Murray, London, 1798.

[17] L. Euler, *Introductio in Analysin Infinitorum,* 1748. *English Translation: John Blanton, Introduction to Analysis of the Infinite*, volume I and II. Springer-Verlag, Berlin, 1988.

[18] P.F. Verhulst, *Notice sur la loi que la population pursuit dans son accroisement.* Corresp. Math. Phys. **10** (1838), 113–121.

[19] P. Verhulst, *Recherches mathématiques sur la loi d'accroisement de la population.* Nouv. Mém. de l'Académie Royale des Sci. et Belles-Lettres de Bruxelles **18** (1845), 1–41.

[20] P. Verhulst, *Deuxième mémoire sur la loi d'accroisement de la population.* Mém. de l'Académie Royale des Sci., des Lettres et de Beaux-Arts de Belgique **20** (1847), 1–32.

[21] P.E. Phillipson and P. Schuster, *Modeling by Nonlinear Differential Equations. Dissipative and Conservative Processes*, volume 69 of *World Scientific Series on Nonlinear Science A.* World Scientific, Singapore, 2009.

[22] D. Zwillinger, *Handbook of Differential Equations.* Third edition, Academic Press, San Diego, CA, 1998.

[23] M. Eigen and P. Schuster, *The hypercycle. A principle of natural self-organization. Part B: The abstract hypercycle.* Naturwissenschaften **65** (1978), 7–41.

[24] P. Schuster and K. Sigmund, *Replicator dynamics.* J. Theor. Biol. **100** (1983), 533–538.

[25] R.A. Fisher, *The Genetical Theory of Natural Selection.* Clarendon Press, Oxford, 1930.

[26] J.J. Bull, L. Ancel Myers, and M. Lachmann, *Quasispecies made simple.* PLoS Comp. Biol. **1** (2005), 450–460.

[27] J.J. Bull, R. Sanjuán, and C.O. Wilke, *Theory for lethal mutagenesis for viruses.* J. Virology **81** (2007), 2930–2939.

[28] W.M. Hirsch and S. Smale, *Differential Equations, Dynamical Systems, and Linear Algebra.* Academic Press, New York, 1974.

[29] S. Shahshahani, *A new mathematical framework for the study of linkage and selection.* Mem. Am. Math. Soc. **211** (1979), 34.

[30] P. Schuster and K. Sigmund, *Dynamics of evolutionary optimization.* Ber. Bunsenges. Phys. Chem. **89** (1985), 668–682.

[31] W.J. Ewens, *Mathematical Population Genetics*, volume 9 of *Biomathematics Texts.* Springer-Verlag, Berlin, 1979.

[32] J.B. Schnog, A.J. Duits, F.A.J. Muskiet, H. ten Cate, R.A. Rojer, and D.P.M. Brandjes, *Sickle cell disease: A general overview.* The Netherlands Journal of Medicine **62** (2004), 364–374.

[33] G.A. Barabino, M.O. Platt, and D.K. Kaul, *Sickle cell biomechanics.* Annu. Rev. Biomed. Eng. **12** (2010), 345–367.

[34] G. Price, *Fisher's "Fundamental Theorem" made clear.* Ann. Hum. Genet. **36** (1972), 129–140.

[35] A.W. Edwards, *The fundamental theorem of natural selection.* Biological Reviews **69** (1994), 443–474.

[36] S. Okasha, *Fisher's fundamental theorem of natural selection – A philosophical analysis.* Brit. J. Phil. Sci. **59** (2008), 319–351.

[37] F. Schlögl, *Chemical reaction models for non-equilibrium phase transitions.* Z. Physik **253** (1972), 147–161.

[38] G. von Kiedrowski, B. Wlotzka, J. Helbig, M. Matzen, and S. Jordan, *Parabolic growth of a self-replicating hexanucleotide bearing a 3′-5′-phosphoamidate linkage.* Angew. Chem. Internat. Ed. **30** (1991), 423–426.

[39] B.R.M. Stadler, P.F. Stadler, G.P. Wagner, and W. Fontana, *The topology of the possible: Formal spaces underlying patterns of evolutionary change.* J. Theor. Biol. **213** (2001), 241–274.

[40] C.K. Biebricher, M. Eigen, and J. William C. Gardiner, *Kinetics of RNA replication.* Biochemistry **22** (1983), 2544–2559.

[41] J.D. Watson and F.H.C. Crick, *A structure for deoxyribose nucleic acid.* Nature **171** (1953), 737–738.

[42] H.F. Judson, *The Eighth Day of Creation. The Makers of the Revolution in Biology.* Jonathan Cape, London, 1979.

[43] R.K. Saiki, D.H.G. amd Susanne Stoffel, S.J. Scharf, R. Higuchi, G.T. Horn, K.B. Mullis, and H.A. Erlich, *Primer-directed enzymatic amplification of DNA with a thermostable DNA polymerase.* Science **239** (1988), 487–491.

[44] C.K. Biebricher, M. Eigen, and J. William C. Gardiner, *Kinetics of RNA replication: Plus-minus asymmetry and double-strand formation.* Biochemistry **23** (1984), 3186–3194.

[45] C.K. Biebricher, M. Eigen, and J. William C. Gardiner, *Kinetics of RNA replication: Competition and selection among self-replicating RNA species.* Biochemistry **24** (1985), 6550–6560.

[46] C.J. Thompson and J.L. McBride, *On Eigen's theory of the self-organization of matter and the evolution of biological macromolecules.* Math. Biosci. **21** (1974), 127–142.

[47] B.L. Jones, R.H. Enns, and S.S. Rangnekar, *On the theory of selection of coupled macromolecular systems.* Bull. Math. Biol. **38** (1976), 15–28.

[48] R.W. Hamming, *Coding and Information Theory.* 2nd edition, Prentice-Hall, Englewood Cliffs, NJ, 1986.

[49] E. Seneta, *Non-negative Matrices and Markov Chains.* Second edition, Springer-Verlag, New York, 1981.

[50] P. Schuster and J. Swetina, *Stationary mutant distribution and evolutionary optimization.* Bull. Math. Biol. **50** (1988), 635–660.

[51] M. Eigen, *Selforganization of matter and the evolution of biological macromolecules.* Naturwissenschaften **58** (1971), 465–523.

[52] J.W. Drake, *Rates of spontaneous mutation among RNA viruses.* Proc. Natl. Acad. Sci. USA **90** (1993), 4171–4175.

[53] E. Domingo, ed., *Virus entry into error catastrophe as a new antiviral strategy.* Virus Research **107** (2005), 115–228.

[54] J.W. Drake, B. Charlesworth, D. Charlesworth, and J.F. Crow, *Rates of spontaneous mutation.* Genetics **148** (1998), 1667–1686.

[55] T. Wiehe, *Model dependency of error thresholds: The role of fitness functions and contrasts between the finite and infinite sites models.* Genet. Res. Camb. **69** (1997), 127–136.

[56] C.O. Wilke, J.L. Wang, C. Ofria, R.E. Lenski, and C. Adami, *Evolution of digital organisms at high mutation rates leads to survival of the flattest.* Nature **412** (2001), 331–333.

[57] M. Kimura, *The Neutral Theory of Molecular Evolution.* Cambridge: University Press, UK, 1983.

[58] G.P. Wagner and P. Krall, *What is the difference between models of error thresholds and Muller's ratchet.* J. Math. Biol. **32** (1993), 33–44.

[59] N. Takeuchi and P. Hogeweg, *Error-thresholds exist in fitness landscapes with lethal mutants.* BMC Evolutionary Biology **7** (2007), 15.

[60] H. Tejero, A. Marín, and F. Moran, *Effect of lethality on the extinction and on the error threshold of quasispecies.* J. Theor. Biol. **262** (2010), 733–741.

[61] K. Lucas and P. Roosen (eds.), *Emergence, Analysis, and Evolution of Structures. Concepts and Strategies across Disciplines.* Understanding Complex Systems, Springer-Verlag, Berlin, 2010.

[62] S.A. Kauffman, *The Origins of Order. Self-Organization and Selection in Evolution.* Oxford University Press, New York, 1994.

[63] W. Fontana and P. Schuster, *Continuity in evolution. On the nature of transitions.* Science **280** (1998), 1451–1455.

[64] P. Schuster, *Prediction of RNA secondary structures: From theory to models and real molecules.* Reports on Progress in Physics **69** (2006), 1419–1477.

Peter Schuster
Institut für Theoretische Chemie, Universität Wien
Währingerstraße 17, A-1090 Wien, Austria

and

Santa Fe Institute, 1399 Hyde Park Road
Santa Fe, New Mexico 87501, USA
e-mail: pks@tbi.univie.ac.at

Some Mathematical Models in Evolutionary Genetics

Reinhard Bürger

Abstract. Since most traits of evolutionary or economic importance are determined by several or many genes, a proper understanding of the evolution of such traits requires the study of multilocus models. Accordingly, we present the basic models and the most fundamental results about the evolutionary dynamics of a population in which selection acts on many gene loci. First, important aspects of the classical case, when selection acts on a single diploid locus with multiple alleles, are highlighted. Then the general model with selection on a finite number of recombining multiallelic loci is treated, with the focus on asymptotic and convergence results, including generalizations of Fisher's Fundamental Theorem. Extensions dealing with migration or frequency-dependent selection are briefly outlined. Finally, the selection response and the evolution of (multivariate) quantitative traits is investigated.

Mathematics Subject Classification (2000). 92D15, 92D10.

Keywords. Population genetics, quantitative genetics, multilocus models, selection, recombination, migration.

1. Introduction

As explained beautifully in Warren Ewens' chapter [1], mathematical models played a decisive role in reconciling Mendelian genetics with Darwin's theory of evolution by natural selection. Out of this successful enterprise during the early 20th century, the field of population genetics emerged. Population genetics is concerned with the study of the genetic composition of populations. This composition may be changed by segregation, selection, mutation, recombination, mating, migration, and other genetic, ecological, and evolutionary mechanisms. Population

---

This work was supported by a grant of the Vienna Science and Technology Fund (WWTF) and by grant P21305 of the Austrian Science Foundation (FWF). I thank Thomas Nagylaki, Simon Aeschbacher, and Stephan Peischl for comments on the manuscript.

genetics is the field in which these mechanisms, their interactions, and their evolutionary consequences are investigated. Because Darwinian evolution is impossible without selection and inheritance, population genetics forms the main corner stone of evolutionary theory. The founding fathers of population genetics were R.A. Fisher, S. Wright, and J.B.S. Haldane. They not only developed almost all the basic theory and employed it to empirical findings and data, but they also initiated numerous experiments to test their theories. Much of their work was highly mathematical, far beyond what the majority of their fellow geneticists and biologists could understand. Almost as a by-product, their work laid the foundations for modern statistics and for important developments in stochastic processes.

One of the first crucial insights, an immediate consequence of the Hardy-Weinberg law, was that Mendelian inheritance preserves genetic variation on which natural selection can act. Early work culminated in Fisher's Fundamental Theorem of Natural Selection (FTNS) [2] which he formulated as

> "The rate of increase in fitness of any organism at any time is equal to its genetic variance in fitness at that time."

Although it was argued convincingly that this statement has been misinterpreted for many decades (see Ewens' chapter [1]), the 'classical' interpretation has led to deep insights into the evolutionary process. Fisher's Fundamental Theorem not only implies that evolution is impossible in the absence of genetic variation, but it gave rise to important quantitative predictions about the response to selection. It is the purpose of this chapter to present some of the most fundamental results about the evolutionary dynamics of a population subject to selection.

I shall first treat the classical case when selection acts on a single diploid locus at which an arbitrary number of alleles can occur. Then I turn to generalizations that include recombination and selection at multiple loci, and point out extensions that include migration or frequency-dependent selection. Eventually, the selection response and the evolution of (multivariate) quantitative traits is studied. In this case, the theory is formulated in terms of directly measurable quantities and thus is relatively easily amenable to experimental and empirical testing. Special cases of this theory have been successfully employed for many decades in animal and plant breeding.

Throughout this chapter, we treat deterministic models, i.e., systems of difference or differential equations. Hence, we assume that populations are effectively infinite, so that stochasticity can be ignored. Of course, it has been known since the early days that reproduction and selection have inherent stochastic effects. Incorporation of these effects, in particular of random genetic drift, produces models that are formulated in terms of (usually Markovian) stochastic processes and adds considerable mathematical complexity. The study of such models has been an active research area since it was initiated by Fisher and Wright, which nowadays, especially in the context of the inference of past evolutionary events from molecular data, seems more relevant than ever. Keeping in mind the most important effects of random genetic drift (the reduction of genetic variance by an approximate factor of

$1-1/(2N)$ per generation and the eventual fixation or loss of alleles), deterministic models are valuable guides to understanding the consequences of the interaction of two or more evolutionary mechanisms that would not be amenable to analysis otherwise. They also yield close approximations for short-term or medium-term predictions of evolutionary change.

2. A single locus

In general, I prefer to formulate models for populations with discrete, nonoverlapping generations because this allows their derivation from a well-defined life cycle based on biological principles. Continuous-time models are sometimes closer to biological reality and frequently easier to treat mathematically. However, their derivation often involves mathematically inconsistent assumptions (see below).

Throughout, we assume that organisms are diploid, that the genotype frequencies are the same in both sexes, that adults mate at random with respect to considered gene loci, and that selection acts on juveniles through differential viabilities that are constant. As a consequence of random mating, zygotes are in Hardy-Weinberg proportions, and it is sufficient to consider gamete frequencies (instead of genotype frequencies) among zygotes. For a detailed treatment including proofs and more complete references, we refer to [3, Chapters I.9 and I.10].

2.1. Discrete, non-overlapping generations

Suppose that at a given gene locus the I alleles $A_1, \dots, A_I$ occur. We denote the frequency of allele A_i by p_i and the fitness of genotype A_iA_j by $w_{ij} = w_{ji} \geq 0$. Then the (marginal) fitness of allele A_i and the mean fitness of the population are

$$w_i = \sum_j w_{ij} p_j \quad \text{and} \quad w = \sum_i w_i p_i = \sum_{i,j} w_{ij} p_i p_j \,, \tag{2.1}$$

respectively. The frequency p_i' of allele A_i in the next generation is given by

$$p_i' = p_i \frac{w_i}{\overline{w}} \quad \text{for } i = 1, \dots, I \,. \tag{2.2}$$

This recursion preserves the relation[1] $\sum_i p_i = 1$ and describes the evolution of allele frequencies at a single autosomal locus in a diploid population. It describes a discrete dynamical system on the simplex

$$\mathsf{S}_I = \{p = (p_1, \dots, p_I)^T \in \mathbb{R}^I : \sum_i p_i = 1,\, p_i \geq 0,\, i = 1, \dots, I\} \,. \tag{2.3}$$

For multiplicative fitnesses, i.e., if constants v_i exist such that $w_{ij} = v_i v_j$ for every i and j, (2.2) reduces to the selection dynamics of a haploid species. It has the same structure as (2.2) but the linear term w_i is replaced by the constant v_i, and $\bar{w}$ by $\overline{v}$.

---

[1]Throughout, sums or products without ranges indicate summation over all admissible indices, e.g., $\sum_i = \sum_{i=1}^I$.

The equilibria of (2.2) are the solutions of the system

$$p_i(w_i - \bar{w}) = 0\,, \quad i = 1, \dots, I\,. \tag{2.4}$$

Obviously, the monomorphic states ($p_i = 1$ for some allele A_i) are always equilibria. If equilibria are isolated, then at most $2^I - 1$ equilibria exist. If an internal equilibrium (i.e., $p_i > 0$ for every i) exists, it is uniquely determined. If an internal equilibrium is (asymptotically) stable, then the (asymptotic) stability is global. This follows from (2.5) below because $\bar{w}$ is concave then.

The selection dynamics (2.2) has the important property that mean fitness is nondecreasing along trajectories. More precisely, it has been shown [4, 5] that

$$\Delta\bar{w} \geq 0 \quad \text{and} \quad \Delta\bar{w} = 0 \quad \text{only at equilibria}\,, \tag{2.5}$$

where $\Delta\bar{w} = \bar{w}' - \bar{w}$. A particularly elegant proof can be found in [6]. Fisher not only stated that mean fitness is nondecreasing but that its rate of change is equal to the additive genetic variance in fitness,

$$\sigma_{\mathrm{A}}^2 = 2\sum_i p_i(w_i - \bar{w})^2\,. \tag{2.6}$$

In general, σ_{A}^2 is strictly smaller than the total genetic variance, which is defined by $\sigma_{\mathrm{G}}^2 = \sum_{i,j} p_i p_j (w_{ij} - \bar{w})^2$, and $\sigma_{\mathrm{A}}^2 = \sigma_{\mathrm{G}}^2$ holds if there is no dominance.

The classical interpretation of the FTNS has been that

$$\Delta\bar{w} = \sigma_{\mathrm{A}}^2/\bar{w}\,, \tag{2.7}$$

at least approximately. Unless there is no dominance, (2.7) does generally not hold exactly. Importantly, error estimates of the degree of failure of the classical interpretation of the FTNS have been derived. Let

$$s = (\max_{i,j} w_{ij} - \min_{i,j} w_{ij})/\min_{i,j} w_{ij} \tag{2.8}$$

denote the biggest selection coefficient. It measures the maximum strength of selection and, hence, the maximum response to selection,

$$\max_i \frac{|\Delta p_i|}{p_i} \leq s\,, \tag{2.9}$$

where Δp_i is the change in gene frequency across one generation (which, as a consequence of the Hardy-Weinberg law, equals the change caused by selection).

Nagylaki [7] proved

$$\Delta\bar{w} = \frac{\sigma_{\mathrm{A}}^2}{\bar{w}}(1 + E)\,, \tag{2.10}$$

where

$$|E| \leq \tfrac{1}{2}s\,, \tag{2.11}$$

and $E = 0$ in the absence of dominance. In particular, for weak selection ($s \ll 1$), the following asymptotic version of the FTNS is valid:

$$\Delta\bar{w} = \frac{\sigma_{\mathrm{A}}^2}{\bar{w}} + O(s^3)\,. \tag{2.12}$$

If, in addition and without loss of generality, $\max_{i,j} w_{ij}$ is normalized to 1, then $\bar{w} \approx 1$ and $\Delta\bar{w} \approx \sigma_{\mathrm{A}}^2$. Because typical selection coefficients among segregating alleles rarely exceed 10%, and often are much smaller, $\Delta\bar{w} \approx \sigma_{\mathrm{A}}^2$ can be expected to provide a good approximation to reality if viability selection acts on a single locus. In [8] it was proved that (2.12) holds if fertility and viability selection both are weak.

Before we turn to generalizations and limitations, we briefly treat the continuous-time model for which some nice and important mathematical properties can be derived with great ease. In particular, we present alternative representations of the allele-frequency dynamics that provide deeper insight into the evolutionary dynamics and also in they way how the selection response depends on the variational properties of a population.

2.2. Continuous time

The preferable way to derive the continuous-time version of the selection dynamics (2.2) is to rescale fitnesses and time according to

$$w_{ij} = 1 + \epsilon m_{ij} \quad \text{and} \quad t = \lfloor \tau/\epsilon \rfloor , \tag{2.13}$$

where m_{ij} is called the Malthusian fitness (or parameter) of A_iA_j and ϵ can be interpreted as the selection intensity or the generation time. Let $m_i = \sum_j m_{ij}p_j$ and $\bar{m} = \sum_{i,j} m_{ij}p_ip_j$ denote the marginal fitness of allele A_i and the mean fitness of the population, respectively. Then, setting $q_i = q_i(\tau) = p_i(t)$, (2.2) is rearranged such that the limit of $(q_i(\tau+\epsilon) - q_i(\tau))/\epsilon$ as $\epsilon \to 0$ can be computed. Returning to the notation p_i and t instead of q_i and τ, one obtains the weak-selection approximation

$$\dot{p}_i = p_i(m_i - \bar{m}) , \quad i = 1, \ldots, I . \tag{2.14}$$

This is a dynamical system on the simplex S_I which has the same equilibria as the corresponding discrete-time system.

A proper derivation of (2.14) from biological principles requires the inclusion of age structure. However, when birth and selection occur continuously, Hardy-Weinberg proportions no longer hold and the model needs to be formulated in terms of genotype frequencies. Equation (2.14) is obtained only if a stable age distribution is assumed and Hardy-Weinberg proportions are (inconsistently) imposed (see [9] for a lucid treatment).

For the dynamics (2.14), a one-line calculation shows that

$$\dot{\bar{m}} = \sigma_{\mathrm{A}}^2 , \tag{2.15}$$

where $\sigma_{\mathrm{A}}^2 = 2\sum_i p_i(m_i - \bar{m})$ is the additive genetic variance in Malthusian fitness. According to the classical view, this has been the quintessence of the FTNS.

The selection dynamics (2.14) can be represented in other ways which provide interesting perspectives and insights. To this aim, we define the indicator variable

$$f_i(A_kA_l) = \begin{cases} 1 & \text{if } k = l = i\,, \\ \frac{1}{2} & \text{if } k \neq l \text{ and } k = i \text{ or } l = i\,, \\ 0 & \text{otherwise}\,, \end{cases} \tag{2.16}$$

where i is fixed and k and l are independent and have probability distribution p. Thus, f_i measures the frequency of allele A_i in a given genotype. It has expectation p_i. The associated $I \times I$ covariance matrix is designated $\mathbf{G}_p = (g^{ij})$, where

$$g^{ij} = \mathrm{Cov}(f_i, f_j) = \tfrac{1}{2}p_i(\delta_{ij} - p_j)\,, \tag{2.17}$$

and δ_{ij} is the Kronecker delta.

Then the allele-frequency dynamics (2.14) can be written as a generalized gradient system on S_I [10, 11],

$$\dot{p} = \mathbf{G}_p \nabla \bar{m} = \mathbf{G}_p \left(\frac{\partial \bar{m}}{\partial p_1}, \ldots, \frac{\partial \bar{m}}{\partial p_n} \right)^T . \tag{2.18}$$

Hence, with respect to the associated metric, $\dot{p}$ is perpendicular to the level surfaces of $\bar{m}$. We also point out that the matrix $\mathbf{G}_p$ plays a central role in the diffusion approximation of the multiallelic Wright-Fisher model. There, it occurs not only in the drift coefficients but also as the matrix of diffusion coefficients in the generator. As a consequence, the stationary distribution can be determined explicitly for general selection and a certain class of mutation matrices [3, Appendix E]

For a diallelic locus, the single frequency $p = p_1$ is sufficient to describe the state of the system. Then (2.18) simplifies to Wright's [12] result

$$\dot{p} = \frac{p(1-p)}{2} \frac{\mathrm{d}\bar{m}}{\mathrm{d}p} . \tag{2.19}$$

Finally, if we define the indicator variable $m(A_kA_l) = m_{kl}$, a simple calculation establishes the covariance formulation [13]

$$\dot{p}_i = \mathrm{Cov}(f_i, m) \tag{2.20}$$

which, in fact, holds under much more general circumstances [14, 15]. Both representations, (2.18) and (2.20), have important analogs in quantitative genetics that will be treated below.

2.3. Limitations

If additional evolutionary forces are allowed to act, then mean fitness does often not increase, not even in the continuous-time approximation. For instance, if mutation is included, then already with three alleles stable limit cycles can exist [16, 17]. The examples require that mutation rates and selection coefficients are of similar magnitude. For a detailed and very general treatment of mutation-selection models, including a continuum of possible alleles, see [3, Chapters 3, 4, 6] Under frequency-dependent selection, mean fitness may be decreasing along all trajectories [18] (see [19, 20] for a detailed investigation). Further, non-random mating, fertility selection, and migration may lead to a decrease of mean fitness and to complex dynamics. We study multiple loci and recombination in more detail below.

3. Multiple loci

For two recombining loci under selection, the existence of stable limit cycles has been established for continuous-time [21, 22] and discrete-time models [23, 24]. Therefore, in general, mean fitness does not increase. Essentially, the demonstration of limit cycles requires that selection coefficients and recombination rates are of similar magnitude and that fitnesses of loci interact non-additively, i.e., there is epistasis. However, for weak selection as well as for weak epistasis, we shall see that the evolutionary dynamics is simple and extensions of the single-locus results on the increase of mean fitness are available.

There are several reasons why the complexity of multilocus models exceeds that of single-locus models by far. First, it is not sufficient to use the allele frequencies at the different loci. Instead, the evolutionary dynamics needs to be formulated in terms of gamete frequencies. This is so because selection generates nonrandom associations, called linkage disequilibria, among the alleles at different loci. Recombination breaks up these associations to a certain extent but changes gamete frequencies in a quite complex way. If all gamete frequencies are the products of the frequencies of the constituent alleles, then the population is said to be in linkage equilibrium. If this is the case, then the dynamics is simplified greatly and no complex behavior can occur. For reasons of continuity, one also expects simple dynamical behavior in quasi-linkage equilibrium, i.e., close to linkage equilibrium. These considerations can be made precise in a way outlined below. First, however, we need to introduce the multilocus selection model. Essentially, our formulation follows [25] and [26].

3.1. The multilocus selection model

As before, we consider a diploid, randomly mating population with discrete, nonoverlapping generations, in which the two sexes need not be distinguished. Selection acts through differential viabilities on juveniles, which are time and frequency independent (although this can be relaxed to a certain extent). The linkage map is arbitrary but mutation and random genetic drift are ignored.

The genetic system consists of $L \geq 1$ loci. At locus n there are $I_n \geq 2$ alleles, $A^{(n)}_{i_n}$ $(i_n = 1, \dots, I_n)$. We use the multi-index $i = (i_1, \dots, i_L)$ as an abbreviation for the gamete $A^{(1)}_{i_1} \dots A^{(L)}_{i_L}$ and write $I = \prod_n I_n$ for the number of gametes. We use the letters i, j, l for gametes and k, n for loci.

Let $p_i = p_i(t)$ represent the frequency of gamete i among zygotes in generation t, and $p = (p_1, \dots, p_I)^T$ the vector of all gamete frequencies. The frequency of allele $A^{(n)}_{i_n}$ among gametes is

$$p^{(n)}_{i_n} = \sum\nolimits_{i|i_n} p_i\,, \tag{3.1}$$

where the sum runs over all multi-indices i with the nth component fixed as i_n. Let w_{ij} denote the fitness of genotype ij. We designate the marginal fitness of gamete i and the mean fitness of the population by

$$w_i = w_i(p) = \sum_j w_{ij} p_j \quad \text{and} \quad \bar{w} = \bar{w}(p) = \sum_{i,j} w_{ij} p_i p_j \,, \tag{3.2}$$

respectively.

After selection the frequency of the genotype ij is $p_i p_j w_{ij}/\bar{w}$. The frequency of gamete i in the next generation, i.e., after recombination and reproduction, is

$$p_i' = \frac{1}{\bar{w}} \sum_{j,l} R_{i,jl} p_j p_l w_{jl}, \tag{3.3}$$

where $R_{i,jl}$ is the probability that during gametogenesis the paternal haplotypes j and l produce a gamete i by recombination.

The complications introduced by recombination are disguised by the terms $R_{i,jl}$ which depend on the recombination frequencies among all subsets of loci. To obtain an analytically useful representation of the dynamics (3.3), more effort is required.

Let $\{\mathsf{K}, \mathsf{N}\}$ be a nontrivial decomposition of the set L of all loci, i.e., K and its complement $\mathsf{N} = \mathsf{L} \setminus \mathsf{K}$ are each proper subsets of L and contain at least one locus. (The decompositions $\{\mathsf{K}, \mathsf{N}\}$ and $\{\mathsf{N}, \mathsf{K}\}$ are identified.) We designate by c_K the probability of reassociation of the genes at the loci in K, inherited from one parent, with the genes at the loci in N, inherited from the other.

We designate the recombination frequency between loci k and n, where $k < n$, by c_{kn}. It is given by

$$c_{kn} = \sum_{\mathsf{K} \in \mathsf{L}_{kn}} c_\mathsf{K} \,, \tag{3.4}$$

where $\mathsf{L}_{kn} = \{\mathsf{K} \subset \mathsf{L} : k \in \mathsf{K} \text{ and } n \in \mathsf{N}\}$. We assume that all pairwise recombination rates c_{kn} are positive. Hence,

$$c_{\min} = \min_{k,n:k<n} c_{kn} > 0 \,. \tag{3.5}$$

We define

$$D_i = \frac{1}{\bar{w}} \sum_j \sum_\mathsf{K} c_\mathsf{K} \left(w_{ij}\, p_i p_j - w_{i_\mathsf{K} j_\mathsf{N}, j_\mathsf{K} i_\mathsf{N}}\, p_{i_\mathsf{K} j_\mathsf{N}}\, p_{j_\mathsf{K} i_\mathsf{N}} \right) , \tag{3.6}$$

where $\sum_\mathsf{K}$ runs over all (different) decompositions $\{\mathsf{K}, \mathsf{N}\}$ of L, and $i_\mathsf{K} j_\mathsf{N}$ is the gamete consisting of alleles $A_{i_k}^{(k)}$ and $A_{j_n}^{(n)}$ at loci in K and N, respectively. D_i is a measure of linkage disequilibrium in gamete i. Then the recursion equations describing the evolution of gamete frequencies are given by ([25] and [3, p. 56])

$$p_i' = p_i \frac{w_i}{\bar{w}} - D_i \,. \tag{3.7}$$

Obviously, this is a much more complicated dynamical system (on S_I) than the single-locus selection dynamics (2.2).

Let

$$\Lambda_0 = \left\{p : p_i = p_{i_1}^{(1)} \cdot \dots \cdot p_{i_L}^{(L)}\right\} \subseteq \mathsf{S}_I \tag{3.8}$$

denote the *linkage-equilibrium manifold* (also called the Wright manifold). If there is no position effect, i.e., if $w_{ij} = w_{i_K j_N, j_K i_N}$ for every i, j, and K, then $D_i = 0$ for every $p \in \Lambda_0$. Hence,

$$\Lambda_0 \subseteq \{p : D = 0\}\,, \tag{3.9}$$

where $D = (D_1, \dots, D_I)^T$ is the vector of all linkage disequilibria. In the absence of selection equality holds in (3.9).

3.2. The additive genetic variance

Below, and in evolutionary genetics in general, the so-called additive genetic variance, σ_{A}^2, plays a central role. Its definition is complicated because it is based on the concept of the additive (or average) effect $\alpha_{i_n}^{(n)}$ of an allele $A_{i_n}^{(n)}$. In words, the additive effects provide the best linear approximation to the deviations of the fitnesses from mean fitness. More precisely, the $\alpha_{i_n}^{(n)}$ are obtained by minimizing the quantities

$$(w_{ij} - \bar{w}) - (\alpha_i + \alpha_j)\,, \tag{3.10}$$

where

$$\alpha_i = \sum_n \sum_{i_n} \alpha_{i_n}^{(n)}\,, \tag{3.11}$$

in the sense of a least-squares approximation. If all genotype frequencies are positive, the vector $(\alpha_{i_n}^{(n)})$ is obtained as the unique solution of an $I \times I$-dimensional linear system (e.g., [3, Chap. II.3]).

It is important to note that, in general, the additive effects are not constant but depend on the frequency distribution p. In the present case, where we assume Hardy-Weinberg proportions, they depend only on the allele frequencies and the pairwise linkage disequilibria. Essentially, this procedure was introduced by Fisher [27] and is the basis of an analysis of variance (see also [1]).

Given the additive effects and assuming Hardy-Weinberg proportions, the additive genetic variance is defined as

$$\sigma_{\mathrm{A}}^2 = 2 \sum_i p_i \alpha_i^2\,. \tag{3.12}$$

The additive genetic variance may evolve as a consequence of changes in allele frequencies or additive effects (see Section 4.4). The latter may change as a consequence of changes in the (pairwise) linkage disequilibria, even if allele frequencies remain constant. In the absence of epistasis and of dominance, the additive effects are constant and σ_{A}^2 is equal to the total genetic variance.

In general, the total genetic variance can be written as the sum of the additive genetic variance and the residual variance. The latter can be decomposed further into variance components arising from nonadditive effects within loci (dominance), nonadditive effects between loci (epistasis), and all kinds of interactions.

For detailed and much more general treatments we refer to [3, Chapter II.3] and especially to [28, Chapter 6] and [25].

3.3. Weak epistasis

Weak epistasis means that the fitness scheme has the form

$$w_{ij} = \sum_n u^{(n)}_{i_n j_n} + \epsilon e_{ij}\,, \tag{3.13}$$

where ϵ, the strength of epistasis, is sufficiently small and $|e_{ij}| \le 1$. We assume $u^{(n)}_{i_n j_n} > 0$.

If $\epsilon = 0$, then there is no (additive) epistasis. We review this case first. Let s be defined as in (2.8), where now ij is a multilocus genotype. Thus, s is the biggest multilocus selection coefficient. We call a point $p \in \mathsf{S}_I$ a selection equilibrium if it satisfies $p_i(w_i - \bar{w}) = 0$ for every i. The set of all points which are a selection equilibrium for $\epsilon = 0$ is denoted by F.

We summarize the main results that hold for (3.7) if epistasis is absent.

Theorem 3.1.

(a) *Mean fitness is nondecreasing,* $\bar{w}' \ge \bar{w}$ [29], *and* $\bar{w}' = \bar{w}$ *holds exactly on the set* F *of selection equilibria* [26].

(b) $\Delta\bar{w} = \sigma^2_{\mathrm{A}}/\bar{w} + O(s^3)$ *as* $s \to 0$ [30].

(c) $\Delta\bar{w} = \dfrac{\sigma^2_{\mathrm{A}}}{\bar{w}}(1+E)$, *where* $|E| \le \frac{1}{2}s$, *and* $E = 0$ *if dominance is absent* [7].

(d) *Every trajectory of* (3.7) *converges to an equilibrium point* [31].

(e) *A point* p *is an equilibrium of* (3.7) *if and only if it is both a selection equilibrium for each locus (with fitnesses given by the* $u^{(n)}_{i_n j_n}$*) and it is in linkage equilibrium* [26, 32].

We point out that (c) implies that mean fitness remains constant if there is no additive genetic variation. Nevertheless, gamete frequencies may change due to linkage disequilibrium.

If fitnesses are multiplicative instead of additive, then mean fitness may decrease and (with two diallelic loci) two asymptotically stable equilibria in linkage disequilibrium, one with $D > 0$ the other with $D < 0$, may coexist [33, 34]. In particular, the linkage-equilibrium manifold Λ_0 is in general not attracting. However, in contrast to the case of additive fitnesses, Λ_0 is forward invariant. For a review, see [3, Chap. II.1].

The next theorem summarizes the results that can be extended to weak epistasis.

Theorem 3.2 ([26]). *If, for* $\epsilon = 0$, *every equilibrium of* (3.7) *is hyperbolic, as is generic, then for sufficiently small* ϵ *the following hold:*

(a) *Every equilibrium is within* $O(\epsilon)$ *of the corresponding equilibrium (on* $D = 0$*) with* $\epsilon = 0$ *and has the same stability properties. Unstable boundary equilibria may leave the state space if perturbed.*

(b) *Every trajectory converges to an equilibrium point.*
(c) *If p is bounded away from the set* F, *then* $\Delta\overline{w} > 0$.

The mathematical essence of this result is that it is a global perturbation result which not only establishes local stability properties of a perturbed equilibrium but proves that the perturbed dynamics is qualitatively the same as the much simpler dynamics with $\epsilon = 0$. If $\epsilon > 0$, it is no longer true that mean fitness is nondecreasing. Nagylaki et al. [26] gave an example in which for every $\epsilon > 0$, $\Delta\overline{w} \leq 0$ holds near and at the set F, and equality holds only at equilibrium. However, mean fitness increases until evolution has nearly come to a halt, which occurs when a neighborhood of F has been reached.

General results such as those above are biologically relevant because epistasis is wide spread, and sign and magnitude can vary widely. They provide the mathematical justification to investigate specific models by perturbation methods.

3.4. Weak selection

To study weak selection, we assume

$$w_{ij} = 1 + s\omega_{ij}\,, \tag{3.14}$$

where the selection intensity s is a nonnegative, sufficiently small parameter, and the selection coefficients satisfy $|\omega_{ij}| < 1$. We speak of weak selection (or strong recombination) if s is small relative to the minimum two-locus recombination rate $c_{\min}$ (3.5).

In the absence of selection ($s = 0$), the dynamics (3.7) is degenerate because every point $p \in \Lambda_0$ is an equilibrium. In addition, Λ_0 is invariant and globally attracting at a geometric rate [25, 35]. For generic conditions, the rate of approach is $1 - c_{\min}$.

For weak selection, the theory of normally hyperbolic manifolds implies the existence of a smooth invariant manifold Λ_s close to Λ_0, which is globally attracting at a geometric rate for (3.7). The manifold Λ_s is characterized by an equation of the form $D = s\zeta(\rho, s)$, where ζ is a smooth function of the vector ρ of all gene frequencies. Thus, on Λ_s, and more generally, for any initial values, after a long time, $D(t) = O(s)$. It follows that on Λ_s, linkage disequilibria are of order s and, in fact, change very slowly, i.e., $\Delta D(t) = O(s^2)$. Therefore, Λ_s is called the quasi-linkage equilibrium manifold [26].

Let us define the gametic, allelic, and average selection coefficients at linkage equilibrium by

$$\omega_i = \sum_j \omega_{ij} \prod_n p_{i_n}^{(n)}\,, \quad \omega_{i_n}^{(n)} = \sum_{i|i_n} \omega_i \prod_{k:k\neq n} p_{i_k}^{(k)}\,, \quad \overline{\omega} = \sum_i \omega_i \prod_n p_{i_n}^{(n)}\,, \tag{3.15}$$

respectively. Then the dynamics in a neighborhood of Λ_s can be shown to be a perturbation of the so-called *weak-selection limit*,

$$\dot{p}_{i_n}^{(n)} = p_{i_n}^{(n)}\left(\omega_{i_n}^{(n)} - \overline{\omega}\right)\,, \tag{3.16}$$

which 'lives' on $\mathsf{S}_{I_1} \times \cdots \times \mathsf{S}_{I_L}$, the natural parameterization of the linkage-equilibrium manifold Λ_0.

The main results can be summarized as follows:

Theorem 3.3 ([25], [26]). *Suppose that, as is generic, all equilibria of* (3.16) *are hyperbolic and s is sufficiently small.*

(a) *The set of equilibria of* (3.7) *contains only isolated points and every equilibrium is within an $O(s)$ neighborhood of the corresponding equilibrium of* (3.16) *on* Λ_0.
(b) *There is a one-to-one correspondence between these equilibria. Every pair has the same stability properties.*
(c) *Every trajectory of the multilocus dynamics* (3.7) *converges to an equilibrium point.*
(d) *The so-called Asymptotic Fundamental Theorem of Natural Selection holds: If $t \geq t^* \sim 2 \ln s / \ln(1 - c_{\min})$, then*

$$\Delta \bar{w} = \frac{\sigma_{\mathrm{A}}^2}{\bar{w}} + O(s^3) \quad as \quad s \to 0\,. \tag{3.17}$$

(e) *If $t \geq t^*$ and p is bounded away from the set of equilibria, then* $\Delta \bar{w}(p) > 0$.

These results have several important consequences. (i) They show that no complex dynamical behavior can occur. (ii) Under weak selection the equilibrium and stability structure can be inferred from the much simpler weak-selection limit (3.16). The latter is not only a generalized gradient system, analogous to (2.18), but its dimension is (much) lower than that of the full dynamics. If greater accuracy than provided by the weak-selection limit is needed, quasi-linkage equilibrium approximations can sometimes be developed (e.g., [36]). (iii) The mean fitness may decrease during the short period, of order $\ln(1/s)$ generations, of approach to Λ_s. Because this initial period is short, relatively little change in allele frequencies occurs during it. After a long time, $\bar{w}$ may also decrease close to equilibrium. For intermediate times, $\bar{w}$ increases. This is during the period when most of the (evolutionary important) gene frequency-change occurs.

Finally, let us assume that both selection and recombination are weak, i.e., for small $\epsilon \geq 0$ and constants m_{ij} and r_{K}, we posit

$$w_{ij} = 1 + \epsilon m_{ij} \quad \text{and} \quad c_{\mathsf{K}} = \epsilon r_{\mathsf{K}} \tag{3.18}$$

for all gametes i, j and subsets $\mathsf{K} \subseteq \mathsf{L}$. Then, proceeding similarly as in Section 2.2, the continuous-time dynamics

$$\dot{p}_i = p_i(m_i - \bar{m}) - \sum_{\mathsf{K}} r_{\mathsf{K}} \left[p_i - p_{i_{\mathsf{K}}}^{(\mathsf{K})} p_{i_{\mathsf{N}}}^{(\mathsf{N})} \right] \tag{3.19}$$

is derived, where $p_{i_{\mathsf{K}}}^{(\mathsf{K})}$ designates the marginal frequency of the gamete with components i_k for the loci $k \in \mathsf{K}$ fixed. Theorems 3.2 and 3.3 also hold for (3.19).

3.5. Generalizations

Without giving much detail, we mention that some of the above results can be generalized to models that include migration among subpopulations or frequency-dependent selection in a panmictic population.

If epistasis and migration are weak, then statements (a) and (b) of Theorem 3.2 remain valid [37, Theorem 5.4]. If selection is weak relative to migration and recombination, then statements (a), (b), (c), and (e) of Theorem 3.3 hold subject to the following modifications [37, Theorems 4.3 and 4.8]: (i) Λ_0 is replaced by the manifold on which linkage equilibrium holds and gamete frequencies are spatially homogeneous; (ii) the weak-selection limit describes the change of spatially averaged allele frequencies. The single-locus case was treated in [38].

For a multilocus model, in which frequency-dependent selection is caused by intraspecific competition for a continuum of resources, a weak-selection limit was derived and analyzed in [39]. It was developed further by Schneider [40, 41], who applied it to study long-term evolution under disruptive frequency-dependent selection. Not surprisingly, the conclusions drawn on the basis of a proper population-genetic analysis partially differ from those derived from an adaptive dynamics approach which, as is usual in ecological modeling, ignores intraspecific variation.

4. Quantitative traits

Quantitative traits are characters that vary (almost) continuously and can be measured on a metric scale. Typical examples include body weight or height, abdominal bristle number in *Drosophila*, milk yield, oil content in maize, and, more generally, many morphological, physiological, or economically important traits. Also fitness can be considered as a quantitative trait. Quantitative traits have genetic and environmental components. They have a complex genetic basis in the sense that they are determined by many genes, most of which have small effects although some may have very large effects. In addition, epistasis, pleiotropy (which means that one gene affects more than one trait), and genotype-environment interaction are common (e.g., [3, 42, 43]). An important feature is that, on an appropriate scale, quantitative traits are often normally distributed. In view of their multifactorial determination and the Central Limit Theorem this is not too surprising.

In the models treated below we assume that there is no genotype-environment interaction and that the environmental contribution E can be treated as white noise, i.e., as a Gaussian random variable with mean zero. Thus, we can write the phenotypic value P as

$$P = G + E\,, \tag{4.1}$$

where the genetic and environmental components, G and E, are independent. Unless stipulated otherwise, we assume that G is determined additively, i.e., trait effects can be assigned directly to alleles and G is the sum of the effects of the paternal and maternal alleles of an individual at the contributing loci. Then all genetic variance is additive (Section 3.2) and the phenotypic variance can be written as

$$\sigma_{\mathrm{P}}^2 = \sigma_{\mathrm{A}}^2 + \sigma_{\mathrm{E}}^2 \,. \tag{4.2}$$

In general, the phenotypic variance can be decomposed into a genetic, an environmental, and an interaction component. As indicated in Section 3.2, the genetic variance can be decomposed further into additive, dominance, epistatic, and various interaction components.

Of fundamental importance in quantitative genetics is the (narrow sense) heritability, h^2. It is defined as the ratio of additive genetic to total phenotypic variance:

$$h^2 = \frac{\sigma_{\mathrm{A}}^2}{\sigma_{\mathrm{P}}^2} \,. \tag{4.3}$$

The heritability can be estimated from correlations among relatives [28]. Most traits, in most populations, show substantial heritabilities (typically between 20% and 50%), and a few patterns have been identified (see, e.g., [43]). Thus, much trait variation is inherited, in fact more than can be explained by simple universal mechanisms (see below).

What distinguishes quantitative-genetic models from many other models in biology is that they are formulated in terms of quantities that are measurable (and not only in the lab or in breeding programs but also, with greater effort, in nature). The basis for this theory was laid by Fisher and Wright. Over the decades the additive model has received widespread empirical support in the sense that cases where the additive-genetic component of variance is not the dominating term in the total genetic component occur rarely. This does, however, not imply that epistasis is irrelevant because functional epistasis typically generates additive effects, hence contributes to the additive variance.

4.1. The breeder's equation

With the help of the heritability, the response to selection in breeding experiments can be predicted. Let $\overline{P}$ and $\overline{P}_{\mathrm{s}}$ denote the mean phenotype before and after selection, respectively. Because the environment contributes only white noise, we have $\overline{P} = \overline{G}$. The so-called *breeder's equation* states that the selection response across generations is

$$\Delta\overline{P} = h^2(\overline{P}_{\mathrm{s}} - \overline{P}) \,. \tag{4.4}$$

It can be derived without recourse to detailed genetic assumptions, simply by regressing offspring values on mid-parent values. This is based on the assumptions that there are no non-genetic causes for resemblance of relatives and that the regression is indeed linear ([28]; see [44] for discussion). Nevertheless, systematic departures from (4.4) apparently require quite extreme assumptions [45]. The importance of the breeder's equation arises from the fact that it allows predictions of the selection response from readily measurable quantities.

4.2. The Secondary Theorem of Natural Selection

Assuming a linear offspring-parent regression, Robertson [46, 47] extended the breeder's equation to predict the correlated response caused by overall (artificial and natural) selection:

$$\Delta\overline{G} = \mathrm{Cov}_{\mathrm{A}}(G,W)/\overline{W}\,. \tag{4.5}$$

Here, $W = W(G)$ denotes the fitness of individuals with genotypic value G, and $\mathrm{Cov}_{\mathrm{A}}(G,W)$ is the additive genetic covariance of G and W (for precise definitions, see [25] or [3, Chap. II.3]). This is sometimes called the Secondary Theorem of Natural Selection and is closely related to the Li-Price equation (2.20).

Nagylaki [25] extended this result and derived an asymptotic version from first principles and in great generality. He admitted arbitrary dominance and epistasis, both at the level of the trait and of fitness, but assumed that selection is weak relative to recombination ($s \ll c_{\min}$ in the notation of Section 3). He proved

$$\Delta\overline{G} = \mathrm{Cov}_{\mathrm{A}}(G,W)/\overline{W} + O(s^2)\,, \quad t \geq t^*\,, \tag{4.6}$$

where t^* is defined as in Theorem 3.3. If $G = W$, then (3.17) is obtained.

These are important results that provide deep insight into the way how quantitative traits respond to selection. However, they can not necessarily be used to predict evolutionary trajectories for several or many generations because $\mathrm{Cov}_{\mathrm{A}}(G,W)$ and $\overline{W}$ may and will change under selection. Moreover, if there is epistasis and either selection is strong or linkage disequilibria are large, changes in the linkage disequilibria can induce large changes in $\overline{G}$ even if $\sigma_{\mathrm{A}}^2 = 0$ (e.g., [25, 48]).

4.3. Lande's equation and multivariate evolution

In nature, and often also in breeding programs, selection acts on several or many traits. To study the evolution of multivariate phenotypes, Lande [49] developed a theory for their evolutionary dynamics. The phenotype of an individual is characterized by a vector of measurements of K quantitative traits, $P = (P_1,\ldots,P_K)^T$. It is assumed to be determined by an additive genetic component G and an environmental component E such that $P = G + E$, where the mean of E vanishes. Hence, the mean vectors satisfy $\overline{P} = \overline{G}$. Lande's central, empirically testable, assumption is that the distributions of G (sometimes called the breeding values) and of E are both multivariate normal and independent. If the corresponding covariance matrices are denoted by $\mathbf{P}$, $\mathbf{G}$, and $\mathbf{E}$, then $\mathbf{P} = \mathbf{G} + \mathbf{E}$ holds. If the fitness of individuals with phenotype P is $W(P)$ and $f(P)$ denotes the (Gaussian) density of phenotypes, the mean fitness of the population is $\overline{W} = \int f(P)W(P)\mathrm{d}P$, which is a function of $\overline{P}$ and $\mathbf{P}$.

Lande showed that the change of the mean phenotype between generations is

$$\Delta\overline{P} = \mathbf{G}\nabla\ln\overline{W} = \mathbf{G}\left(\frac{\partial\ln\overline{W}}{\partial\overline{P}_1},\ldots,\frac{\partial\ln\overline{W}}{\partial\overline{P}_K}\right)^T\,, \tag{4.7}$$

where $\nabla\ln\overline{W}$ is called the selection gradient. In the univariate case, this can be shown to be equivalent to Robertson's equation (4.5). Also the analogy to the

Svirezhev-Shahshahani gradient (2.18) is notable. Moreover, it can be shown that (4.7) implies

$$\Delta \ln \overline{W} \geq 0\,, \tag{4.8}$$

and mean fitness is (strictly) increasing except at equilibria.

Lande's theory integrated quantitative-genetic models and methods into evolutionary genetics and had a huge impact on evolutionary biology. It has received innumerable applications and is of great heuristic and predictive value. Equation (4.7) also reveals that selection on ignored, correlated traits may greatly obscure the selection response of the trait(s) under consideration [50].

Lande also established the powerful integrative concept of an adaptive landscape for phenotypic traits. (A related concept for genotypes had been introduced by Wright in 1931.) The adaptive landscape summarizes the selection pressures that act on a *population* and direct its evolution. Its height represents the mean fitness of the population as a function of the mean phenotype. Its slope and curvature determine the strength of directional and stabilizing selection, respectively. Equation (4.7) shows that the population mean evolves upwards on this surface, toward an adaptive peak. The precise direction is given by $\mathbf{G}$ (the so-called $\mathbf{G}$ matrix) and is usually not perpendicular to the level surfaces. Ecological change can be captured by models of peak movement (for a review, see [51]).

The strength and attractiveness of Lande's theory arises from the simple and intuitive nature of his equation and from the fact that no detailed genetic information is necessary (though measuring W and $\mathbf{G}$ in natural populations is usually cumbersome and laborious). The derivation of Lande's equation does not specify the mechanisms that maintain variation and, in order to predict the evolutionary response for more than one generation, a constant $\mathbf{G}$-matrix has to be assumed. In particular, this assumption of constancy has been criticized severely because, in general, $\mathbf{G}$ evolves and predicting its evolution is difficult [52].

In the meantime, the stability properties of the $\mathbf{G}$-matrix have been investigated using individual-based simulations of explicit multilocus models that include recurrent pleiotropic mutation. For a constant and for a moving peak, conditions have been identified when a bivariate $\mathbf{G}$-matrix (or some if its characteristics such as its total size, or shape, or orientation) remains approximately stable despite random genetic drift (e.g., [53, 54], see [51] for a comprehensive review). Also substantial effort has been devoted to derive the evolution of the whole distribution of genotypic values under general selection scenarios. Due to the inherent complexity, these approaches were mainly restricted to univariate traits and we discuss them below.

4.4. The evolutionary dynamics derived from genetic principles

The derivation of the evolutionary dynamics of a quantitative trait from genetic principles is a complex enterprise that faces severe difficulties. The focus has been on equations for the mean, the variance, and the higher moments under the assumption that the trait is determined additively by finitely many loci [36, 55, 56, 57]. A comprehensive account of the theory is given in [3, Chap. V].

Kirkpatrick et al. [58] developed further generalizations and implemented a computer algebra package to handle the recursions.

Here, we present only two results that complement the above theory and can be formulated easily. We represent the fitness function acting on the genotypic values G as a Taylor series,

$$W(G) = \sum_{j=0}^{\infty} s_j G^j \,. \tag{4.9}$$

(Since E has a normal distribution, the coefficients of $W(G)$ are straightforwardly calculated from the series expansion of $W(P)$.) No assumptions about the distribution of allele frequencies, of genotype frequencies, or of breeding values are made.

The distribution of G evolves as a result of the interaction of selection, mutation, recombination, and random mating. For simplicity, we assume that the distribution of mutational effects is symmetric with respect to the mean. Let us denote by M_j^0 the jth moment around zero of the distribution of G and by C_j its jth cumulant; thus, $C_1 = \overline{G}$ and $C_2 = \sigma_{\mathrm{A}}^2$. Because of their additivity property, cumulants are particularly useful to describe the effects of selection on an additive trait [57].

Then the response of the mean phenotype can be written in each of the following ways:

$$\begin{aligned}
\Delta\overline{G} &= \frac{1}{\overline{W}} \operatorname{Cov}_{\mathrm{A}}(G, W) && (4.10\mathrm{a})\\
&= \frac{1}{\overline{W}} \sum_j s_j M_{j+1}^0 - \overline{G} && (4.10\mathrm{b})\\
&= s_1\sigma_{\mathrm{A}}^2 + s_2(C_3 + 2\overline{G}\sigma_{\mathrm{A}}^2) + s_3(C_4 + 3\overline{G}C_3 + 3\sigma_{\mathrm{A}}^4 + 3\overline{G}^2\sigma_{\mathrm{A}}^2) + \cdots && (4.10\mathrm{c})\\
&= \sigma_{\mathrm{A}}^2 \frac{\partial \ln \overline{W}}{\partial \overline{G}} + C_3 \frac{\partial \ln \overline{W}}{\partial \sigma_{\mathrm{A}}^2} + C_4 \frac{\partial \ln \overline{W}}{\partial C_3} + \cdots \,. && (4.10\mathrm{d})
\end{aligned}$$

Explicit formulas instead of the dots can be given. Obviously, the first equation is Robertson's (4.5). If the distribution of G is Gaussian, the last equation yields the univariate version of Lande's (4.7) because then all cumulants of order three and higher vanish. The terms $\partial \ln \overline{W} / \partial C_k$ are called (higher-order) selection gradients.

Equations (4.10) show that, in general, the evolutionary dynamics depends not only on the variance, but also on the higher moments or cumulants of the distribution of breeding values. Importantly, however, the response of the mean does not depend on genetic details, such as number of loci, the distribution of allelic effects at particular loci, or the linkage map.

This changes dramatically when the dynamics of the higher moments is studied. We confine our attention to the additive genetic variance. Its change depends not only on all pairwise recombination rates and the mutation distribution at individual loci, but also on all pairwise linkage disequilibria and even on the pairwise associations between loci at homologous gametes within an individual that are generated by selection (see [3, p. 194]). Under the assumption of quasi-linkage equilibrium, which is fulfilled if selection is weak relative to recombination and $t \geq t^*$ (Result 3.3), the change of σ_{A}^2 across generations is given by

$$\Delta\sigma_{\mathrm{A}}^2 = s_1 C_3 + s_2 \left(C_4 + 2\overline{G}C_3 + 4\sum_{n=1}^{L} \kappa_{nn} \right) + \cdots + 2\sum_n \mu_n \gamma_n^2 + O(s^2)\,. \quad (4.11)$$

Here, κ_{nn} designates the second cumulant (= variance) of the distribution of allelic effects at locus n, μ_n is the haploid mutation rate of locus n, γ_n^2 is the variance of mutational effects at locus n, and s may be defined as $s = \sum_j |s_j|$.

Equation (4.11) shows that, unless selection is linear ($s_j = 0$ if $j \geq 2$), the response of the variance depends on higher cumulants and on genetic details. Even for equivalent loci, when $\sum_{n=1}^{L} \kappa_{nn} = \frac{1}{L}\sigma_{\mathrm{A}}^4$, the number L of loci needs to be known to predict the variance. (The total input of mutational variance per generation, $2\sum_n \mu_n \gamma_n^2$, is relatively easy to estimate from mutation-accumulation experiments.) These complications underly and explain the skepticism that has been articulated, especially by Turelli, against the use of Lande's (or any other existing) theory for inferences about long-term evolution. Nevertheless, some positive results have been obtained (e.g., [45, 53, 54, 59, 60] and [3, Chap. VII]).

The above results are based on the assumption that allelic effects contribute additively, i.e., without dominance and epistasis, to the trait. However, because the fitness function may and usually will be nonlinear, dominance and epistasis in fitness is included and not necessarily weak.

Little is known about the genotype-phenotype map, except that it is exceedingly complex. The quantitative genetics approach to treat epistasis as a statistical deviance from additivity has been highly successful, but makes epistasis more a population property than a property of the functional interaction among genes. A systematic approach to develop a population-genetic framework for studying the role of functional epistasis has been pioneered in [61]; see [62] for a progenitor. Based on this and a very similar model, interesting consequences for the evolution of quantitative traits were derived in [63, 64].

4.5. Maintenance of genetic variation

So far, we have not said anything about how genetic variation is maintained. As outlined in Ewens' chapter [1], this problem already plagued Darwin. Although Mendelian inheritance solves part of the problem, a satisfactory quantitative resolution has not yet been achieved. Many facts and open questions contribute to this failure. First, there is plenty of genetic variation in quantitative traits. For instance, heritabilities are typically in the range of 20%–50%, and short-term and long-term selection responses can be enormous. Second, many forms of selection, especially the commonly observed stabilizing selection and most forms of directional selection, tend to deplete genetic variation. Third, mutation is the ultimate source of genetic variability, and recombination disperses variability greatly, but models of mutation-selection balance can, in general, account only for a fraction of the empirically observed heritabilities. However, a final answer will not be possible before we understand the nature of quantitative genetic variation, especially the roles of regulatory versus structural genes much better (see [3] for a comprehensive treatment of this topic, and [43] for a more recent review).

High levels of genetic variation can be maintained under various forms of balancing selection, which may be caused by migration-selection balance, genotype-environment interaction, frequency-dependent selection, or pleiotropy (e.g., [20, 38, 39, 40, 65, 66, 67, 68]). However, none of these explanations is universally applicable, and quantitative predictions require not only information about the genetic basis of the trait(s), but also detailed ecological knowledge about the population.

5. Fundamental topics omitted

There are a number of fundamental topics in evolutionary genetics that have not even been mentioned above. These include speciation (see [69] for a fairly recent account focusing on theory), the evolution of sex and recombination (for a brief and recent review, see [70]), and molecular evolution.

As mentioned in the Introduction, stochastic models have been of paramount importance since the early work of Fisher and Wright. Essential further developments are due to Malécot and Kimura, who established the method of diffusion approximation as a powerful tool for the study of the evolutionary dynamics under the influence of random genetic drift. Kimura [71] also advocated the neutral theory of evolution which, upon modification and extension, has become one of the corner stones of evolutionary and population genetics. This, and much more, work is covered comprehensively by [72].

More recently, the focus in population genetics has changed to study evolution backward in time. This line of work was motivated by the desire to infer the evolutionary history of species from molecular data of extant populations. It has been fueled by the increasing availability of dense maps of genetic markers and sequence data which, under certain conditions, allow the reconstruction of ancestral relationship and past evolutionary events. Its mathematical ancestor is Kingman's [73] coalescent process. For a beautiful treatment consult [74].

References

[1] W. Ewens, *What changes has mathematics made to the Darwinian theory? In* F.A.C.C. Chalub and J.F. Rodrigues (eds.), *The Mathematics of Darwin's Legacy*, 7–26, Birkhäuser, Basel, 2011, *This issue.*

[2] R.A. Fisher, *The Genetical Theory of Natural Selection.* Clarendon Press, Oxford, 1930.

[3] R. Bürger, *The Mathematical Theory of Selection, Recombination and Mutation.* Wiley Series in Mathematical and Computational Biology. Chichester: Wiley., 2000.

[4] P.A.G. Scheuer and S.P.H. Mandel, *An inequality in population genetics.* Heredity (1959), 519–524.

[5] H.P. Mulholland and C.A.B. Smith, *An inequality arising in genetical theory.* Am. Math. Mon. **66** (1959), 673–683.

[6] J.F.C. Kingman, *A mathematical problem in population genetics.* Proc. Camb. Phil. Soc. **57** (1961), 574–582.

[7] T. Nagylaki, *Error-bounds for the fundamental and secondary theorems of natural-selection.* Proc. Natl. Acad. Sci. USA **88** (1991), 2402–2406.

[8] T. Nagylaki, *Evolution under fertility and viability selection.* Genetics **115** (1987), 367–375.

[9] T. Nagylaki and J.F. Crow, *Continuous selective models.* Theor. Popul. Biology **5** (1974), 257–283.

[10] Yu.M. Svirezhev, *Optimality principles in population genetics. In Studies in Theoretical Genetics*, 86–102, Novisibirsk: Inst. of Cytology and Genetics, 1972.

[11] S. Shahshahani, *A new mathematical framework for the study of linkage and selection.* Mem. Am. Math. Soc. **211** (1979), 34.

[12] S. Wright, *Evolution in populations in approximate equilibrium.* J. Genetics (1935), 257–266.

[13] C.C. Li, *Fundamental theorem of natural selection.* Nature **214** (1967), 505.

[14] G.R. Price, *Selection and covariance.* Nature **227** (1970), 520.

[15] S. Lessard, *Fisher's fundamental theorem of natural selection revisited.* Theor. Popul. Biol. **52** (1997), 119–136.

[16] J. Hofbauer, *The selection mutation equation.* J. Math. Biol. **23** (1985), 41–53.

[17] E. Baake and T. Wiehe, *Bifurcations in haploid and diploid sequence space models.* J. Math. Biol. **35** (1997), 321–343.

[18] S. Wright, *Genetics of populations. In Encyclopaedia Britannica, 14th ed. revised, vol 10*, 111–115, 1948.

[19] K.A. Schneider, *Maximization principles for frequency-dependent selection I: the one-locus two-allele case.* Theor. Popul. Biol. **74** (2008), 251–262.

[20] K.A. Schneider, *Maximization principles for frequency-dependent selection II: the one-locus multiallele case.* J. Math. Biol. **61** (2010), 95–132.

[21] E. Akin, *The Geometry of Population Genetics*, volume 31 of *Lecture Notes in Biomathematics.* Springer-Verlag, Berlin, 1979.

[22] E. Akin, *Cycling in simple genetic systems.* J. Math. Biol. **13** (1982), 305–324.

[23] A. Hastings, *Stable cycling in discrete-time genetic models.* Proc. Natl. Acad. Sci. USA **78** (1981), 7224–7225.

[24] J. Hofbauer and G. Iooss, *A Hopf-bifurcation theorem for difference-equations approximating a differential-equation.* Monatsh. Math. **98** (1984), 99–113.

[25] T. Nagylaki, *The evolution of multilocus systems under weak selection.* Genetics **134** (1993), 627–647.

[26] T. Nagylaki, J. Hofbauer, and P. Brunovský, *Convergence of multilocus systems under weak epistasis or weak selection.* J. Math. Biol. **38** (1999), 103–133.

[27] R.A. Fisher, *The correlation between relatives on the supposition of Mendelian inheritance.* Trans. of the Roy. Soc. of Edinburgh **52** (1918), 399–433.

[28] M.G. Bulmer, *The Mathematical Theory of Quantitative Genetics.* Oxford: Clarendon Press, 1980.

[29] W.J. Ewens, *Mean fitness increases when fitnesses are additive.* Nature **221** (1969), 1076.

[30] T. Nagylaki, *Rate of evolution of a character without epistasis.* Proc. Natl. Acad. Sci. USA **86** (1989), 1910–1913.

[31] L.A. Kun and Yu.I. Lyubich, *Convergence to equilibrium in a polylocus polyallele population with additive selection.* Probl. Inform. Transmiss. **16** (1980), 152–161.

[32] Yu.I. Lyubich, *Mathematical Structures in Population Genetics*, volume 22 of *Biomathematics.* Springer-Verlag, Berlin, 1992.

[33] S. Karlin and M.W. Feldman, *Simultaneous stability of $D = 0$ and $D \neq 0$ for multiplicative viabilities at two loci.* Genetics **90** (1978), 813–825.

[34] A. Hastings, *Simultaneous stability of $D = 0$ and $D \neq 0$ for multiplicative viabilities at two loci: an analytical study.* J. Theor. Biol. **89** (1981), 69–81.

[35] Yu.I. Lyubich, *Basic concepts and theorems of evolutionary genetics of free populations.* Russ. Math. Surv. **26** (1971), 51–123.

[36] M. Turelli and N.H. Barton, *Dynamics of polygenic characters under selection.* Theor. Popul. Biol. **38** (1990), 1–57.

[37] R. Bürger, *Multilocus selection in subdivided populations I. Convergence properties for weak or strong migration.* J. Math. Biol. **58** (2009), 939–978.

[38] T. Nagylaki and Y. Lou, *Evolution under multiallelic migration-selection models.* Theor. Popul. Biol. **72** (2007), 21–40.

[39] R. Bürger, *A multilocus analysis of intraspecific competition and stabilizing selection on a quantitative trait.* J. Math. Biol. **50** (2005), 355–396.

[40] K.A. Schneider, *A multilocus-multiallele analysis of frequency-dependent selection induced by intraspecific competition.* J. Math. Biol. **52** (2006), 483–523.

[41] K.A. Schneider, *Long-term evolution of polygenic traits under frequency-dependent intraspecific competition.* Theor. Popul. Biol. **71** (2007), 342–366.

[42] T.F.C. Mackay, *Quantitative trait loci in Drosophila.* Nat. Rev. Genet. **2** (2001), 11–20.

[43] N.H. Barton and P.D. Keightley, *Understanding quantitative genetic variation.* Nat. Rev. Genet. **3** (2002), 11–21.

[44] A. Gimelfarb and J.H. Willis, *Linearity versus nonlinearity of offspring-parent regression: An experimental-study of Drosophila melanogaster.* Genetics **138** (1994), 343–352.

[45] M. Turelli and N.H. Barton, *Genetic and statistical analyses of strong selection on polygenic traits: What, me normal?* Genetics **138** (1994), 913–941.

[46] A. Robertson, *A mathematical model of the culling process in dairy cattle.* Anim. Prod. **8** (1966), 95–108.

[47] A. Robertson, *The spectrum of genetic variation.* *In* R. Lewontin (ed.), *Population Biology and Evolutions*, 5–16, Syracuse University Press, EUA, 1968.

[48] A. Gimelfarb, *Genotypic variation for a quantitative character maintained under stabilizing selection without mutations: Epistasis.* Genetics **123** (1989), 217–227.

[49] R. Lande, *Quantitative genetic-analysis of multivariate evolution, applied to brain-body size allometry.* Evolution **33** (1979), 402–416.

[50] R. Lande and S.J. Arnold, *The measurement of selection on correlated characters.* Evolution **37** (1983), 1210–1226.

[51] S.J. Arnold, R. Bürger, P.A. Hohenlohe, B.C. Ajie, and A.G. Jones, *Understanding the evolution and stability of the G-matrix.* Evolution **62** (2008), 2451–2461.

[52] M. Turelli, *Phenotypic evolution, constant covariances, and the maintenance of additive variance.* Evolution **42** (1988), 1342–1347.

[53] A.G. Jones, S.J. Arnold, and R. Bürger, *Stability of the G-matrix in a population experiencing pleiotropic mutation, stabilizing selection, and genetic drift.* Evolution **57** (2003), 1747–1760.

[54] A.G. Jones, S.J. Arnold, and R. Bürger, *Evolution and stability of the G-matrix on a landscape with a moving optimum.* Evolution **58** (2004), 1639–1654.

[55] N.H. Barton and M. Turelli, *Adaptive landscapes, genetic-distance and the evolution of quantitative characters.* Genetical Research **49** (1987), 157–173.

[56] N.H. Barton and M. Turelli, *Natural and sexual selection on many loci.* Genetics **127** (1991), 229–255.

[57] R. Bürger, *Moments, cumulants, and polygenic dynamics.* J. Math. Biol. **30** (1991), 199–213.

[58] M. Kirkpatrick, T. Johnson, and N. Barton, *General models of multilocus evolution.* Genetics **161** (2002), 1727–1750.

[59] R. Bürger and R. Lande, *On the distribution of the mean and variance of a quantitative trait under mutation-selection-drift balance.* Genetics **138** (1994), 901–912.

[60] J.P. Reeve, *Predicting long-term response to selection.* Genet. Res. **75** (2000), 83–94.

[61] T.F. Hansen and G.P. Wagner, *Modeling genetic architecture: A multilinear theory of gene interaction.* Theor. Popul. Biol. **59** (2001), 61–86.

[62] L.A. Zhivotovsky and S. Gavrilets, *Quantitative variability and multilocus polymorphism under epistatic selection.* Theor. Popul. Biol. **42** (1992), 254–283.

[63] J. Hermisson, T.F. Hansen, and G.P. Wagner, *Epistasis in polygenic traits and the evolution of genetic architecture under stabilizing selection.* Amer. Nat. **161** (2003), 708–734.

[64] N.H. Barton and M. Turelli, *Effects of genetic drift on variance components under a general model of epistasis.* Evolution **58** (2004), 2111–2132.

[65] M. Turelli and N.H. Barton, *Polygenic variation maintained by balancing selection: Pleiotropy, sex-dependent allelic effects and $G \times E$ interactions.* Genetics **166** (2004), 1053–1079.

[66] T. Nagylaki, *Evolution under the multilocus Levene model without epistasis.* Theor. Popul. Biol. **76** (2009), 197–213.

[67] R. Bürger, *Multilocus selection in subdivided populations II. Maintenance of polymorphism under weak or strong migration.* J. Math. Biol. **58** (2009), 979–997.

[68] R. Bürger, *Evolution and polymorphism under the multilocus Levene model without epistasis.* Theor. Pop. Biol. **78** (2010), 123–138.

[69] S. Gavrilets, *Fitness landscapes and the origin of species.* Princeton Univ. Press, Princeton, NJ, 2004.

[70] S.P. Otto, *The evolutionary enigma of sex.* Am. Nat. **174** (2009), S1–S14.
[71] M. Kimura, *The Neutral Theory of Molecular Evolution.* Cambridge: University Press, UK, 1983.
[72] W.J. Ewens, *Mathematical Population Genetics. I: Theoretical Introduction. 2nd ed.* Interdisciplinary Mathematics 27. New York, NY: Springer., 2004.
[73] J.F.C. Kingman, *The coalescent.* Stoch. Process. Appl. **13** (1982), 235–248.
[74] J. Wakeley, *Coalescent Theory: An Introduction.* Greenwood Village, Colorado, USA, 2008.

Reinhard Bürger
Department of Mathematics
University of Vienna
Nordbergstrasse 15
A-1090 Wien, Austria
e-mail: `reinhard.buerger@univie.ac.at`

Extinction, Persistence, and Evolution

Peter Jagers

Abstract. Extinction can occur for many reasons. We have a closer look at the most basic form, extinction of populations with stable but insufficient reproduction. Then we move on to competing populations and evolutionary suicide.

Mathematics Subject Classification (2000). Primary 92B05; Secondary 60J80.

Keywords. Branching process, extinction, survival, population dynamics, evolution.

1. Extinction is Omnipresent

Even Malthus observed that overall growth of a population does not preclude frequent extinction of families and other subpopulations. After establishing and elaborating his fundamental idea of geometric or exponential growth of unchecked populations, he referred to the city of Berne, where 379 out of the 487 bourgeois families died out in two centuries, 1583 to 1783 [1]. Likewise, it was the observed extinction of known family names in 19th Century France and England, that prodded first Bienaymé and then Galton to formulate the family extinction problem in mathematical terms. In Galton's famous wording, mirroring time and environment, [2] Problem 4001: "A large nation, of whom we will only concern ourselves with adult males, N in number, and who each bear separate surnames colonize a district. Their law of population is such that, in each generation, a_0 per cent of the adult males have no male children who reach adult life; a_1 have one such male child; a_2 have two; and so on up to a_5 who have five. Find (1) what proportion of their surnames will have become extinct after r generations; and (2) how many instances there will be of the surname being held by m persons."

Both Malthus and Galton approached the problem from the point of view of society and human populations. In biology, the prevalence and importance of

Research supported by the Swedish Research Council under its grant to the Gothenburg Stochastic Centre. This article version was written while the author was visiting Monash University, supported by the Australian Research Council grant ARC DP0881011.

extinction was noted by Galton's cousin Darwin in the Origin of Species [3]: "Extinction and natural selection go hand in hand". Another catchy phrase ascribed to Darwin is "extinction is the motor of evolution". Paleontologists have estimated that the overwhelming majority of all species (more than 99%!) that ever existed are now extinct [4].

Extinction can occur for many reasons. The basic form, in the absence of competitors and in a stable environment, is the consequence of a combination of variation between individuals, and too low average reproduction. It is often referred to as being due to *demographic stochasticity.* In posing his Problem 4001 [2], Galton seems to have identified it as the natural null hypothesis as distinct from conjectures about extinction for some specified reason. In this vein it can well be referred to as the *intrinsic* form of extinction. In nature, extinction is of course often the consequence of *competition.* In natural history the rôle of *catastrophes* has been ardently debated. Finally, Darwin and later Haldane [5] noted the possibility that natural selection may favour individual traits which turn out harmful to the whole population, *evolutionary!suicide*, like the peacock's tail, [6].

2. Intrinsic extinction

Today, analysis of intrinsic extinction in clonally reproducing single-type populations in terms of simple, generation counting Galton-Watson processes is a textbook matter. Note that the question of extinction or not can be posed within a simple model lacking a realistic time structure: a population dies out if and only if it has an empty generation. We need to know just the distribution of the total number of children of an individual, neither birth times nor life spans are relevant. (In order to study times and paths to extinction in a realistic fashion, more general models are needed [7]. Within the present simple setup, "time to extinction" actually means "the number of generations until dying out". Only if generations do not overlap, will the two coincide.) Thus, the basic extinction theorem resolves the matter:

Theorem 2.1 (Branching or Malthusian Dichotomy). *Let Z_n be the size of the nth generation in a simple branching process with $Z_0 = 1$. Write $q_n = \mathbb{P}(Z_n = 0)$. If p_k is the probability of k children and f its probability generating function, then*

$$q_{n+1} = \sum_k p_k q_n^k = f(q_n), \qquad q_n \uparrow q = \mathbb{P}(extinction),$$

and

$$q = f(q),\ q < 1 \Leftrightarrow m = f'(1) > 1,$$

barring the trivial exception $p_1 = 1$. If the process does not die out, then Z_n grows like m^n precisely under the famous $x \log x$-condition:

$$\sum_k p_k k \log k < \infty.$$

The basic, first part of this was known to Bienaymé as early as in 1845, for a modern proof, see [8], e.g., it hinges upon all individuals reproducing independently and according to the same probability law. Strangely, Galton and Watson overlooked the logical possibility of unlimited growth when responding to Galton's Problem for Solution [9]:

"All the surnames tend to extinction and this result might have been anticipated, for a surname lost can never be recovered". This result must not be confounded with the extinction of the male population, for in every (supercritical) case we have an indefinite increase of the male population."

They had noticed that 1 solves the equation $f(x) = x$, but did not observe that if $m > 1$, there is another root between zero and one, and that the latter yields the correct extinction probability. Since Bienaymé's work remained unknown in England, their oversight was only corrected by Haldane, [10], and ultimately the Danish actuary Steffensen, half a century later, [11].

It is tempting to muse over the blunder. One may note a defensive tone in the quote from their paper, as though the authors had a hunch that the proof did not provide a firm ground for the conclusion. Further, we should keep in mind that the problem was approached precisely because of the ubiquity of extinction. So, in a sense, they found what they were after.

The real truth is that in nature and history we often meet with supercritical ($m > 1$) populations, where due to strong convexity of f, q is close to one, albeit strictly smaller. Here is an illustration, which also describes how simple Galton-Watson processes can catch the extinction risk of more general populations. Yearly survival probabilities for North Atlantic harbour seals are something like 0.6 for he first year, 0.8 for the second, and 0.95 for later years, possibly slightly lower after 30 years of age or so. The first three years no children are born. The fourth year, the probability of a daughter is 0.2, and then it is 0.45 per year. Though it requires some computation it is not difficult to determine the probabilities $p_0, p_1, p_2, p_3, \dots$ that a female gets 0, 1, 2, 3,... daughters throughout her life. Theorem 2.1 then yields the extinction probability as $q = 0.65$. The mean number of daughters is $f'(1) = 3$, so this is actually a quickly growing population, but one where 65% of the "families" die out.

Further, these were times different from ours. Organic, pre-fascist theories of states and peoples prevailed, presuming among other things that populations like nations led a life independently of their members, and why should not life spans of collectives have an upper bound like those of individuals? After all, Watson was also of the clergy, though there is no evidence that ideas like that of a last judgement should have influenced his science.

But how come that their oversight was not corrected quickly? A cynical answer would be that this was a mathematical result, and probably nobody outside the little world of mathematics cared about its implications, whereas mathematicians were not so concerned with its sensibility. But the latter would certainly not hold true for people like Galton himself, and the former seems also not to have been the case. According to Heyde and Seneta [12], "its implications were strongly doubted" at the time of publication.

One (almost) contemporary and non-mathematical criticism, is by a Swedish historian or political scientist, Pontus Fahlbeck. He was a commoner who married a baroness and became the author of a monumental two-volume treatise on the Swedish aristocracy [13]. There he gives a correct, verbal description of the relation between growth of the whole versus frequent extinction of separate family lines, and writes, somewhat condescendingly it may seem: "Galton, who with characteristic curiosity considered the question, has tried to investigate to what extent families ... must die out, with the help of a competent person." Fahlbeck then recounts examples considered by Galton, showing that "the tendency is the extinction of all". (The account is not completely lucid.) This is followed by a sequel of questions, and a reassuring answer: "If this course of events is based on a mathematical law, then it should be as necessary, or not? And what then about our general conclusions, that no necessity forces extinction? Is there not in this a contradiction, which if both arguments are right (i.e., Fahlbeck's verbal argument and Galton's and Watson's mathematical ditto), as they undoubtedly are, leads to what philosophers call an antinomy? However, mathematical calculations, as applied to human matters, may seem unrelenting but are actually quite innocuous. The necessity lies buried in them like an electrical current in a closed circuit, it cannot get out and has no power over reality." (pp. 133–135, my translation).

3. But bounded populations do die out!

The alternative to extinction in branching processes is exponential growth. This is an important result, relevant for real populations in a short to semi-long perspective, while reproduction retains its character of free branching. But on a bounded globe nothing can go on growing forever. And population size stabilisation, as presumed by deterministic mathematical population theory, is unrealistic, since a bounded population subject to individual variation in reproduction including a risk of no offspring, will ultimately die out. This is extremely generally summarised in the following theorem, which shows that no environmental feedbacks or other interactions can ensure stable population size. "In the long run, we are all dead", as Lord Keynes said [14].

Theorem 3.1 (General Dichotomy). *Consider non-negative (not necessarily integer-valued) random variables* $X_1, X_2, \ldots$. *Assume 0 absorbing (i.e.,* $X_n = 0 \Rightarrow X_{n+1} = 0$*) and suppose that for any* x *there is a* $\delta > 0$ *such that* $\mathbb{P}(\exists n; X_n = 0 | X_1, \ldots, X_k) \geq \delta$*, if only* $X_k \leq x$*. Then, with probability one, either there is an* n *such that all* $X_k = 0$ *for* $k \geq n$ *or* $X_k \to \infty$*, as* $k \to \infty$*. If* $\mathbb{E}[X_n]$ *remains bounded, it follows that* X_n *must turn zero, almost surely.*

The main difference between this and the branching case of individuals reproducing independently, is that in the general case growth need not be exponential. An example is the well-known linear growth occurring in PCR, the polymerase chain reaction, [15]. A direct consequence of the dichotomy is that no population, whose expected size is bounded, can persist.

The simple proof is reproduced here.

Proof. Let $D = \{\exists n; X_n = 0\}$ be the event of extinction. By Lévy's theorem, or more generally martingale convergence,

$$\mathbb{P}(D|X_1, \ldots, X_k) \to 1_D, \qquad k \to \infty,$$

since D is measurable with respect to the σ-algebra generated by all the $X_i, i = 1, 2, \ldots$. If the outcome is such that X_k does not tend to infinity, then it comes under some level x infinitely often. The conditional extinction probability on the left-hand side exceeds δ, and hence so must 1_D. But $1_D > 0 \Rightarrow 1_D = 1$. □

In a certain sense, Galton and Watson were thus right, after all: no true population, i.e., one that allows variation in reproduction between individuals and remains bounded (in expectation only) can persist. This gives rise to questions about (a) time to extinction and (b) quasi-stationary states before extinction. As mentioned, (a) has been discussed in very general branching models in [7]. Here we shall describe a simple model with competition, extinction due to competition, final extinction, and quasi stationarity [16].

4. A simple model with competition

The keyword here is not realism but simplicity. All individuals live one time unit (season). At death they either beget no children or two. The probability of the latter event depends upon population size N and a population characteristic *carrying capacity*. While talking of just one morph, we shall denote the latter by K, and assume that the probability of successful division is $K/(K+N)$. The probability of getting no children is thus $1 - K/(K+N) = N/(K+N)$. Reproduction is clonal, and besides the dependence upon population size independent between individuals. When discussing several morphs, we shall write their carrying capacities aK, bK, etc. This makes it possible to discuss varying K and relative carrying capacities. We shall also refine reproduction probabilities into a distinction between competition within your own morph and between morphs, and finally introduce mutation probabilities.

4.1. One single morph

In other words, the population studied here is a binary, population-size-dependent Galton-Watson branching process. It starts from a positive integer number $Z_0 = z$. Let ξ_{nj} be the number of children of individual j in generation n (taking the value zero or two). The population size is then recursively given by

$$Z_{n+1} = \sum_{j=1}^{Z_n} \xi_{nj},$$

with

$$\mathbb{P}(\xi_{nj} = 2|Z_n) = \frac{K}{K + Z_n}, \quad \mathbb{P}(\xi_{nj} = 0|Z_n) = \frac{Z_n}{K + Z_n}. \tag{4.1}$$

The random variables ξ_{nj} are assumed independent and identically distributed, given the population size Z_n, or indeed the whole past population history, $Z_0, Z_1, \ldots, Z_n$. Since reproduction is identically distributed for all individuals in the same generation and the distribution, given Z_n, is the same for all generations n, we shall often delete the suffices, at least when not referring to several individuals in one context.

Whenever the population size Z_n exceeds K, $K/(K + Z_n) < 1/2$ and the process behaves like a subcritical branching process. For sizes smaller than the carrying capacity, it turns supercritical. It is critical in the unlikely event that the size is precisely K (then necessarily an even integer).

It is easy to check that $\mathbb{E}[Z_n]$ is bounded. It follows from Theorem 3.1 that the extinction probability is

$$\mathbb{P}(Z_n \to 0, \text{as } n \to \infty) = 1.$$

Being supercritical while under the level K, the population tends, however, to increase, with a positive probability, while this is the case, and is prone to reach large values (around K) before ultimate extinction. It seems plausible that it either dies out quickly or else persists for a long time. We proceed to make that precise.

Write $T(a)$ for the first time the population reaches, or passes $a \geq 0$, from below or above, depending upon the starting position. For short, let $T = T(0)$ be the time of extinction. What will be the relation between these two random variables for large a?

Theorem 4.1 (Risk of direct extinction). *Let $0 < d < 1$. Then for any $1 \leq z \leq dK$,*

$$\mathbb{P}_z(T < T(dK)) < d^z.$$

In this, and elsewhere, probability or expectation indexed by z, $\mathbb{P}_z, \mathbb{E}_z$, means that the population starts from size $Z_0 = z$.

Proof. Such assertions are proved by comparison with suitably chosen (not population size dependent) simple Galton-Watson branching processes, about which much is known. In the present case, consider such a binary splitting process with the probability of begetting zero children being $d/(d+1)$. Call it $\hat{Z}_n$. Since $x/(1+x)$ is an increasing function of x, any $k < dK$ yields

$$\frac{k}{K+k} < \frac{dK}{K+dK} = \frac{d}{d+1}.$$

Hence, as long as population size stays below dK, the probability of producing no offspring is smaller than the corresponding probability pertaining to this classical Galton-Watson. Therefore, clearly the probability that our process becomes extinct by time n, without crossing dK, is smaller than the corresponding probability for the binary Galton-Watson process $\hat{Z}_n$. The latter must be smaller than the Galton-Watson probability of ultimate extinction, $\mathbb{P}(\hat{Z}_n \to 0) = \hat{q}$. From Theorem 2.1 we know that $\hat{q}$ is the smallest root of the quadratic equation

$$\frac{d}{d+1} + \frac{1}{d+1}\hat{q}^2 = \hat{q},$$

which is simply d. Hence

$$\mathbb{P}_z(T < T(dK)) = \mathbb{P}_z(Z_n = 0 \text{ for some } n < T(dK)) \le \mathbb{P}_z(\hat{Z}_n \to 0) = d^z. \quad \square$$

Thus, with positive probability the population will not die out but reach sizes at the order of the carrying capacity. If $\hat{m} = 2/(d+1)$ denotes the reproduction mean of the minorizing Galton-Watson $\{\hat{Z}_n\}$ considered above, we know that $\hat{Z}_n \approx z\hat{m}^n$ if it does not die out, by the Branching Dichotomy. Hence

$$dK \approx Z_{T(dK)} \ge \hat{Z}_{T(dK)} \approx z\hat{m}^{T(dK)},$$

and approximately

$$T(dK) \le (\log(dK) - \log z)/\log \hat{m}.$$

Since always

$$T(dK) \ge (\log(dK) - \log z)/\log 2,$$

we can conclude that $\log dK$ is indeed the right order of time it takes for our process to reach dK, if it does not first die out.

But sooner or later it will. So the question arises: when? In the case of direct extinction, say not reaching a level $k < K$, the time to extinction will be like that of a conditioned branching process. (Recall that supercritical branching turns subcritical, when conditioned on extinction.) If population size manages to climb up to the vicinity of the carrying capacity, it will stay there for a long time.

Theorem 4.2 (Upper Survival Bound). *Whatever the starting number z, carrying capacity K, and time (generation) n,*

$$\mathbb{P}_z(T > n) \le \left(1 - \mathrm{e}^{-K}\right)^n \le \exp\left\{-n\mathrm{e}^{-K}\right\}$$

and $\mathbb{E}_z[T] \le \mathrm{e}^K$.

Proof. Write

$$Q_n := \mathbb{P}_z(T > n).$$

The elementary inequality

$$\left(\frac{k}{K+k}\right)^k = \left(\frac{1}{1+Kk^{-1}}\right)^k \ge \mathrm{e}^{-K}$$

yields

$$\begin{aligned}\mathbb{P}_z(T \le n+1) &= \mathbb{P}_z(T \le n) + \sum_{k=1}^{\infty} \mathbb{P}_z(Z_n = k)\left(\frac{k}{K+k}\right)^k \\ &\ge \mathbb{P}_z(T \le n) + \mathrm{e}^{-K} Q_n.\end{aligned}$$

Hence,

$$Q_{n+1} \le Q_n - \mathrm{e}^{-K} Q_n,$$

and the asked for upper bounds on the probabilities follow by induction and another elementary inequality, $0 < 1-u < \mathrm{e}^{-u}$ for $0 < u < 1$. The second follows by summation:

$$\mathbb{E}_z[T] = \sum_n \mathbb{P}_z(T > n) \le \sum_n \left(1 - \mathrm{e}^{-K}\right)^n = \mathrm{e}^K. \qquad \square$$

Actually, this upper bound describes persistence fairly well.

Theorem 4.3 (Exit Downwards). *For any $0 < d < 1$ write $c = \frac{d(1-d)^2}{8(1+d)}$. Then for any K and $z \ge dK$,*

$$\mathbb{P}_z(Z_1 > dK) \ge 1 - \mathrm{e}^{-cK}.$$

Further, for any $n > 1$

$$\mathbb{P}_z(T(dK) > n) > (1 - \mathrm{e}^{-cK})^n,$$

and

$$\mathbb{E}_z\left[T(dK)\right] > \mathrm{e}^{cK}.$$

Proof. The proof uses an elegant inequality for the binomial distribution, established by Janson (see [17]): For any natural n, $0 < p < 1$ and $r > 0$

$$\mathbb{P}(\mathrm{Bin}(n,p) \le np - r) \le \mathrm{e}^{-r^2/(2np)}.$$

But if population size is $Z_0 = z$, $dK \le z < K$ and $p(z) = K/(K+z)$, the next generation is $Z_1 = 2\mathrm{Bin}(z, p(z))$ and $r = zp(z) - dK/2 > dKp(K) - dK/2 = 0$, so that

$$\mathbb{P}_z(Z_1 \le dK) = \mathbb{P}_z(Z_1 \le 2zp(z) - r) \le \mathrm{e}^{-r^2/(2zp(z))}.$$

In this the exponent is

$$\frac{r^2}{2zp(z)} = \frac{(2zp(z) - dK)^2}{8zp(z)} = K\frac{(f(x)-d)^2}{4f(x)}$$

in terms of $x = z/K$ and $f(x) = 2x/(1+x)$. Note that for $d < 1$, $z > dK$, $x > d$ and $f(x) > f(d) > d$. The following function is increasing and gives an inequality for $x > d$

$$\frac{(x-d)^2}{4x} > \frac{(f(d)-d)^2}{4f(d)} = \frac{d(1-d)^2}{8(1+d)} = c.$$

Hence, for any $x > d$ ($z > dK$)

$$\mathbb{P}_z(Z_1 \le dK) \le \mathrm{e}^{-cK}.$$

This proves the first assertion.

For the second use induction on n to show that for any $z \ge dK$

$$\mathbb{P}_z(T(dK) > n) > (1 - \mathrm{e}^{-cK})^n.$$

For $n = 1$ this is the first statement, which is proved. Assume it has been established for $n \geq 1$. By the Markov property,

$$\begin{aligned}\mathbb{P}_z(T(dK) > n+1) &= \mathbb{P}_z(Z_1 \geq dK, \ldots, Z_{n+1} \geq dK) \\ &- \sum_{k \geq dK} \mathbb{P}_z(Z_1 - k, Z_2 \geq dK, \ldots, Z_{n+1} \geq dK) \\ &= \sum_{k \geq dK} \mathbb{P}(Z_2 \geq dK, \ldots, Z_{n+1} \geq dK | Z_1 = k)\mathbb{P}_z(Z_1 = k) \\ &= \sum_{k \geq dK} \mathbb{P}_k(T(dK) > n)\mathbb{P}_z(Z_1 = k).\end{aligned}$$

By the assumption of induction, this is

$$\geq (1 - \mathrm{e}^{-cK})^n \mathbb{P}_z(Z_1 \geq dK) \geq (1 - \mathrm{e}^{-cK})^{n+1},$$

as required.

The final assertion of the theorem follows by

$$\mathbb{E}_z\left[T(dK)\right] = \sum_n \mathbb{P}_z(T(dK) > n). \qquad \square$$

With the corresponding upwards excursion result, we can conclude that prevalence around the carrying capacity is of the order e^{cK}, [16]: For any $\delta > 0$ and starting points $z \geq dK$

$$\mathbb{P}_z(\mathrm{e}^{(c-\delta)K} < T < \mathrm{e}^{(c+\delta)K}) \to 1, K \to \infty.$$

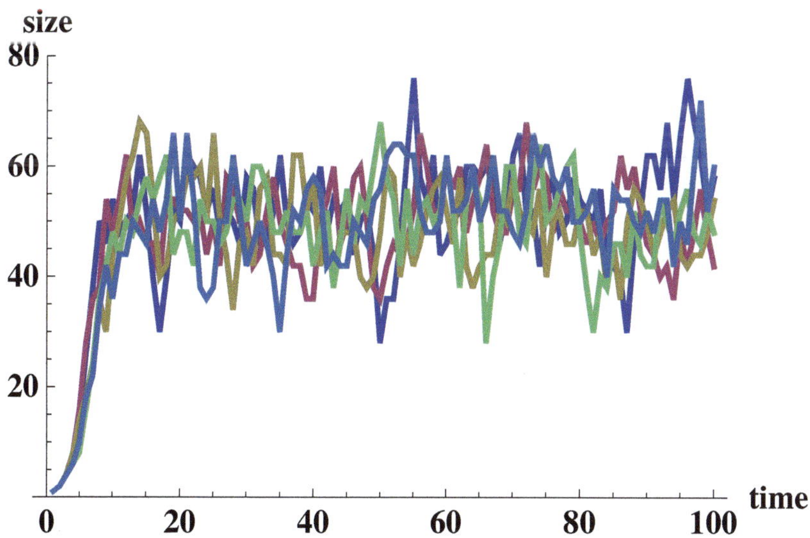

FIGURE 4.1. Five population developments with $K = 50$.

While the population oscillates around its carrying capacity, its size will seem to follow a stationary distribution:

Theorem 4.4. *Consider the distribution of* $X_n = Z_n/K$, *given that* $X_n > 0$, *for fixed* K. *As* $n \to \infty$, *this converges weakly to a proper distribution function, called the quasi-stationary distribution. As then* $K \to \infty$, *the latter concentrates all probability mass at the point* 1.

This was shown in [18]. Indeed, existence of quasi-stationary distributions is a consequence of the Krein-Rutman theory of positive operators.

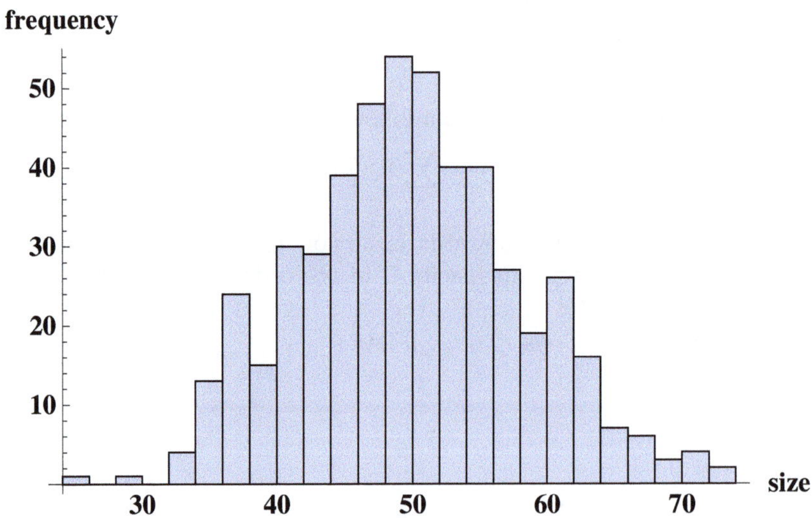

FIGURE 4.2. Histogram of a population size for the last 500 of 10.000 generations with $K = 50$.

It may seem as though this simple model, with its long persistence and pseudo-stabilisation would rather illustrate persistence than extinction. The picture does change, however, if mutation is introduced, and ensuing competition between the fresh mutant and the pseudo-established resident. Then, either the mutant will never grow up to its carrying capacity but disappear quickly, the mutant may take over the habitat, and the old resident thus disappear, or in rare cases both will survive and an evolutionary branching has occurred [19]. The latter case was discussed in [16]. Here, let us only note that if a resident has carrying capacity a_1K and the mutation risk is p, then the waiting time until a mutation is approximately geometrically distributed with the parameter p^{a_1K}. If the mutant has the carrying capacity a_2K, we shall assume that probabilities are further determined by the *competition coefficient* γ through the formulas

$$\mathbb{P}\left(\xi_{nk}^{(1)} = 0 \,|\, Z_n^{(1)}, Z_n^{(2)}\right) = \frac{Z_n^{(1)} + \gamma Z_n^{(2)}}{a_1 K + Z_n^{(1)} + \gamma Z_n^{(2)}},$$
$$\mathbb{P}\left(\xi_{nk}^{(1)} = 2 \,|\, Z_n^{(1)}, Z_n^{(2)}\right) = \frac{a_1 K}{a_1 K + Z_n^{(1)} + \gamma Z_n^{(2)}},$$

and

$$\mathbb{P}\left(\xi_{nk}^{(2)} = 0 \,|\, Z_n^{(1)}, Z_n^{(2)}\right) = \frac{\gamma Z_n^{(1)} + Z_n^{(2)}}{a_2 K + \gamma Z_n^{(1)} + Z_n^{(2)}},$$
$$\mathbb{P}\left(\xi_{nk}^{(2)} = 2 \,|\, Z_n^{(1)}, Z_n^{(2)}\right) = \frac{a_2 K}{a_2 K + \gamma Z_n^{(1)} + Z_n^{(2)}},$$

superscripts referring to the resident and mutant respectively.

If

$$a_2 > \gamma a_1, \tag{4.2}$$

the mutant population will start supercritically, and either die out or else start growing at a geometric rate, until it approaches its carrying capacity. This is the case of possible invasion. It is illustrated in the following figure, which shows five runs of a population system with $\gamma = 0.7$, $a_1 K = 40$, and $a_2 K = 70$. In three cases the resident prevailed, the invader dying out very quickly, in two runs the invader took over.

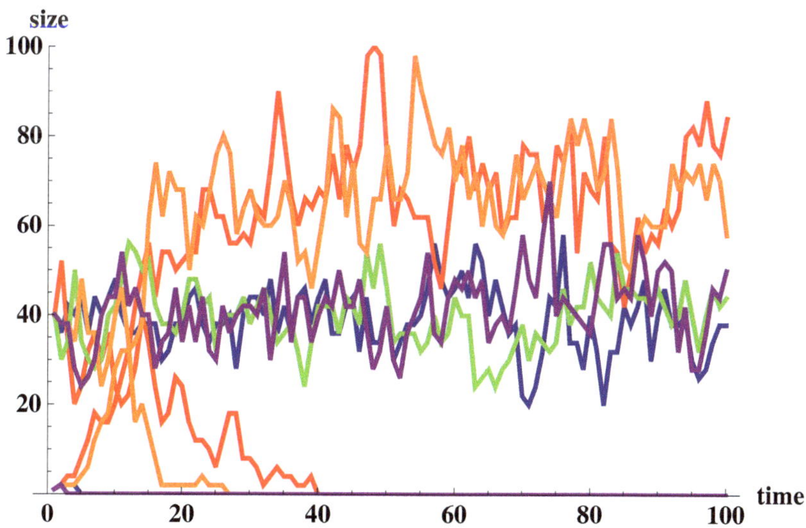

FIGURE 4.3. Five competitive population evolutions, $K = 100, \gamma = 0.7$, $a_1 = 0.4$, and $a_2 = 0.7$. In two of them the invader takes over.

Generally the probability of the mutant not dying out but establishing itself will be $1 - \gamma a_1/a_2$, approximately for large K, and given a_1 and a_2. Thus, it also depends upon how the latter are chosen, the classical approach being the Bugge Christiansen and Loeschke model [20]. If we assume, for the sake of illustration, that the establishment probability is constant and equal to r, we can conclude that for large K we will observe a resident extinction rate of Kpr, until the whole population dies out, after a time span of the order e^{cK}. In this, the mutation probability may well depend upon carrying capacity and satisfy inequalities like

$$\frac{e^{-cK}}{K} \ll p \ll \frac{1}{K \ln K},$$

needed to guarantee on one side that an invader will have time to establish itself before the next mutation, on the other that mutations will occur while the population is around.

5. Evolutionary suicide

A question more interesting than the lifespan of single morphs may be that of the whole population, under presence of mutations. A sequel of mutations increasing the carrying capacity, say from aK to bK, $a < b$, will in general increase the survival time. However, even the simple pattern introduced allows for the intriguing development, known as evolutionary suicide. The simplest possible case may be the following: When a mutant appears its carrying capacity will be a times the resident carrying capacity, $0 < a < 1$, starting from a carrying capacity K. A mutant individual encroaches upon the resident living-space as would a non-mutant, but mutants are not disturbed by the non-mutants.

This would yield conditional reproduction probabilities of the form

$$\mathbb{P}\left(\xi_{nk}^{(1)} = 0 \,|Z_n^{(1)}, Z_n^{(2)}\right) = \frac{Z_n^{(1)} + Z_n^{(2)}}{a_1K + Z_n^{(1)} + Z_n^{(2)}},$$

$$\mathbb{P}\left(\xi_{nk}^{(1)} = 2 \,|Z_n^{(1)}, Z_n^{(2)}\right) = \frac{a_1K}{a_1K + Z_n^{(1)} + Z_n^{(2)}},$$

and

$$\mathbb{P}\left(\xi_{nk}^{(2)} = 0 \,|Z_n^{(1)}, Z_n^{(2)}\right) = \frac{Z_n^{(2)}}{a_2K + Z_n^{(2)}},$$

$$\mathbb{P}\left(\xi_{nk}^{(2)} = 2 \,|Z_n^{(1)}, Z_n^{(2)}\right) = \frac{a_2K}{a_2K + Z_n^{(2)}},$$

superscripts referring to the resident and mutant as before, and $a_2 = aa_1$.

Clearly, as a rule the resident ($Z_n^{(1)} \approx a_1K$) is subcritical and the mutant ($Z_n^{(2)} < a_2K$) supercritical initially. Unless the latter dies out quickly, it will thus invade and replace the old resident. However, the new carrying capacity is lower, and subsequent such mutations force the extinction of the whole population, since

$a^n K \to 0$. This is an extremely simple example of evolutionary suicide due to asymmetry [21]. For a general overview of this phenomenon from a non-stochastic viewpoint, see [6]. I hope to be able to come back to a more substantial analysis in stochastic terms.

Acknowledgment

Section 4 of this article reports joint work with P. Haccou, F.C. Klebaner, S. Sagitov, and V.A. Vatutin [16].

References

[1] T.R. Malthus, *An Essay on the Principle of Population, as it Affects the Future Improvement of Society, with Remarks on the Speculations of Mr Godwin, M. Condorcet and Other Writers.* John Murray, London, 1798.

[2] F. Galton, *Problem 4001: On the extinction of surnames.* Educational Times **26** (1873), 17.

[3] C. Darwin, *On the Origin of Species by Means of Natural Selection or the Preservation of Favoured Races in the Struggle for Life.* John Murray, London, 1859.

[4] D. Raup, *Extinction: Bad Genes or Bad Luck?* NW. W. Norton & Company, USA, 1991.

[5] J.B.S. Haldane, *The Causes of Evolution.* Longmans Green, London, 1932.

[6] K. Parvinen, *Evolutionary suicide.* Acta Biotheor. **53** (2005), 241–264.

[7] P. Jagers, F.C. Klebaner, and S. Sagitov, *On the path to extinction.* Proc. Natl. Acad. Sci. USA **104** (2007), 6107–6111.

[8] P. Haccou, P. Jagers, and V.A. Vatutin, *Branching Processes. Variation, Growth, and Extinction of Populations*, volume 5 of *Cambridge Studies in Adaptive Dynamics.* Cambridge University Press, UK, 2005.

[9] F. Galton and H.W. Watson, *On the probability of extinction of families.* EJ. Anthropol. Inst. **4** (1874), 138 144.

[10] J.B.S. Haldane, *A mathematical theory of natural and artificial selection, Part V: Selection and mutation.* Proceed. of the Cambridge Phil. Soc. **23** (1927), 838–844.

[11] F.H. Steffensen, *Om sandsyndligheden for at afkommet uddør.* Matematisk Tidskrift **B** (1930), 19–23.

[12] C.C. Heyde and E. Seneta, *I.J. Bienaymé, Statistical Theory Anticipated.* Springer-Verlag, New York, 1977.

[13] P. Fahlbeck, *Sveriges adel, statistisk undersökning öfver de å riddarhuset introducerade ätterna, I and II (The Swedish Nobility, a Statistical Investigation of the Families of the House of Nobility).* Gleerups, Sweden, 1898, 1902.

[14] J.M. Keynes, *A Tract on Monetary Reform.* MacMillan, London, 1923.

[15] P. Jagers and F. Klebaner, *Random variation and concentration effects in PCR.* J. Theor. Biol. **224** (2003), 299–304.

[16] F.C. Klebaner, S. Sagitov, V.A. Vatutin, P. Haccou, and P. Jagers, *Stochasticity in the adaptive dynamics of evolution: the bare bones (to appear)*, invited submission. J. Biol. Dynamics.

[17] S. Janson, *Large deviation inequalities for sums of indicator variables* (1994), technical report, Dep. Mathematics, Uppsala University **34**.

[18] F.C. Klebaner and J. Lazar, *On the quasi-stationary distribution in randomly perturbed dynamical systems with a single attracting point. In* R. Wilson, D. Pra Murthy, and S. Osaki (eds.), *Proc. 2nd Australia-Japan Workshop on Stochastic Models in Engineering, Technology, and Management*, 348–354, The University of Queensland, Australia, 1996.

[19] S.A.H. Geritz, É. Kisdi, G. Meszena, and J.A.J. Metz, *Evolutionarily singular strategies and the adaptive growth and branching of the evolutionary tree.* Evol. Ecol. **12** (1998), 35–57.

[20] F.B. Christiansen and V. Loeschcke, *Evolution and intraspecific exploitative Competition. 1. One-Locus theory for small additive gene effects.* Theor. Popul. Biol. **18** (1980), 297–313.

[21] M. Gyllenberg and K. Parvinen, *Necessary and sufficient conditions for evolutionary suicide.* Bull. Math. Biol. **63** (2001), 981–993.

Peter Jagers
Mathematical Sciences
Chalmers University of Technology
and the University of Gothenburg
SE-412 96 Gothenburg, Sweden
e-mail: `jagers@chalmers.se`

Group Theory in Homogeneous Populations (Rescuing Darwin from the mud)

Peter Taylor

Abstract. Considerable recent work on the evolution of behaviour has been set in structured populations. An interesting "cancellation" result is known for structures, such as lattices, cycles and island models, which are homogeneous in the sense that the population "looks the same" from every site. In such populations all proximate or immediate fitness effects on others (for example, payoffs in a game or contest) play no role in the evolution of the behaviour. The altered competitive effects of such behaviour exactly cancel the proximate fitness effects. In mathematics, the internal symmetry which drives this result is powerfully described by the theory of mathematical groups and recent work has used this theory to clarify and extend a number of existing results. I review this body of work here.

Mathematics Subject Classification (2000). 05C57, 91A43, 92D50.

Keywords. Population structure, graph, group theory, homogeneous, selection, relatedness, allele frequency.

1. Introduction

It is a pleasure and a privilege to have been asked to talk about the Mathematics of Charles Darwin on this very day, the 150th birthday of the publication of the "Origin," particularly among such distinguished colleagues. The invitation prompted me to think about my early encounters with Darwin and I realized that one of these occurred 50 years ago, no doubt very close to the 100th birthday. I was in late high school and was interested in science and had discovered *The Origin of Species* [1] in my father's collection of the Harvard Classics. I read bits of it, but not very much. My father, on noticing the book in my possession, told me I must talk to a biologist he knew at the museum, where he was working, and in spite of my mild resistance, he set up a meeting a few days hence. When the time came he

This work was partially supported by a grant from the Natural Sciences and Engineering Research Council of Canada. I am grateful to the organizers of conference on The Mathematics of Darwin's Legacy for hosting me at this marvelous meeting.

drove me into town, took me firmly up the curling flight of stone steps and knocked at his colleague's door. And there he left me saying he would meet me in the lobby.

Well there we were, both of us not quite sure what I was doing there. He asked me uneasily what I wanted and I didn't dare reply with the truth – that there was nothing I wanted but my father had made me come. I told him I had been reading *The Origin of Species* and had found it interesting. "That's perhaps not the best place to start," he replied gently, and for the rest of the mercifully short interview he gave me a list of books I might read "next" and told me a bit about them in words that meant little to me. And here I am 50 years later. It would be interesting now for me to have a copy of that list, but I am certain I never consulted it again.

A couple of years later it was physics I chose at university and then, after a year or two, I settled on mathematics, I believe, for its structure, its beauty, its simplicity, and its independence from the world. Another 15 years would pass before I encountered biology again, and that was in a remarkable series of lectures delivered by John Maynard Smith at a special symposium organized by the Canadian Mathematics Society. John talked about something called "evolutionary game theory" and I suddenly realized that the structure and beauty I had sought in turning, many years ago, to mathematics, was what Charles Darwin had given us in biology, that the theory of evolution opened up to us a whole new way of asking questions about why organisms behave this way rather than that, and thereby gave us powerful new tools for understanding this behaviour. It was a striking revelation.

The impact on me was even greater as I had recently turned to a study of the applications of game theory to economics and had been troubled by its unsettling assumptions of rationality and purpose. But now I could see where game theory might really belong, in biology where the actors were not us crazy humans, but genes, where rationality was not needed, and where purpose was not a precondition but appeared miraculously as a consequence of the unfolding of the evolutionary game.

The 40 years since the work of Maynard Smith and Price have given us an enormous spectrum of rich interactions between mathematics and biology. A wonderful account of many of these is found in Joel Cohen's remarkable 2004 essay called *Mathematics Is Biology's Next Microscope, Only Better; Biology Is Mathematics' Next Physics, Only Better* [2]. Cohen discusses an impressive range of areas in which mathematics and biology interact and in particular presents five biological challenges that could stimulate and benefit from major innovations in mathematics, and then five mathematical challenges that would contribute to the progress of biology. When I heard about this symposium, I went back to this essay looking for ideas and one sentence I read was this: 'Charles Darwin was right when he wrote that people with an understanding "of the great leading principles of mathematics ... seem to have an extra sense" ' [2, page 2017].

Reading that, I started to wonder what else Darwin might have thought of mathematics, so I looked on the web and I found this: "A mathematician is a blind man in a dark room looking for a black cat which isn't there". I had heard this famous remark before but had never known that it was credited to Darwin. And another: "I suppose you are two fathoms deep in mathematics, and if you are, then God help you, for so am I, only with this difference, I stick fast in the mud at the bottom and there I shall remain". This second quote quite challenged me and prompted me to choose for this talk a slice of mathematical biology that I found simple and beautiful, and that might just have had the power to rescue Darwin from the mud, if only he could have seen it.

2. Homogeneous population structures

For the past twenty years I have been interested in the effects of population structure on the direction of evolution, more specifically on allele frequency change. To illustrate the impact of structure, imagine an interaction between two individuals i and j which changes the fitness of each of them. Does the size of these fitness changes tell us all we need to know to work out the evolutionary effects? In general, no. The point is that each of the two fitness effects will in general have an impact on the fitness of other individuals k. For example if the offspring of i have some tendency to compete with the offspring of an individual k, then a change in i's fitness might affect the fitness of k. Thus, a full accounting of evolutionary change requires knowledge of the fitness effects on all individuals in the population.

To have some terminology, let's call the first set of effects proximate, and classify the second set under the general label of ecological feedback. The proximate effects are typically those that are observed in the field or specified in a behavioural model. For example in a two-party altruistic interaction, these are the cost incurred by the actor and the benefit gained by the recipient. The feedback effects are those that derive from the resulting altered competitive pressures and depend on the population structure and the nature of population regulation. For example if the proximate effects are on fecundity, the feedback effects might be a change in mortality among the offspring; if the proximate effects are on adult survival, the feedback effects might be altered offspring recruitment.

To have a specific scenario to work with, I will suppose here that the proximate effects are on breeder fecundity and the feedback effects are on offspring survival deriving from altered competitive pressures.

The proximate effects are typically known (measured or assumed) but the feedback effects are much harder to get hold of and typically require detailed knowledge of offspring dispersal and recruitment patterns. That's the story told by the population structure.

One of my interests has been to identify simple classes of population structures for which the feedback effects can be very simply described. Here I will look at what would appear to be the very simplest such structures, those for which the population "looks the same" to every individual. The term "homogeneous" has been used in the literature with many different meanings, but I will use it here to capture this notion and I will define it more precisely below. This notion of homogeneity might seem rather special or restricted, but some approximate version of it is often assumed. Theoretical models which are "open" in the sense that no population structure is actually given, typically work with a randomly chosen "focal" individual, and for this to make sense, the environment of this individual must be, in some sense, generic. This is effectively an assumption of homogeneity.

To keep things simple, I suppose that the population consists of a fixed collection of breeding sites, each occupied by a single asexual haploid individual. Instead of using the index i to keep track of individuals (who are being born and dying), I will use it to index the sites. Indeed I will assume that the set of sites is permanent and there is no change in population size.

One can depict these breeding sites as a set of nodes with arcs or edges between nodes to represent the relationship among them. For us, these relationships will be of two types, the rate at which offspring disperse from one node to another and the effect of the behaviour of the breeder at one node on the fitness of the breeder at another. I define the *dispersal probability* $d(i,j)$ to be the probability that an offspring born at site i competes to breed at site j and, counting only those offspring who attain a breeding site, I assume that $\sum_j d(i,j) = 1$. I let F_i denote the fecundity of the breeder at site i. In general, F_i will depend on the behaviour z_j (defined as the level of a behavioural trait) at many different sites j and I define the *fecundity effect* of j on i to be the partial derivative $\partial F_i/\partial z_j$.

The notion of homogeneity can be specified in terms of these relationships. To say that the population looks the same from node i as from node j is to say that an individual who could perceive only the dispersal probabilities and the fitness effects could not tell whether it was situated on node i or on node j. In [2] the notion of isomorphism is used to describe this. An isomorphism T (iso $\sim$ same; morph $\sim$ structure) is a bijection of the node set which preserves the dispersal probabilities and the fecundity effects, i.e., for any i and j,

$$\begin{cases} d\left(T(i),T(j)\right) = d(i,j) \ , \\ \dfrac{\partial F_{T(i)}}{\partial z_{T(j)}} = \dfrac{\partial F_i}{\partial z_j} \ . \end{cases} \tag{2.1}$$

The population structure is called *transitive* if, for every pair of nodes i and j, there is an isomorphism T for which $T(i) = j$. The transitive structures are precisely those that I am calling homogeneous.

3. The main result

There is a surprising, even extraordinary result which obtains in this homogeneous case. Suppose an actor at site j carries an allele A which causes her to give a fecundity benefit b to the breeder at site i different from j (that's the proximate effect). We want to measure the selective effect of this behaviour and by this we mean the effect on the population frequency of the allele A. Now as we have discussed, the altered fecundity of i will change the competitive environment in a neighbourhood of i and as a result the fitness of a number of other individuals will also be affected. Some of these others may carry the allele A, so to calculate the change in frequency of A, we will need to work out the effects of the action on all those whose fitness is affected as well as the probability that they will carry A. This can be a big job but it can, after some work, be obtained from knowledge of the population structure.

Now here's the remarkable result – if the structure is homogeneous, this substantial calculation doesn't have to be done [3, 4, 5, 6, 7, 8]. In this case, all these effects on the frequency of A, the proximate effect, b, on i and the resulting competitive effects on any number of others, will all cancel out *so that the net effect on the frequency of A will be zero.* Some who carry A will have increased fitness and others who carry A will have decreased fitness, and the net result is that on average, the decreases will exactly balance the increases. If you like, the ecological feedback effects of the fecundity change of i will exactly neutralize the proximate effect.

Assumptions. Right away I have to declare that there are a number of significant assumptions that are needed for this result to hold. A first class of assumptions is standard for inclusive fitness methods to be valid [9, 10] – fitness effects have to be additive and small. Thus, if the behaviour of several of my neighbours affects my fitness, the net effect must be the sum of the individual effects. Also the analysis is done to first order in the fitness deviations and this will give accurate results only if these deviations are small compared to baseline fitness. A second class of assumptions are more technical and have to do with the way in which generations succeed one another (overlapping or not), whether proximate effects are on fecundity, as we have assumed here, or survival, and whether offspring dispersal is symmetric, i.e., $d(i,j) = d(j,i)$ [8]. Finally, the population needs to be large or a small correction is required. In a finite population, the average effect on the non-focal individuals must be subtracted (equation (7.2)).

I emphasize that this "cancellation result" holds for the effects of the interaction on all other breeders, but not for the actor herself. As a result of this, the direction of change in frequency of the allele A is determined by the sign of the proximate effect of the actor on her own fecundity. If the act is costly to the actor (say it incurs a survival cost c) the allele frequency decreases, and if beneficial, the allele frequency increases. This simple result is indeed surprising, almost unbelievable, and I now discuss its interesting history.

4. History of the result

In 1992 I had a call from David Wilson. He had constructed a simulation to test Hamilton's Rule [9] on a two-dimensional lattice (Fig. 5.2c) and what he was finding quite surprised (and intrigued) him. No matter how large he made the benefit b bestowed by an altruistic allele, he could not get altruism to be selected. In his simulations, it always eventually died out and that appeared to flatly contract Hamilton's Rule [3].

Let's go back to Hamilton. Consider an allele A which causes the bearer (the actor) to bestow a fitness benefit b on a "neighbour" at a personal fitness cost c. Then if we want to calculate the resultant direction of frequency change of A, we will need to know the probability that the neighbour is an A-individual and (in this simple haploid case) that is a measure of what is called the relatedness of the actor to the neighbour [10, 11]. This idea, that some notion of relatedness ought to have something to do with allele-frequency change was recognized long ago, most famously by Haldane [12] who while walking beside a flooded river, considered the question of whether to jump in to save a drowning child and decided that the closer was his relatedness to the child, the more likely he ought to jump in. Haldane, of course, was able to think in genetic terms, but even Darwin [1], before he knew of Mendel's work, knew that what counted was the "blood" [13] rather than the individual. But it was Hamilton's genius that put forward the simple quantification of this idea. Hamilton's Rule states that the allele will be selectively favoured when $bR > c$ where R is the relatedness of the actor to the recipient.

Now back to Wilson's population. It had the structure of a lattice and each generation offspring dispersed to neighbouring sites, so there was certainly a significant probability that a neighbour would be an offspring or grandchild or niece etc. and we ought to have a positive R. Thus if b is big enough, we should have $bR > c$ and Hamilton's Rule [9] ought to give a selective advantage to the altruistic trait. Hence Wilson's (and my!) dismay.

The discussion above clarifies the dilemma. Wilson's b and c were in fact changes in *fecundity* – they represented a small increase (to the recipient) and decrease (to the actor) in offspring number. As we have seen, that's only the first part of the story. To translate these changes in fecundity into changes in fitness, we need to know the effect of these extra offspring on the fitness of others who might live nearby and who might also share genes with the actor. And when this is done all the effects of the b-gift cancel, leaving only the effects of the actor's fecundity decrement c. This certainly reduces the fitness of the actor and any positive effects on others (from reduced competition) cannot be strong enough to turn that negative effect around. Wilson's lattice population is an example of a homogeneous population structure and the main result applies.

What becomes of Hamilton's Rule $bR > c$? It has a number of versions which are valid. It holds when b and c represent the total fitness effect on recipient and actor and the fitness of no other individual is affected. There is one special case in which it holds with b and c representing only the proximate effects on fecundity, and that is when the resulting competitive effects (of the fecundity changes) are randomly distributed in the population as a whole. This might typically be the case in a randomly mixed population.

But in the usual situation the fitness effects of a proximate fitness transaction will be felt by a number of individuals, in this case a generalization of Hamilton's

Rule would have the form $\sum_i w_i R_i > 0$ where w_i is the overall fitness effect on individual i and R_i is the relatedness of i to the actor.

5. Mathematical groups

Different treatments of the main result, at various levels of generality, have appeared [3, 4, 5, 6, 7, 8, 14, 15] but the most recent of these [8] presents the result in a particularly elegant mathematical framework, and to celebrate Darwin's anniversary, I discuss that here.

It turns out that a powerful description of structural homogeneity can be obtained using the language and notation of the theory of mathematical groups. That's hardly surprising – group theory arose as a need for a formal structure to study the geometry of objects, such as regular polyhedra, that have a significant amount of internal symmetry.

First of all, instead of working with the individuals in the population (who are ephemeral – they die and are replaced by others) we work with the collection of breeding sites, each occupied by a single adult breeder. To have a picture we represent them as the nodes of a graph and use the arcs between nodes to represent the relationship between them, capturing both the dispersal probabilities and the fitness interactions (Fig. 5.1).

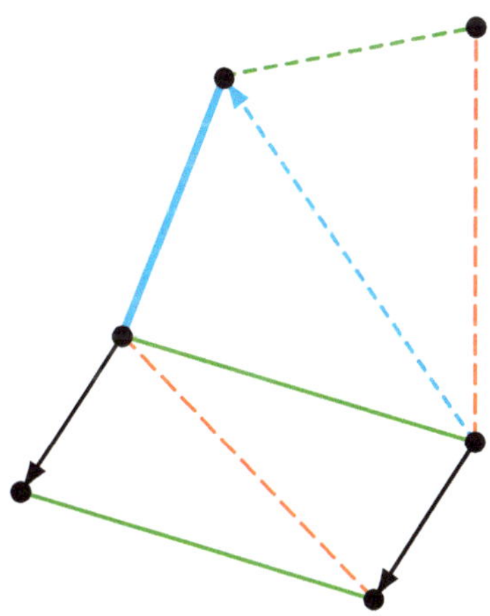

FIGURE 5.1. A directed graph. The different arcs represent different dispersal probabilities and/or different fitness interactions between the nodes.

So far we have described what is simply called a directed graph. What we don't yet have is the condition that the population "look the same" from every node. An elegant way to obtain that is to suppose that the set of nodes can be given the structure of a group.

We begin with a formal definition. A group G is a set of elements i with a closed binary operation (which we represent multiplicatively) which satisfies the following three axioms:

1. There is an identity element e with the property $ei = ie = i$ for all i.
2. Every element i has an inverse (denoted i^{-1}) such that $ii^{-1} = i^{-1}i = e$.
3. The operation is associative: $i(jk) = (ij)k$ for all i, j and k.

Now the wonderful thing about groups is that we have a natural transitive set of bijections which can serve as our isomorphisms, and these are the group multiplications. Indeed, given two elements j and k of the group, multiplication on the left by $i = kj^{-1}$ is a bijection T of the group which maps j into k. Indeed, $T(j) = ij = (kj^{-1})j = k(j^{-1}j) = ke = k$. Thus if our breeding sites are the elements of a group, we can use these left multiplications as a natural transitive set of maps preserving our two critical relationships – offspring dispersal and fecundity effects. That is, for any i, j and k, we specify:

$$\begin{cases} d(j,k) = d(ij,ik) \ , \\ \dfrac{\partial F_j}{\partial z_k} = \dfrac{\partial F_{ij}}{\partial z_{ik}} \ . \end{cases} \tag{5.1}$$

It turns out that the homogeneous population structures that have appeared in the theoretical literature can all be given a group structure in a natural way so that the group multiplication provides the isomorphisms satisfying equation (2.1).

For example how do we put a group structure on the examples depicted in Fig. 5.2? The answer is that in every case this comes from the geometry of our representation. In each case we choose an arbitrary node to be the identity element e. Then for (a) and (b) we can simply use the rotations about the centre identifying each node with the angle required to rotate the identity to that node. Group multiplication, of course, is composition. For (c) we use the horizontal and vertical translations if the population is infinite, but in the case of a finite lattice, we fold the array into a torus by identifying boundaries (right with left and top with bottom) and use a 2-parameter family of rotations. In (d) we again have a 2-parameter family of rotations, first the four rotations through 90° that rotate each island within itself, and secondly the three rotations through 120° which cycle the islands. All possible products of these give us the twelve elements of the group.

6. Can every homogeneous population be given a group structure?

That's all fine for these standard examples which we can easily draw, but the question arises as to whether any homogeneous population structure can be represented as a group. Note first that the converse of this holds and was mentioned above – any group-structured population for which the invariance equations (5.1) hold is homogeneous and the left multiplications give us a transitive set of isomorphisms (equation (2.1)). But suppose that we have a population structure with a

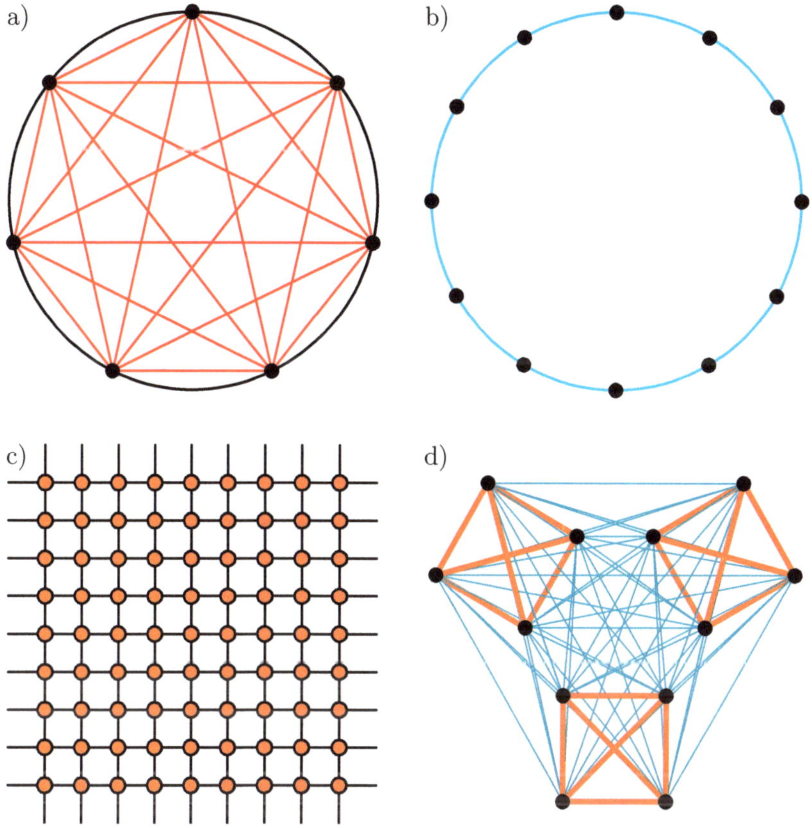

FIGURE 5.2. Some examples of standard homogeneous populations structures. (a) A random-mixing population with 7 breeding sites. (b) A cycle. Interactions with neighbours. Dispersal to neighbours or at random. (c) A lattice. Interactions with neighbours. Dispersal to neighbours or at random. (d) An island structure with 3 demes of size 4. Interactions at random within deme. Dispersal at random within deme or at random in population.

transitive set of isomorphisms (equation (2.1)). Can we put a group structure on the set of nodes so that the left multiplications are isomorphisms (equation (5.1))?

Well here is an idea. Take a random node and label it e. Now take any other node i. The homogeneity property tells us that the population should "look the same" from i as it does from e. Thus, for any node j, there should be a node which "looks the same" from i as j looks from e. *We could call that node* ij and this in fact would define the group multiplication operation on the node set. That seems at first to work nicely, at least the three group axioms (above) seem to hold.

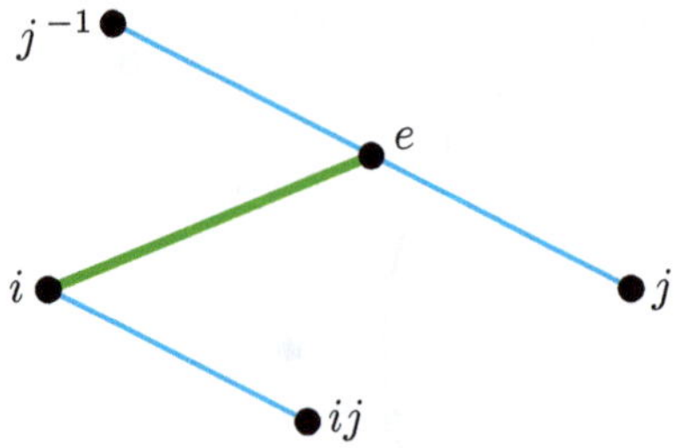

FIGURE 6.1. ij is the node that looks the same from i as j looks from e.

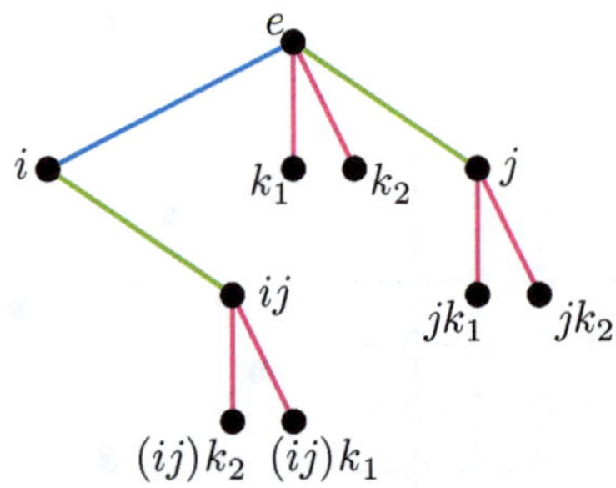

FIGURE 6.2. Possible difficulties with associativity. The nodes k_1 and k_2 bear the same relationship to e and there is no consistent way to distinguish them. Of course any diagram gives us a temporary right-left distinction. In the realization here we have used one ordering attached to j and another attached to ij. In this case, the verification that $(ij)k = i(jk)$ for either of the k's, will fail.

Except they don't. If you argue carefully, you run across a problem with the associative axiom 3.

The problem arises when there are several candidates bearing the same relation to a fixed node. This would not be the case in the structures of Figures 5.2 (b) and (c), but it could be an issue in (a) and (d) (Figure 6.2).

Typically, of course, we can use other aspects of the population structure (such as geometry) to pick out a consistent set of nodes playing the role of k, but the question we started with is whether this can always be done. Are there homogeneous populations which cannot be given a group structure?

It turns out that this is unknown, indeed it is closely related to an open problem in the theory of mathematical groups [8]. Certainly any homogeneous population I have ever seen (or imagined) can be given a group structure, but that is the best I can do. For sure this question is of much more mathematical than biological interest, but it is intriguing none the less.

7. Finite and infinite populations

Our main result takes a slightly different mathematical form in a finite and an infinite population and I look at that now. Recall that the result says that the effects of allele action on the fecundity of others do not "at the end of the day" cause a change in the average frequency of the allele. We conclude that the selective change in the average frequency will only be determined by, and must have the same sign as, the effect of the behaviour on the actor's own fecundity – if the actor suffers a cost, the overall frequency of the allele must decrease; if she gives herself a net benefit, the overall frequency of the allele will increase.

Actually there is one clarification or adjustment needed. The factor to pay attention to is not the effect of the behaviour on the actor's own fecundity, but rather the effect relative to the average population-wide effect. In an infinite population (or even a very large one), this correction is negligible, as a single actor could have only a negligible average effect. But in a small finite population, this normalization factor has to be included. Formally, the conditions for the sign of the selective change in the average allele frequency are written [8]:

$$\text{Infinite population:} \qquad \Delta\bar{x} \equiv \frac{\partial F_e}{\partial z_e}\,, \tag{7.1}$$

$$\text{Finite population:} \qquad \Delta\bar{x} \equiv \frac{\partial F_e}{\partial z_e} - \mathbb{E}_{i\neq e}\left(\frac{\partial F_i}{\partial z_e}\right)\,, \tag{7.2}$$

where "≡" means "has the same sign as". In a finite population we normalize by subtracting the average effect of the behaviour on the fecundity of all *other* breeders in the population.

8. Examples

To illustrate the result, I present three brief examples of altruistic behaviour.

Example. Suppose we have a finite population with a deviant trait which provides a cost-free public good which increases the fecundity of everyone by the same amount. Such a trait should have no effect on the frequency of the allele causing it. Since focal behaviour has the same effect on everyone, $\partial F_i/\partial z_e = \partial F_e/\partial z_e$ for all i, and equation (7.2) gives us $\Delta\bar{x} = 0$ as expected.

Example. Suppose we have an infinite island population with demes of size n, and a cost-free deviant trait which provides a fecundity benefit to all deme-mates, but not to self. Then there is no fecundity effect on the actor and equation (7.1) gives us $\Delta\bar{x} = 0$. Do we in fact expect no change in allele frequency? In this population, an individual's fitness depends on the number of its deme-mates which are deviant. If we suppose the demes are randomly formed, then the average number of deviant deme-mates should be the same for a deviant and a normal individual and there should indeed be no change in allele frequency, confirming the result. Now suppose that the demes are not randomly formed, but some offspring stay at home. Then we expect that a deviant individual will have more deviant deme-mates on average than a normal individual and will have higher average fitness. But this will also be the case for the deviant individual's deme-mates so that the competitive pressures at home will be higher than average. Equation (7.1) implies that these two opposing factors must exactly cancel.

Example. Altruism. Consider an altruistic trait in which individuals give fecundity benefits to various other individuals at total cost c. Then focal fecundity might have the form $F_e = -cz_e + \sum_i b_i z_i$ where b_i is the focal benefit received from site i when there is an altruist at that site. To move from there to a calculation of $\Delta\bar{x}$ we need the offspring dispersal patterns, both to calculate the competitive effects (which are needed for focal fitness w_e) and to get the focal relatedness coefficients. However, in a homogeneous population, equations (7.1) and (7.2) tell us that none of that is needed:

$$\text{Infinite population:} \qquad \Delta\bar{x} \equiv \frac{\partial F_e}{\partial z_e} = -c\,, \tag{8.1}$$

$$\text{Finite population:} \qquad \Delta\bar{x} \equiv \frac{\partial F_e}{\partial z_e} - \mathbb{E}_{i\neq e}\left(\frac{\partial F_e}{\partial z_i}\right) = -c - \bar{b}\,, \tag{8.2}$$

where $\bar{b}$ is the average value of the b_i over all non-focal individuals. The finite population equation was obtained in [6] under the assumption that the actor gives b to a single other individual so that $\bar{b} = b/(N-1)$ where N is population size. These equations make it clear that altruism can never be selected. In an infinite population, spite [16, 17] can also never be selected, but it can be selected in a finite population if the average harm done by a focal actor to other individuals in the population exceeds the focal cost.

What does the group-formalism do for us that the old notion of transitivity did not? First of all it does allow us to strengthen a number of the results previously obtained working with the notion of transitivity. But secondly, and of more mathematical significance, it provides simpler more elegant proofs of a number of previous results. In many ways it provides the right "natural" setting for the type of homogeneity we are looking for.

9. Limitations

I end with a warning that these results apply only to a restricted class of behaviours. They do not generally apply in a class- or age-structured population, nor to ploidies other than 1, though the results do extend to a sexual diploid population if males and females are treated the same. In particular, they do not apply to sex ratio traits, or to sex-specific behaviour. Secondly, while the trait is supposed to affect the fecundities F_i, it cannot affect the offspring dispersal probabilities $d(i,j)$. In particular, it does not apply in models of optimal dispersal. It seems to apply most readily in models of cooperation and competition.

References

[1] C. Darwin, *On the Origin of Species by Means of Natural Selection or the Preservation of Favoured Races in the Struggle for Life.* John Murray, London, 1859.

[2] J.E. Cohen, *Mathematics is biology's next microscope, only better; biology is mathematics' next physics, only better.* PLoS. Biol. **2** (2004), 2017–2023.

[3] D.S. Wilson, G.B. Pollock, and L.A. Dugatkin, *Can altruism evolve in purely viscous populations?* Evol. Ecol. **6** (1992), 331–341.

[4] P.D. Taylor, *Altruism in viscous populations – an inclusive fitness model.* Evol. Ecol. **6** (1992), 352–356.

[5] P.D. Taylor, *Inclusive fitness in a homogeneous environment.* Proc. R. Soc. Lond. Ser. B-Biol. Sci. **249** (1992), 299–302.

[6] P.D. Taylor, T. Day, and G. Wild, *Evolution of cooperation in a finite homogeneous graph.* Nature **447** (2007), 469–472.

[7] A. Grafen and M. Archetti, *Natural selection of altruism in inelastic viscous homogeneous populations.* J. Theor. Biol. **252** (2008), 694–710.

[8] P.D. Taylor, T. Lillicrap, and D. Cownden, *Inclusive fitness analysis on mathematical groups* (2010), *Submitted.*

[9] W.D. Hamilton, *Genetical evolution of social behaviour I.* J. Theor. Biol. **7** (1964), 1.

[10] R.E. Michod and W.D. Hamilton, *Coefficients of relatedness in sociobiology.* Nature **288** (1980), 694–697.

[11] A. Grafen, *A geometric view of relatedness.* Oxford Surveys in Evolutionary Biology **2** (1985), 28–89.

[12] J.B.S. Haldane, *Population genetics.* New Biol. **18** (1955), 34–51.

[13] L.A. Dugatkin, *Inclusive fitness theory from Darwin to Hamilton.* Genetics **176** (2007), 1375–1380.

[14] D.C. Queller, *Genetic relatedness in viscous populations.* Evol. Ecol. **8** (1994), 70–73.

[15] S.A. West, I. Pen, and A.S. Griffin, *Conflict and cooperation – cooperation and competition between relatives.* Science **296** (2002), 72–75.

[16] W.D. Hamilton, *Selection of selfish and altruistic behaviour in some extreme models. In* J.F. Eisenberg and W.S. Dillon (eds.), *Man and Beast: Comparative Social Behaviour*, 57–91, Smithsonian Press, USA, 1971.

[17] A. Gardner and S.A. West, *Spite and the scale of competition.* J. Evol. Biol. **17** (2004), 1195–1203.

Peter Taylor
Department of Mathematics and Statistics
Queen's University
Kingston
Ontario K7L 3N6 Canada
e-mail: `peter.taylor@queensu.ca`

Evolutionary Dynamics of Collective Action

Jorge M. Pacheco, Francisco C. Santos,
Max O. Souza and Brian Skyrms

Abstract. In the natural world, performing a given task which is beneficial to an entire group requires the cooperation of several individuals of that group who often share the workload required to perform the task. The mathematical framework to study the dynamics of collective action is game theory. We study the evolutionary dynamics of cooperators and defectors in a population in which groups of individuals engage in N-person, non-excludable public goods games. We analyze the N-person Prisoner's dilemma (NPD), where the collective benefit increases proportional to the cost invested, and the N-person Snowdrift game (NSG), where the benefit is fixed but the cost is shared among those who contribute. We impose the existence of a threshold which must be surpassed before collective action becomes successful, and discuss the evolutionary dynamics in infinite and finite populations. In infinite populations, the introduction of a threshold leads, in both dilemmas, to a unified behavior, characterized by two interior fixed points. The fingerprints of the interior fixed points are still traceable in finite populations, despite evolution remaining active until the population reaches a monomorphic end-state. As the group size and population size become comparable, we find that spite dominates, making cooperation unfeasible.

Mathematics Subject Classification (2000). Primary 92D50; Secondary 91A22.

Keywords. Evolution of Cooperation, Collective Coordination, Threshold Games, Population Dynamics.

1. Introduction

The last decades have witnessed the discovery of key insights into the emergence and sustainability of cooperation at different levels of organization [1, 2, 3, 4, 5, 6, 7, 8, 9, 10, 11, 12, 13, 14, 15, 16, 17]. Special attention has been paid to two-person dilemmas such as the Prisoner's Dilemma (PD) [18, 19], the Snowdrift Game (SG) [20] and the Stag-Hunt game (SH) [16], which constitute powerful metaphors to

---

This work was supported by FCT Portugal.

describe conflicting situations often encountered in the natural and social sciences [5, 16]. Many real-life situations, however, are associated with collective action based on joint decisions made by a group often involving more than 2 individuals. This is the case, for instance, in the upper primates, where problems of collective action are recurrent [2, 21]. These types of problems are best dealt-with in the framework of N-person games [22, 23, 24, 25, 26, 27, 28]. The crossover from two-person games to N-person games brings along additional difficulties, similar to what one observes in the physical sciences when moving from the study of interactions between two particles and those involving many particles. The impact of this additional complexity in the context of biology has been well captured by the words of W.D. Hamilton [29]:

> "The theory of many person games may seem to stand to that of two-person games in the relation of sea-sickness to a headache."

The prototypical example of a Public Goods Game (PGG) is captured by the so-called N-person PD (NPD). It involves a group of N individuals, who can be either Cooperators (C) or Defectors (D). Cs contribute a cost "c" to the public good, whereas Ds refuse to do so. After all individuals are given the chance to contribute, the accumulated contribution is multiplied by an enhancement factor "F", and the total amount is equally shared among all individuals of the group. In other words, if there are k Cs in a group of N individuals, Ds end up with kFc/N, whereas Cs only get $kFc/N - c$, that is, in mixed groups Cs are always worse off than Ds.

Group hunting provides an excellent example of this type of setting. From lionesses in Etosha National Park, Namibia [30], to Chimpanzees in the Tai forest [31] and African wild dogs [32], group hunting, being ubiquitous, usually requires, to be effective, the joint action of at least a minimum number of animals. Of course, the more individuals participate, the more effective the hunting will be. In animals, other collective actions, such as lions defending a kill against a pack of hyenas, can also be seen as generalized Stag Hunt games [33]. In human affairs we also find collective action problems that can be viewed as generalized Stag hunts, not only in literal hunts such as the whale hunts discussed in [34], but also in international relations [35] and macroeconomics [36].

Despite their abundance, N-person generalizations of the Prisoner's Dilemma and the Stag-Hunt games do not exhaust the spectrum of collective action dilemmas encountered in the natural and social phenomena. Indeed, generalized snowdrift games appear all too often. In the standard SG, two individuals are driving on a road which is blocked by a snowdrift. To proceed with their journey home, the snow must be removed. Three possibilities occur: No-one shovels, and hence no-one gets home: The two drivers cooperate and shovel, and both get home, each one sharing the workload of shoveling the snow. If only one driver decides to shovel, both get home despite one driver incurring the entire cost of snow shoveling. If we define the benefit of getting home as b and the cost of shoveling as c, then if both drivers cooperate and shovel, each gets $b - c/2$. If both defect, no one gets anything: 0. If one cooperates and the other defects, the Cooperator (C) gets $b - c$ while the defector (D) gets b. Assuming, as usual, that the benefit is greater than the cost, we get a payoff ranking characteristic of a chicken, hawk-dove or snowdrift dilemma [6]. The generalization of this game to a public goods game involving N players is straightforward. To remain with the previous example, we can imagine that the snowdrift occurs at a cross-road where N drivers meet. Again, all want to go home (getting all the same benefit b), but perhaps not all are willing to shovel. If all shovel, then each gets $b - c/N$. But if only k individuals shovel (C), they get $b - c/k$ whereas those who defect by refusing to shovel get home for free and get b.

Similar to group hunting, however, it is often the case that no common benefit is produced unless its cost is shared by a minimum threshold of cooperating individuals. In keeping with the metaphor introduced above, the fact that individuals have a finite capacity of clearing the snow, may lead to the requirement of a minimum threshold of people to cooperate (shovel) so that the road is cleared.

The existence of thresholds in NSG abounds. For example, not all Amish need to participate in the construction of a church for the church to be built (see, e.g., the movie *Witness*, directed by P. Weir (1985)). Yet, the more contribute the better, since the effort to be invested by each member of the construction group will be smaller. On the other hand, the cost of building a church cannot be provided by a single individual. In this example, the public good is the church. Note, further, that the size of the church, or the benefits of having one, do not necessarily increase with the number of individuals that worked on it. Similar settings apply whenever individuals act collectively to setup sandbag levees to prevent river flooding.

Hence, as with the NPD, the need for collective coordination in the NSG introduces a behavioral tension common to conventional coordination games [15, 16]: if the others do their work, it might be profitable to do it as well; otherwise you definitely gain from *opting out*.

Mathematically, this means that for a given group of size N, we define a threshold $1 \leq M \leq N$ such that only when the number k of Cs in the group is at least M $(k \geq M)$ a public good is achieved. In all cases, a cost c must be paid before a common benefit b is produced. For the NPD, the benefit increases with the cost invested. For the NSG, the benefit is fixed but the cost is shared among those that contribute. In Table 1 we summarize the payoffs of Cs and Ds in any case (as usual in N-person games, $k = 0$ means no cost is expended and no benefit is produced).

We shall assume a population of size Z, from which groups of size N are randomly sampled. Let us first study the conventional limit in which $Z \to \infty$, under deterministic replicator dynamics. Subsequently, we shall consider stochastic dynamics in finite populations. The fitness of individuals is determined by their payoff collected when engaging in N-person PGG, requiring at least $0 < M < N$ individuals to produce any public good at all. We shall find that requiring a minimum threshold of cooperators to produce a benefit leads to the appearance of both coexistence and coordination features in an otherwise defector dominance game (NPD), and to coordination features in an otherwise coexistence game (NSG).

Game	*N*PD		*N*SG	
Strategy	C	D	C	D
$1 \leq k < M$	$-c$	0	$-\frac{c}{M}$	0
$M \leq k$	$\frac{Fkc}{N} - c$	$\frac{Fkc}{N}$	$b - \frac{c}{k}$	b

TABLE 1. Payoff values $\Pi_{\rm C}$ and $\Pi_{\rm D}$ for the NPD and NSG.

Hence, we obtain a richer evolutionary dynamics scenario in infinite populations, which, at least qualitatively, brings about a unified picture of N-person games with a threshold. We find that this scenario remains qualitatively valid whenever we remove the approximation of assuming infinite populations, although the stochastic dynamics only ends whenever a monomorphic composition of the population is reached. Nonetheless, for small populations and/or group sizes spanning nearly the entire population, we observe the "spite" effect first noted by Hamilton in 1970, and which works against cooperation [37].

2. Evolutionary dynamics of PGGs in infinite populations

Let us assume a very large population, a fraction x of which is composed of Cs, the remaining fraction $(1-x)$ being Ds. Let groups of N individuals be sampled randomly from the population. Such a random sampling leads to groups whose composition follows a binomial distribution. The fitness of Ds is given by

$$f_{\rm D} = \sum_{k=0}^{N-1} \binom{N-1}{k} x^k (1-x)^{N-1-k} \Pi_{\rm D}(k), \tag{2.1}$$

whereas the average fitness of Cs is given by

$$f_{\rm C} = \sum_{k=0}^{N-1} \binom{N-1}{k} x^k (1-x)^{N-1-k} \Pi_{\rm C}(k+1), \tag{2.2}$$

$\Pi_{\rm C}$ and $\Pi_{\rm D}$ are defined in Table 1 for each of the games. The evolutionary dynamics is given by the replicator equation [4],

$$\dot{x} = x(1-x)(f_{\rm C} - f_{\rm D}) \tag{2.3}$$

following that there exists an interior fixed point, x^*, whenever $Q(x^*) = f_{\rm C}(x^*) - f_{\rm D}(x^*) = 0$.

2.1. *N*-person PD with thresholds in infinite populations

For the NPD, with a given threshold M, the payoff of Defectors and Cooperators can be explicitly written as (see Table 1) $\Pi_{\rm D} = (kFc/N)\theta(k-M)$ and $\Pi_{\rm C} = \Pi_{\rm D} - c$,

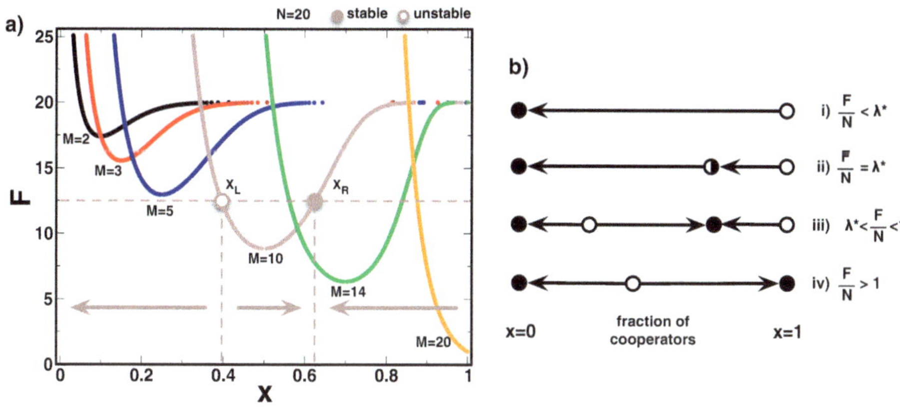

FIGURE 2.1. a) Interior fixed points of the replicator equation for N-person PD games with coordination threshold. The curves provide the location of the critical values of the fraction of cooperators (x_L,x_R) at which $f_C = f_D$. For each value of F (defining a horizontal line), the critical values are given by the intersection of this line with each curve (one curve for given fixed M and $N = 20$). Scenarios with none, one and two interior fixed points are possible as detailed in the right panel. b) Dynamics of N-person PD in infinite populations with coordination threshold. Empty circles represent unstable fixed points; full circles represent stable fixed points and arrows indicate the direction of evolution by natural selection.

respectively, where the Heaviside step function $\theta(x)$ is equal to 1 whenever $x \geq 0$ and equal to 0 otherwise. The introduction of a threshold ($M > 1$) leads to a symmetry breaking of the sampling, which does not allow a closed form expression for the fitness. Thus, the determination of the possible interior equilibrium points, i.e., the zeros of $Q(x)$ has to be done numerically. However, a great deal of information can be obtained without solving explicitly for $Q(x) = 0$. Indeed, as shown in [38], introducing Π_C and Π_D above in Eqs. (2.1) and (2.2) leads to

$$\begin{aligned} Q(x) &= f_C(x) - f_D(x) \\ &= c\left(\frac{F}{N} - 1\right) \\ &\quad - c\frac{F}{N}(1-x)^{N-M} \sum_{k=0}^{M-1} \binom{N-1}{k} (1 - M\delta_{k,M-1})\, x^k (1-x)^{M-1-k}. \end{aligned}$$

In what follows, we shall strictly assume that $N \geq 2$. For most of the time, we shall assume that $1 < M \leq N$; the degenerate cases can be handled as well, and the reader is referred to [38] for details. Let

$$R(x) = \sum_{k=M}^{N-1} \binom{N-1}{k} x^k (1-x)^{N-1-k} + M \binom{N-1}{M-1} x^{M-1} (1-x)^{N-M}$$
$$= x^{M-1} \left(\sum_{k=M}^{N-1} \binom{N-1}{k} x^{k-M+1} (1-x)^{N-1-k} + M \binom{N-1}{M-1} (1-x)^{N-M} \right). \tag{2.4}$$

Since,

$$1 = 1^{N-1} = (x + 1 - x)^{N-1} = \sum_{k=0}^{N-1} \binom{N-1}{k} x^k (1-x)^{N-1-k},$$

we have that

$$Q(x) = -c\,(1 - \lambda R(x))\,, \tag{2.5}$$

with $\lambda = F/N$.

Lemma 2.1. *The polynomial R defined above satisfies*

1. $R(0) = 0$;
2. $R(1) = 1$;
3. $R(x) > 0, \quad x \in (0,1)$;
4. *Let $x^* = M/N$. Then we have that $R'(x) > 0$ for $0 \leq x < x^*$, and $R'(x) < 0$ for $x^* < x < 1$. In particular, $R'(x^*) = 0$, and x^* is a point of maximum of R with $R(x^*) > 1$.*

Proof. First, notice that 1., 2. and 3. are straightforward from the form of the polynomial $R(x)$; cf. (2.4).

To prove 4., we let $k = N - 1 - k'$, and on noting that

$$\binom{N-1}{N-1-k'} = \binom{N-1}{k'},$$

we may write

$$R(x) = x^{M-1} \left[\sum_{k'=0}^{N-M-1} \binom{N-1}{k'} x^{N-M-k'} (1-x)^{k'} + M \binom{N-1}{M-1} (1-x)^{N-M} \right]$$
$$= x^{N-1} \left[\sum_{k'=0}^{N-M-1} \binom{N-1}{k'} \left(\frac{1-x}{x} \right)^{k'} + M \binom{N-1}{M-1} \left(\frac{1-x}{x} \right)^{N-M} \right].$$

Let

$$z = \frac{1-x}{x}.$$

Then, we have that

$$z' = -\frac{1}{x^2} = -\frac{1}{x}(z+1)$$

Thus,

$$R(x) = x^{N-1}p(z), \quad \text{with} \quad p(z) = \sum_{i=0}^{N-M} a_i z^i,$$

where

$$a_i = \binom{N-1}{i}, \quad 0 \le i < N-M \quad \text{and} \quad a_{N-M} = M\binom{N-1}{M-1}$$

We now compute R':

$$\begin{aligned}
R'(x) &= (N-1)x^{N-2}p(z) - x^{N-2}p'(z)(z+1) \\
&= x^{N-2}\left[(N-1)p(z) - (z+1)p'(z)\right] \\
&= x^{N-2}\left[(N-1)\sum_{i=0}^{N-M} a_i z^i - \sum_{i=1}^{N-M} i a_i z^i - \sum_{i=1}^{N-M} i a_i z^{i-1}\right] \\
&= x^{N-2}\left[(N-1)a_0 - a_1 + (N-1)\sum_{i=1}^{N-M} a_i z^i - \sum_{i=1}^{N-M} i a_i z^i - \sum_{i=2}^{N-M} i a_i z^{i-1}\right].
\end{aligned}$$

Since $a_0 = 1$ and $a_1 = N-1$, and writing $i = i+1$ in the last sum, we find that

$$\begin{aligned}
R'(x) &= x^{N-2}\left[(N-1)\sum_{i=1}^{N-M} a_i z^i - \sum_{i=1}^{N-M} i a_i z^i - \sum_{i=1}^{N-M-1} (i+1)a_{i+1} z^i\right] \\
&= x^{N-2}S(z),
\end{aligned}$$

where

$$\begin{aligned}
S(z) = &\sum_{i=1}^{N-M-2} \left[(N-1-i)a_i - (i+1)a_{i+1}\right] z^i \\
&+ \left[M a_{N-M-1} - (N-M)a_{N-M}\right] z^{N-M-1} + (M-1)a_{N-M} z^{N-M}.
\end{aligned}$$

On noting that

$$\binom{L}{j+1} = \frac{L-j}{j+1}\binom{L}{j}, \tag{2.6}$$

we obtain, for $1 \le i < N-M$, that

$$a_{i+1} = \frac{N-1-i}{i+1} a_i.$$

Hence,

$$\sum_{i=1}^{N-M-2} \left[(N-1-i)a_i - (i+1)a_{i+1}\right] z^i = 0.$$

Also, we have

$$M a_{N-M-1} - (N-M)a_{N-M} = M\binom{N-1}{M} - (N-M)\binom{N-1}{M-1},$$

which on calling upon (2.6) yields

$$M\binom{N-1}{M} - (N-M)\binom{N-1}{M-1} = (N-M)\binom{N-1}{M} - (N-M)\binom{N-1}{M-1}$$
$$= -(N-M)(M-1)\binom{N-1}{M-1}.$$

Thus, we write

$$S(z) = z^{N-M-1}\binom{N-1}{M-1}\left[-(N-M)(M-1) + M(M-1)z\right]$$

which yields

$$R'(x) = x^{M-1}(1-x)^{N-M-1}\binom{N-1}{M-1}\left[-(N-M)(M-1) + M(M-1)z\right] \quad (2.7)$$

For $x \in (0,1)$, (2.7) vanishes at

$$z^* = \frac{N-M}{M} = \frac{1-M/N}{M/N}.$$

Since

$$z = \frac{1-x}{x}$$

is one-to-one.

$$x^* = \frac{M}{N}.$$

Also, from (2.7), we see that

1. for $0 < z < z^*$, $R'(x) < 0$;
2. for $z > z^*$, $R'(x) > 0$

Furthermore, $z = (1-x)/x$ is monotonically decreasing and maps $(0,1)$ in $(0,\infty)$ (thus reversing the orientation), which yields that $0 < z < z^*$ corresponds to $x^* < x < 1$ and $z > z^*$ corresponds to $0 < x < x^*$.

This proves 4. □

Using the information provided by Lemma 2.1, we have

Theorem 2.2. *Let $\lambda^* = 1/R(x^*)$. Then we have that $0 < \lambda^* < 1$. Moreover, we have that $Q(x)$ satisfies:*

1. *For $\lambda < \lambda^*$ there are no roots in $(0,1)$;*
2. *For $\lambda = \lambda^*$ there exists one double root at $x = x^*$;*
3. *For $\lambda^* < \lambda \leq 1$ there are two simple roots $\{x_L, x_R\}$, with $x_L \in (0,x^*)$ and $x_R \in (x^*,1]$.*
4. *For $\lambda > 1$ there is only one root in $(0,x^*)$.*

From Theorem 2.2, we can infer the complete evolutionary dynamics of the system. Thus, if $F < \lambda^* N$, no interior equilibrium is possible. For $F = \lambda^* N$, $x = M/N$ is an unstable equilibrium. For

$$\lambda^* < \frac{F}{N} < 1,$$

we have the existence of two equilibria. The leftmost equilibrium is always less than M/N and it is unstable. On the other hand, the rightmost equilibrium is always greater than M/N, and it is stable. The reader is referred to [38] for the detailed proofs.

Overall, the analysis above shows that the properties of $Q(x)$ lead to a very interesting dynamics of the replicator equation, with possibly two interior fixed points (x_L and x_R), as illustrated in Fig. 2.1, for $N = 20$, different values of $1 < M \leq 20$ and variable F. Note, in particular, that the fact that $R'(x_L) > 0$ and $R'(x_R) < 0$ [38] allows us to classify immediately x_L as an unstable fixed point whereas x_R, if it exists, corresponds to a stable fixed point, as illustrated also in Fig. 2.1. Moreover, when $F/N = R(M/N)^{-1}$, M/N is the unique interior and unstable fixed point.

Between these two limiting values of F, and given the nature of the interior fixed points x_L and x_R, one can easily conclude that below x_L all individuals will ultimately forego the public good. Conversely, for all $x > x_L$, the population will evolve towards a mixed equilibrium defined by x_R, corresponding to a stable fixed point of the associated replicator equation (even if, initially, $x > x_R$). Similar to the N-person PD, whenever $F/N < R(M/N)^{-1}$, $f_C(x) < f_D(x)$, for all $x \in (0, 1)$, which means that all individuals will end up foregoing the public good.

2.2. N-person SG with thresholds in infinite populations

For the NSG, we may formally write the payoffs in Table 1 in the form

$$\Pi_D(k) = b\theta(k - M) \tag{2.8}$$

for the payoff of a defector in the group and

$$\Pi_C(k) = \Pi_D(k) - \frac{c}{k}\theta(k - M) - \frac{c}{M}(1 - \theta(k - M)) \tag{2.9}$$

for the payoff of a cooperator in the same group. Under these assumptions, one can show that $Q(x)$ now reads [39]

$$\begin{aligned} Q(x) = & \frac{c}{xN}\left\{N\frac{b}{c}\binom{N-1}{M-1}x^M(1-x)^{N-M}\right. \\ & \left. - \left[1 + \sum_{k=0}^{M-1}\binom{N}{k}x^k(1-x)^{N-k}\left(\frac{k}{M} - 1\right)\right]\right\}. \end{aligned}$$

Although the polynomial Q in this case is quite distinct from the NPD case, we can show similar results for the internal fixed points. More precisely, let $\gamma = c/b$.

We find that it will be more appropriate to study

$$\begin{aligned} p(x,\gamma) =& N\binom{N-1}{M-1}x^M(1-x)^{N-M} \\ & -\gamma\left[1+\sum_{k=0}^{M-1}\binom{N}{k}x^k(1-x)^{N-k}\left(\frac{k}{M}-1\right)\right]. \end{aligned}$$

$p(x,\gamma)$ has the same interior roots as $Q(x)$, and we made the dependence on γ explicit. Notice also that $p(x,\gamma)$ implies the same dynamics for the Replicator equation as that implied by $Q(x)$ in $(0,1)$ up to a time rescaling. We then have the following result

Theorem 2.3. *There exists* $0<\bar{\gamma}$ *and* $0<\bar{x}<1$ *such that, if*

1. $\bar{\gamma}/\gamma<1$, *then the evolutionary dynamics has no interior equilibria.*
2. $\bar{\gamma}/\gamma=1$, *then* $\bar{x}$ *is a unique interior equilibrium.*
3. $\bar{\gamma}/\gamma>1$, *then there are two interior equilibria* $x_{\mathrm{L}}<\bar{x}<x_{\mathrm{R}}$. *Moreover,* x_{L} *is always an unstable equilibrium point, while* x_{R} *is always a stable point.*

In order to prove Theorem 2.3, it turns out that is more convenient to determine what γ will render a given $x\in(0,1)$ an interior point, rather than determining what x are equilibria for a given γ. Let us define

$$\Gamma(x)=\begin{cases} 0, & x=0, \\ N\binom{N-1}{M-1}\dfrac{x^M(1-x)^{N-M}}{1+\sum_{k=0}^{M-1}\binom{N}{k}x^k(1-x)^{N-k}\left(\frac{k}{M}-1\right)}, & 0<x\leq 1. \end{cases} \tag{2.10}$$

Then $\Gamma:[0,1]\to\mathbb{R}$ is continuous in $[0,1]$ and differentiable in $(0,1)$. Also, by solving for γ the equation $p(x,\gamma)=0$, it is straightforward to verify that we have the identity

$$p(x,\Gamma(x))=0. \tag{2.11}$$

Ultimately, $\Gamma(x)$ is responsible for the existence of a cost-to-benefit ratio at which a given interior x can become an equilibrium of the replicator dynamics. The critical value $\bar{x}$ corresponds to the first interior equilibrium which emerges when $c/b=\bar{\gamma}$ and which divides the unit interval into two pieces, in which the stable and unstable equilibria remain confined whenever $c/b<\bar{\gamma}$. The thrust of the argument is to study the number of solutions of $\Gamma(x)=\gamma$, for a given γ, which then can be used to prove Theorem 2.3. In order to achieve our goal, we establish a series of results about Γ. In what follows, we shall assume $N>2$ and $1<M<N$.

Proposition 2.4. *There is a unique* $\bar{x}\in(0,1)$ *such that* $\Gamma'(\bar{x})=0$. *Such* $\bar{x}$ *will be the unique point of global maximum for* Γ.

Proposition 2.5. *Let* $\bar{\gamma}=\Gamma(\bar{x})$, *with* $\bar{x}$ *given above. Then the equation* $\Gamma(x)=\gamma$ *has*

1. *two solutions,* x_{L} *and* x_{R}, *for* $\gamma<\bar{\gamma}$. *Moreover* $x_{\mathrm{L}}\in[0,\bar{x})$ *and* $x_{\mathrm{R}}\in(\bar{x},1]$.
2. *one solution for* $\gamma=\bar{\gamma}$;
3. *no solution for* $\gamma>\bar{\gamma}$.

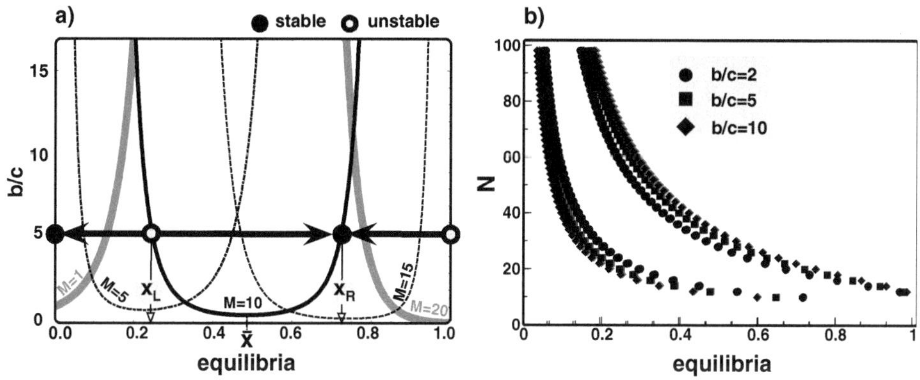

FIGURE 2.2. Equilibria of the N-person Snowdrift Game with threshold. We assume infinite, well-mixed populations, fix the group size at $N = 20$ and vary the threshold M above which cooperation leads to a common benefit b. The total cost involved is c. In a) we show how the occurrence of a threshold leads to the appearance of at most 2 interior fixed points x_{L} and x_{R}, which can be found via the intersection of a horizontal line with the appropriate curve (illustrated for $M = 10$); in this case, the leftmost root is always an unstable fixed point whereas the rightmost corresponds to stable fixed point, as illustrated by the horizontal arrows (see main text for details). For a given M/N, there is a critical value $\bar{\gamma}$ for the critical cost-to-benefit ratio c/b below which the 2 interior roots discussed above always exist. In panel b) we show how these interior fixed points scale with variable group size N for some values of the b/c ratio indicated. For $\bar{\gamma}b < c$ no interior fixed points exist and defectors dominate unconditionally, whereas for $\bar{\gamma}b = c$ the only root corresponds to an unstable fixed point.

Finally, the following asymptotic result allows an approximate determination of $\bar{x}$.

Proposition 2.6. *Let* $x_0 = \frac{M}{N}$ *and assume that*

$$0 < \epsilon = \frac{N - M}{N} \ll 1$$

Then, we have that

$$\bar{x} = x_0 - \frac{x_0^M}{M}\binom{N}{M-1}\epsilon^{N-M+1} + \mathcal{O}(\epsilon^{N-M+2})\,.$$

Therefore, when the threshold is comparable in order to the size of the group, we have that while the critical equilibrium is not quite M/N, it is quite close to it. We refer the interested reader to [39] for detailed proofs.

As in the case of the NPD, the exact position of the roots of $Q(x)$ in the NSG regime may be cumbersome to find analytically, but is easy to compute numerically. Fig. 2.2 pictures the position of the interior roots of $Q(x)$ for a fixed group size of $N = 20$ and variable threshold values of M (right panel).

For each value of M there is a critical benefit-to-cost value b/c above which two interior fixed points emerge. These can be found in Fig. 2.2 by drawing a horizontal line at a fixed b/c – its intersection with the appropriate curve for a given threshold M provides the location of the points.

As shown above and illustrated in Fig. 2.2, one root corresponds to an unstable fixed point (x_{L}) and the other to a stable fixed point (x_{R}) inducing a coexistence between Cs and Ds. This means there is a range of values of x ($x_{\mathrm{L}} < x < x_{\mathrm{R}}$), in which Cs are favored against Ds ($f_{\mathrm{C}}(x) > f_{\mathrm{D}}(x)$). When $x > x_{\mathrm{L}}$, the system will always evolve to the mixed configuration given by x_{R}, and below x_{L} all individuals will end up refusing to contribute to the public good.

3. Evolutionary dynamics of PGGs in finite populations

Let us focus on a well-mixed population of size Z in the absence of mutations. Sampling of individuals is no longer binomial, following a hypergeometric distribution. Consequently, the average fitness of Cs and Ds can now be written as

$$f_{\mathrm{C}}(k) = \binom{Z-1}{N-1}^{-1} \sum_{j=0}^{N-1} \binom{k-1}{j} \binom{Z-k}{N-j-1} \Pi_{\mathrm{C}}(j+1) \tag{3.1}$$

and

$$f_{\mathrm{D}}(k) = \binom{Z-1}{N-1}^{-1} \sum_{j=0}^{N-1} \binom{k}{j} \binom{Z-k-1}{N-j-1} \Pi_{\mathrm{D}}(j) \tag{3.2}$$

respectively.

The fraction of cooperators is no longer a continuous variable, varying in steps of $1/Z$. We adopt a stochastic birth-death process [40] combined with the pairwise comparison rule [41, 42, 43] in order to describe the evolutionary dynamics of Cs (and Ds) in a finite population. Under pairwise comparison, two individuals from the population, A and B are randomly selected for update (only the selection of mixed pairs can change the composition of the population). The strategy of A will replace that of B with a probability given by the Fermi function (from statistical physics)

$$p = \frac{1}{1 + e^{-\beta(f_{\mathrm{A}} - f_{\mathrm{B}})}} \, . \tag{3.3}$$

The reverse will happen with probability $1 - p$. The quantity β, which in physics corresponds to an inverse temperature, controls the intensity of selection: For $\beta \ll 1$ selection is weak, and one recovers the replicator equation in the limit $Z \to \infty$ [41, 42, 43]. For arbitrary β, the quantity corresponding to the right-hand side of the replicator equation, specifying the *gradient of selection*, is given in finite populations by [41, 42, 43]

$$g(k) \equiv T^{+}(k) - T^{-}(k) = \frac{k}{Z}\frac{Z-k}{Z}\tanh\left\{\frac{\beta}{2}[f_{\mathrm{C}}(k) - f_{\mathrm{D}}(k)]\right\} \quad (3.4)$$

The right-hand side of $g(k)$ is similar to the replicator equation, only that the (non-linear) pairwise comparison [41, 42, 43] defined in Eq. 3.3 leads to the appearance of the hyperbolic tangent of the fitness difference, instead of the fitness difference. This has implications in the characteristic evolutionary times, which now depend on β [41, 42, 43], but not in what concerns the roots of $g(k)$. Importantly, the evolutionary dynamics in finite populations will only stop whenever the population reaches a monomorphic state ($k = 0$ or $k = Z$). Hence, the sign of $g(k)$, which indicates the direction of selection, is important in that it may strongly influence the evolutionary time required to reach any of the absorbing states.

3.1. N-person PD with thresholds in finite populations

Whenever $M = 0$ (NPD without the requirement to coordinate to obtain collective benefits) we may write

$$f_{\mathrm{C}}(k) - f_{\mathrm{D}}(k) = c\left[\frac{F}{N}\left(1 - \frac{N-1}{Z-1}\right) - 1\right] \quad (3.5)$$

which is independent of k being, however, population and group size dependent. This means frequency independent selection. In particular, whenever the size of the group equals the population size, $N = Z$, we have that $f_{\mathrm{C}}(k) - f_{\mathrm{D}}(k) = -c$ and cooperators have no chance irrespective of the value of the enhancement factor. This contrasts with the result in infinite, well-mixed populations ($Z \to \infty$), where to play C would be the best option whenever $F > N$. For finite populations, the possibility that group size equals population size leads to the demise of cooperation. Moreover, given the independence of $f_{\mathrm{C}}(k) - f_{\mathrm{D}}(k)$ on k in finite populations, for a given population size, it is straightforward to obtain a critical value of F for which selection is neutral, and above which cooperators will win the evolutionary race. From the equations above this critical value reads $F = N\left(1 - \frac{N-1}{Z-1}\right)^{-1}$.

Let us now discuss the NPD with $1 < M < N \leq Z$. Whenever $N = Z$, the result is easily inferred from the NPD above – all individuals in the population will ultimately forego the public good. This will happen, in finite populations, irrespective of the existence (or not) of a threshold M. However, whenever $N < Z$ the threshold brings about a strong disruption of the finite population dynamics, which we illustrate numerically, given the unappealing look of the analytical equations.

Let us start with the case in which $F > N$, that is, the regime for which we obtain a pure coordination game with a single (unstable) fixed point in the replicator dynamics equation (cf. Fig. 2.1). In finite populations the possible scenarios are depicted in the left panel of Fig. 3.1. Clearly, for small population sizes, cooperators are always disadvantageous. With increasing Z, however, one approaches

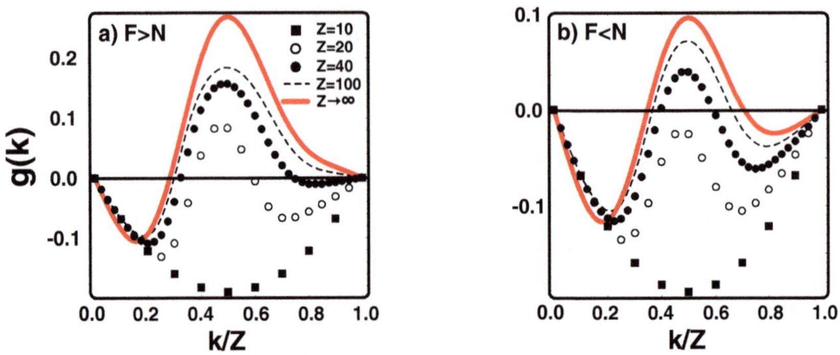

FIGURE 3.1. Behaviour of $g(k)$ for a N-person PD game with coordination threshold $M = 5$ in a population of variable size Z and fixed group size $N = 10$. a) Since $F = 12 > N$, the game becomes a pure coordination game in infinite populations. In finite populations, however, it strongly depends on Z: For $Z = N$, Cs are always disadvantageous and evolutionary dynamics leads mostly to 100% Ds. For $Z = 20$ (and using a terminology which is only correct for $Z \to \infty$), we obtain a profile for $g(k)$ evidencing the emergence of a coordination point and a coexistence point. For increasingly large Z (e.g., $Z = 40$), the coexistence point disappears and we recover the behaviour of the replicator dynamics (see Fig. 2.1: Selection favours Cs above a given fraction k/Z and Ds below that fraction which, in turn, depends on the population size. b) Since $F = 8 < N$, the game exhibits now 2 interior fixed points in infinite populations (red curve). Similar to a), for small Z Cs are disadvantageous for all k. Unlike a), however, now two interior fixed points emerge together for a critical population size, and remain for larger population sizes.

the replicator dynamics scenario (see Fig. 2.1), despite the fact that, e.g., for $Z = 20$, convergence towards the absorbing state at 100% Cs is hindered because Cs become disadvantageous for large k. Indeed, for this population size, Cs are advantageous only in a small neighbourhood of $k/Z = 0.5$, being disadvantageous both for smaller and larger values of k/Z. In other words, and despite the fact that evolution will stop only at $k = 0$ or $k = Z$, the time it takes to reach an absorbing state will depend sensitively on the population size, given the occurrence (or not) of interior roots of $g(k)$.

Whenever $F < N$, yet above the critical limit below which Cs become disadvantageous for all x in Fig. 2.1, we observe that for small population sizes Cs are always disadvantageous, and the two interior fixed points of the replicator dynamics equation only manifest themselves above a critical population size, as illustrated in the right panel of Fig. 3.1.

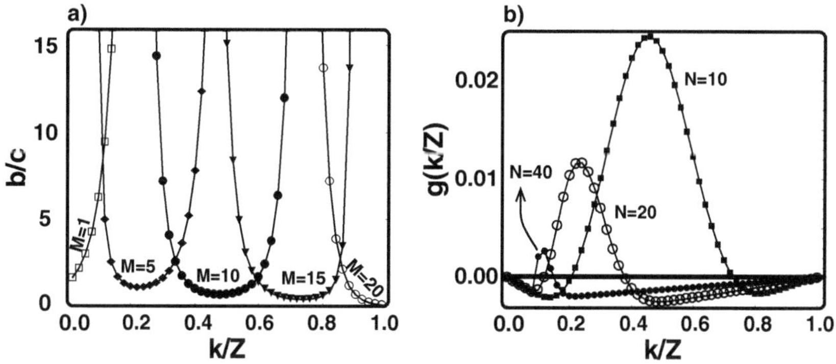

FIGURE 3.2. a) Equilibria of the N-person snowdrift game with threshold in finite populations. Population size is $Z = 50$ and group size is $N = 20$. We vary the threshold M above which cooperation leads to a common benefit b. For each k/Z we show the corresponding b/c at which $g(k) = 0$ (cf. Eq. (13)). Whenever the population size is large compared to group size, selection in finite populations is qualitatively similar to that in infinite populations. b) Effect of group size in the evolution of cooperation.We plot $g(k)$ as a function of the fraction of cooperators k/Z, for $b/c = 5$. We fixed the population size at $Z = 50$ and the threshold at $M = 5$, while varying the group size N. As the group size approaches the population size, the range of values of k/Z for which cooperation is advantageous ($g(k) > 0$) is reduced.

3.2. N-person SG with thresholds in finite populations

In Fig. 3.2a, we show how the qualitative behavior of selection under stochastic dynamics in finite populations mimics closely that already encountered in the previous section (cf. Fig. 2.2), associated with deterministic dynamics in infinite populations. Although the population will always fixate in one of the two *absorbing states* ($k = 0$ and $k = Z$ in the absence of mutations), selection will act to drive the population toward a composition reflecting the rightmost root of $g(k)$, which constitutes the deepest point of the basin of attraction of the evolutionary dynamics.

On the other hand, as the group size approaches the population size the previous basin of attraction is reduced. In Fig. 3.2b we show a typical behavior of $g(k)$ as a function of the fraction of cooperators k/Z for fixed population size $Z = 50$, threshold $M = 5$ and different group sizes N. As N increases, cooperation becomes increasingly unfeasible – in the limit when $N \to Z$, cooperators have no chance and defectors dominate unconditionally. Moreover, for a given b/c ratio, the existence of a finite population analogue of a stable root of $g(k)$ (in infinite populations) occurs for values of the frequency k/Z of cooperators which decrease as N increases. This has been first noted by Hamilton [37] and reflects the occurrence of "spite" which works against cooperation, as illustrated in Fig. 3.2.

4. Discussion

We showed how generalizing the conventional versions of the NPD and NSG dilemmas by introducing thresholds below which collective action is unfeasible leads to the emergence of an entirely new evolutionary scenario. Irrespectively of the game played, in infinite, well-mixed populations, the existence of a threshold opens the possibility for the appearance of two interior fixed points in the replicator equation (x_{L} and x_{R}). The one at lower frequency of cooperators is always an unstable fixed point (coordination), which determines a threshold for cooperative collective action. The other, at higher frequency of cooperators, is a stable fixed point (coexistence), and hence determines the final frequency of cooperators in the population, assuming the coordination threshold is overcome. Moreover, both dilemmas converge to a pure coordination game whenever the coordination threshold approaches the group size.

In the particular case of the NSG with a given threshold M and group size N, there is always a critical cost-to-benefit ratio c/b above which the two interior roots discussed above emerge. The same qualitative behavior can be observed in finite populations. However, as soon as the group size approaches the population size, cooperation becomes increasingly unfeasible.

In the NPD, besides the above-mentioned regime with two interior roots, there are also the possible outcomes of no cooperation or of a pure coordination game, which depends sensitively on the minimum number of cooperators M in a group of N individuals required to produce any public good. In finite populations, the evolutionary dynamics of the NPD game may be profoundly affected, mostly when the population size (Z) is comparable to the group size (N). In this regime, one observes an overlap of the different scenarios observed in infinite populations. Hence, for $Z = N$, cooperators are always disadvantageous, irrespective of the existence or not of a threshold. For $Z > N$, the direction of selection in a finite population is strongly size dependent. For fixed $F > N$, there is a critical value, Z_1, above which the interior roots of $g(k)$ emerge, which constitute the finite-population analogs of x_{L} and x_{R} in infinite populations (cf. Fig. 2.1). Above a second critical value, Z_2, x_{R} disappears, and one ends up with a coordination game. For $M < F < N$ and a small population size, that is, $F < N$ but yet above the critical value $\lambda^* = R(M/N)^{-1}$ defined in section 2.1, cooperators are always disadvantageous; however, above a critical population size (Z_{C}) the interior roots of $g(k)$ emerge simultaneously and the evolutionary dynamics approaches that observed in infinite populations.

5. Conclusions

Unlike two-person games, current models of collective action have typically overlooked the necessity of some form of coordination among individuals, pervasive in biological and social collective dilemmas. From social organization [17] to the salvation of the planet against environmental hazards [44, 45], examples abound where a minimum number of individuals, which does not necessarily equal the entire group, must simultaneously cooperate before any outcome (or public good) is produced.

In this chapter we investigate the predictions of evolutionary game theory in both finite and infinite populations, whenever a minimum threshold of individuals must cooperate simultaneously in a group before any viable public good is achieved. We have concentrated on two of the most important collective dilemmas: the N-person snowdrift game (NSG) [39] and N-person prisoner's dilemma (NPD) [38]. In doing so, we uncover a new framework in which the advantage or not of cooperators depends sensitively on group and population size, as well as on the threshold for collective action. Such interplay leads to rich evolutionary scenarios, impossible to anticipate based on the traditional assumption of infinite populations, providing valuable insights into the variety and complexity of many person social dilemmas, inescapable especially among humans.

In addition, it is noteworthy that irrespectively of the distinctive features of the N-person Prisoner's dilemma (a defector's dominance dilemma) and the N-person Snowdrift game (a coexistence game), the existence of a coordination threshold is able to produce an unifying framework associated with a generalized stag-hunt game [38]. Moreover, the necessity of coordination is shown to increase the equilibrium fraction of cooperators, even if this enhancement comes together with a strong dependence on the initial level of cooperation, since coexistence between cooperators only emerges when a minimum number of cooperators is already present in the population. This result is of particular relevance given that the existence of coordination thresholds constitutes a rule, rather than the exception. Finally, our results reinforce the idea that even minor differences in the nature of collective rewards and/or costs can have a profound effect in the final outcome of evolution.

References

[1] R. Axelrod and W.D. Hamilton, *The evolution of cooperation.* Science **211** (1981), 1390–1396.

[2] R. Boyd and P.J. Richerson, *Culture and the Evolutionary Process.* University of Chicago Press, USA, 1985.

[3] P.E. Hammerstein, *Genetic and Cultural Evolution of Cooperation.* MIT press, Cambridge, MA., USA, 2003.

[4] J. Hofbauer and K. Sigmund, *Evolutionary Games and Population Dynamics.* Cambridge Univ. Press, Cambridge, UK, 1998.

[5] M.W. Macy and A. Flache, *Learning dynamics in social dilemmas.* Proc. Natl. Acad. Sci. USA **99** (2002), 7229–7236.

[6] J. Maynard Smith, *Evolution and the Theory of Games.* Cambridge University Press, Cambridge, UK, 1982.

[7] M.A. Nowak, *Evolutionary Dynamics: Exploring the Equations of Life.* The Belknap Press of Harvard University Press, Cambridge, MA, 2006.

[8] M.A. Nowak, *Five rules for the evolution of cooperation.* Science **314** (2006), 1560–1563.

[9] M.A. Nowak and K. Sigmund, *Evolutionary dynamics of biological games.* Science **303** (2004), 793–799.

[10] H. Ohtsuki, C. Hauert, E. Lieberman, and M.A. Nowak, *A simple rule for the evolution of cooperation on graphs and social networks.* Nature **441** (2006), 502–505.

[11] F.C. Santos and J.M. Pacheco, *Scale-free networks provide a unifying framework for the emergence of cooperation.* Phys. Rev. Lett. **95** (2005), 098104.

[12] F.C. Santos, J.M. Pacheco, and T. Lenaerts, *Evolutionary dynamics of social dilemmas in structured heterogeneous populations.* Proc. Natl. Acad. Sci. USA **103** (2006), 3490–3494.

[13] F.C. Santos, M.D. Santos, and J.M. Pacheco, *Social diversity promotes the emergence of cooperation in public goods games.* Nature **454** (2008), 213–216.

[14] J.M. Pacheco, F.C. Santos, and F.A.C.C. Chalub, *Stern-judging: A simple, successful norm which promotes cooperation under indirect reciprocity.* PLoS Computational Biology **2** (2006), e178.

[15] B. Skyrms, *The stag hunt.* Proceedings and Addresses of the American Philosophical Association **75** (2001), 31–41.

[16] B. Skyrms, *The Stag Hunt and the Evolution of Social Structure.* Cambridge University Press, UK, 2004.

[17] S. Bowles, *Microeconomics: Behavior, Institutions and Evolution.* Princeton University Press, USA, 2003.

[18] M.M. Flood, *Some experimental games, research memorandum RM-789.* RAND Corporation, Santa Monica, CA (1952).

[19] M. Dresher, *The Mathematics of Games of Strategy: Theory and Applications.* Prentice-Hall, Englewood Cliffs, NJ, 1961.

[20] R. Sugden, *The Economics of Rights, Co-operation and Welfare.* Basil Blackell, Oxford, UK, 1986.

[21] C. Boehm, *Hierarchy in the Forest: The Evolution of Egalitarian Behavior.* Harvard University Press, USA, 1999.

[22] G. Hardin, *The tragedy of the commons.* Science **162** (1968), 1243–1248.

[23] T.C. Schelling, *Hockey helmets, concealed weapons, and daylight saving: A study of binary choices with externalities.* J. Conflict Resolution **17** (1973), 381.

[24] R.M. Dawes, *Social dilemmas. social dilemmas. social dilemmas.* Annu. Rev. of Psychol. **31** (1980), 169–193.

[25] R. Boyd and P.J. Richerson, *The evolution of reciprocity in sizable groups.* J. Theor. Biol. **132** (1988), 337–356.

[26] P. Kollock, *Social dilemmas: The anatomy of cooperation.* Annu. Rev. Sociol. **24** (1998), 183–214.

[27] C. Hauert, F. Michor, M.A. Nowak, and M. Doebeli, *Synergy and discounting of cooperation in social dilemmas.* J. Theo. Bio. (2006), 195–202.

[28] C. Hauert, A. Traulsen, H. Brandt, M.A. Nowak, and K. Sigmund, *Via freedom fo coercion: The emergence of costly punishment.* Science **316** (2007), 1905–1907.

[29] W.D. Hamilton, *Biosocial anthropology. In Biosocial anthropology*, 133–155, Malaby Press, London, UK, 1975.

[30] P.E. Stander, *Cooperative hunting in lions – the role of the individual.* Behavioral Ecology and Sociobiology **29** (1992), 445–454.

[31] C. Boesch, *Cooperative hunting roles among tai chimpanzees.* Human Nature – an Interdisciplinary Biosocial Perspective **13** (2002), 27–46.

[32] S. Creel and N.M. Creel, *Communal hunting and pack size in African wild dogs, Lycaon-pictus.* Anim. Behav. **50** (1995), 1325–1339.

[33] J. Maynard Smith and E. Szathmáry, *The Major Transitions in Evolution.* Freeman, Oxford, UK, 1995.

[34] B. Beding, *The stone-age whale hunters who kill with their bare hands.* Daily Mail, 12th April (2008).

[35] R. Jervis, *Cooperation under the security dilemma.* World Politics **30** (1978), 167–214.

[36] J. Bryant, *Coordination theory, the stag hunt and macroeconomics. In* J. Friedman (ed.), *Problems of Coordination in Economic Activity*, 207–225, Kluwer, Dordrecht, The Netherlands, 1994.

[37] W.D. Hamilton, *Selfish and spiteful behaviour in an evolutionary model.* Nature **228** (1970), 1218–1220.

[38] J.M. Pacheco, F.C. Santos, M.O. Souza, and B. Skyrms, *Evolutionary dynamics of collective action in n-person stag-hunt dilemmas.* P. Roy. Soc. B: Biological Sciences **276** (2009).

[39] M.O. Souza, J.M. Pacheco, and F.C. Santos, *Evolution of cooperation under N-person snowdrift games.* J. Theor. Biol. **260** (2009), 581–588.

[40] S. Karlin and H.M.A. Taylor, *A First Course in Stochastic Processes.* 2nd edition, Academic, London, UK, 1975.

[41] A. Traulsen, M.A. Nowak, and J.M. Pacheco, *Stochastic dynamics of invasion and fixation.* Phys. Rev. E: Stat. Nonlin. Soft. Matter. Phys. **74** (2006), 011909.

[42] A. Traulsen, M.A. Nowak, and J.M. Pacheco, *Stochastic payoff evaluation increases the temperature of selection.* J. Theor. Biol. **244** (2007), 349–356.

[43] A. Traulsen, J.M. Pacheco, and M.A. Nowak, *Pairwise comparison and selection temperature in evolutionary game dynamics.* J. Theor. Biol. **246** (2007), 522–529.

[44] M. Milinski, D. Semmann, H.J. Krambeck, and J. Marotzke, *Stabilizing the Earth's climate is not a losing game: Supporting evidence from public goods experiments.* Proc. Natl. Acad. Sci. USA **103** (2006), 3994–3998.

[45] M. Milinski, R.D. Sommerfeld, H.J. Krambeck, F.A. Reed, and J. Marotzke, *The collective-risk social dilemma and the prevention of simulated dangerous climate change.* Proc. Natl. Acad. Sci. USA **105** (2008), 2291–2294.

Jorge M. Pacheco
Departamento de Matemática e Aplicações
Universidade do Minho
4710-057 Braga, Portugal

and

ATP group, CMAF
Complexo Interdisciplinar
Universidade de Lisboa, Portugal
e-mail: jmpacheco@math.uminho.pt

Francisco C. Santos
CENTRIA & Departamento de Informática
Universidade Nova de Lisboa
Caparica, Portugal

and

ATP group, CMAF
Complexo Interdisciplinar
Universidade de Lisboa, Portugal

Max O. Souza
Departamento de Matemática Aplicada
Universidade Federal Fluminense
R. Mário Santos Braga, s/n
Niterói RJ, 24020-140, Brasil

Brian Skyrms
Logic and Philosophy of Science
School of Social Sciences
University of California at Irvine
Irvine, CA 92612, USA

On Kin and Group Selection, and the Haystack Model

Vincent A.A. Jansen

Abstract. Kin and group selection are two different ways to describe the evolution of social behaviour. Although these two explanations are compatible in many cases, they lead to a different perspective on the interpretation of the drivers of the evolution of social behaviour. Here, I will illustrate in a model based on the haystack model, which is often used in the context of group selection, that it allows a kin selection as well as a group selection interpretation. To do so I will analyse a variant of the haystack model in which the local dynamics are specified through a continuous time model. From the description of the dynamics the cost and the benefits of the interaction can be calculated, as well as the relatedness. We also revisit the interpretation of Maynard-Smith, who originally described the model, and show that this interpretation can be found if one assumes strong selection. This shows how the various interpretations of the evolution of social behaviour all can follow from the same model. It also shows how ecological details of the interaction are crucially important in interpreting and understanding the process of evolution.

Mathematics Subject Classification (2000). 91-XX, 91A22, 92Bxx, 92Dxx, 92D15.

Keywords. Game theory; Evolutionary games; Genetics and population dynamics; Problems related to evolution; Population dynamics (general); Ecology.

1. Introduction

The evolution of social behaviour poses a puzzle within the Darwinian paradigm. If the process of adaptation results from selection that benefits individuals with favourable traits, how is it then possible that behaviour that promotes the reproductive output of others at a cost to oneself evolves? Darwin himself was aware of

The research presented inspired and informed by many discussions with John Bryden, Sébastien Lion and Troy Day. This work benefitted from the financial support of the UK Engineering and Physical Sciences Research Council contract EP/D002249/1.

this issue, and commented on it on various occasions. He wrote: "A tribe including many members who... were always ready to aid one another, and to sacrifice themselves for the common good, would be victorious over most other tribes; and this would be natural selection" [1]. In this we recognize what nowadays would be called a group selection argument. In the context of insects with sterile casts he wrote: "... with the working ant we have an insect... absolutely sterile; so that it could never have transmitted successively acquired modifications of structure or instinct to its progeny. It may well be asked how is it possible to reconcile this case with the theory of natural selection?" and he continues: "... selection may be applied to the family, as well as to the individual, and may thus gain the desired end. Thus, a well-flavoured vegetable is cooked, and the individual is destroyed; but the horticulturist sows seed of the same stock, and confidently expects to get nearly the same variety" [2]. This last argument is reminiscent of what could be called a kin selection argument.

Others, notably Fisher [3] and Haldane [4] commented on the evolution of social behaviour, but it was W.D. Hamilton who elaborated and formalised the explanation [5] that later became known as kin selection [6]. Hamilton's understanding appears be based on an intuitive insight, to which he added a formal justification. Hamilton's basic idea follows from the observation that it is not so much the benefit to an individual's fitness that matters, as the benefit to the fitness associated with the gene that conveys an advantage to the behaviour under study. It is entirely feasible that populations have structure, so that most interactions are not with random individuals from the population, but with individuals that are likely to carry the same genes as the individual that displays the behaviour. If this is so, the benefits of a behaviour that are bestowed on another individual benefit the gene that causes the behaviour if this gene is also present in the other individual. Therefore, one need not just consider the fitness consequences of a gene on the individual that carries the gene, but also its effect on the same gene in other individuals that reap the benefit of the interaction.

Hamilton's quantified the idea that a gene would benefit directly through its influence of the carrier it finds itself in, as well as indirectly through increasing the fitness of other carriers of the same gene with which its carrier interacts. A crucial part of the argument is that populations have structure, which leads to assortment of genes. This can, for instance, be through limited dispersal (Hamilton used the term "viscous population" to describe this). The fitness of an altruistic gene is then

$$-C + RB,$$

where C is the cost an individual pays for altruism, B is the benefit that others receive from individuals carrying this gene, and R is a measure of the correlation between the genotype of the acting individual and that of its neighbour. Relatedness is proportional to the probability of a carrier of a gene to encounter this same gene among those that (s)he interacts with, over and above the average frequency of this gene in the population. The coefficient of relatedness is often defined as the regression coefficient that is obtained in a plot of the actors genotype versus the genotype of the receiving individual [7].

Prior to Hamilton's ideas of kin selection being widely accepted, the main explanation for traits from which others benefit was that such traits would be adaptations to living in groups. Groups of similar individuals would benefit from such traits and therefore more of such groups would be formed so that the trait would eventually come to dominate a population. One of the main proponents of this group selection perspective was Wynne-Edwards [8].

It was pointed out by John Maynard-Smith [6] that there are two main problems with the idea of group selection. Firstly, the explanation is limited to situations where groups can be identified. To describe situations in which the reproductive success of relatives of the individual carrying certain traits is enhanced Maynard Smith coined the word kin selection. Secondly, and this was the main point of Maynard Smith' argument, groups which are composed of benevolent individuals are vulnerable to exploitation by individuals who benefit from the advantages of the group without contributing to it. Such exploitation easily arises if there is dispersal between the groups.

To illustrate this idea Maynard Smith used a simple model describing mice living in haystacks. At the beginning of a season a haystack is colonised by a pregnant single female mouse. The female carries an allele which renders her timid or aggressive. Each mouse produces a colony, and within the colony competition aggressive individuals will replace the timid ones through competition. The only colonies that will produce timid mice are those that were founded by a female carrying only homozygous timid mice. Although timid mice are competitively at a disadvantage, they can produce far more offspring in a colony they solely occupy. Maynard Smith showed that if mating is mainly within the colony the timid allele will increase in frequency. Mixing between the colonies will make that the selection is easily favoured towards aggressive mice. The model has since been known as the haystack model.

A similar type of argument is used extensively in the literature on group selection. Within groups individuals, often haploid, interact and influence their own and each other's fitness through interactions. At the end of the interaction time individuals disperse and colonize new groups (such groups are sometimes referred to us "trait groups" [9]). The point that is often put forward with these models is that it is possible that a trait evolves that loses out in all groups to competition because yet can still evolve because of the increased outputs of groups it produces on its own.

Here, I will try to scrutinise these arguments by formulating a model for the haystack scenario, based on a detailed, if fictitious, description of the local dynamics of the model. The model is a variation of a mathematical model which fills in many of the dynamical details of the haystack model [10]. We will analyse these model to demonstrate how the details of the local interaction and the details of the biology impact the findings of the model, and in how far it is justified to interpret these models in the light of group and kin selection.

2. Local dynamics including resource

Our model is akin to the haystack model, in that it consist of isolated patches that are colonized at the beginning of a season by a number of haploid individuals. We assume that within the patches there is a total amount of resource U available. The individuals in the patch can sequester this resource with rate c_i. The sequestered resource is converted into new individuals, and they die and are converted back into resource with rate d.

There are two strains, the numbers in each strain is given by v_i, that compete through the availability of resource. The strains differ in the rate with which they sequester resource, c_i. The strain that sequesters the most resource will also be the most competitive strain.

At the end of the season both type of strains produce dispersers. To produce dispersers the remaining free resources is converted into dispersers. To optimise the number of dispersers the best strategy would be to be prudent with the resources. However, such a strategy backfires if the patch is shared with another strain that sequesters the resource faster and which will be the winner of the competition.

This is a typical scenario used in group selection studies: we have a situation where a strain could lose out in competition, in almost all patches, but where the premium that is gained in patches that are not shared is sufficient to provide a selective advantage.

3. Model description

We will try to find out how evolution proceeds in such a system by analysing the dynamics in some detail. The within patch dynamics are given by

$$\begin{aligned} \dot{u} &= -u(c_1 v_1 + c_2 v_2) + d(v_1 + v_2)\ , \\ \dot{v}_1 &= c_1 u v_1 - d v_1\ , \\ \dot{v}_2 &= c_2 u v_2 - d v_2\ , \end{aligned}$$

where u is the amount of free resource, and v_i is the number of mice of type i. Note that the total amount of resource $U = u + v_1 + v_2$ is constant. Using this, we can simplify the local dynamics to

$$\begin{aligned} \dot{v}_1 &= v_1(c_1(U - v_1 - v_2) - d)\ , \\ \dot{v}_2 &= v_2(c_2(U - v_1 - v_2) - d), \end{aligned} \tag{3.1}$$

and thus within a patch the local dynamics of v_1 and v_2 take the from of a Lotka-Volterra type interaction. It is a well-known result from this type of competition models that the equilibrium which can maintain the highest population density, and in which thus the amount of free resource is most reduced, is stable and the other equilibrium is unstable, in the sense that it can be invaded by the other type. This means that the local dynamics describe a process of competition in which the type with the highest c_i will be the dominant local competitor. Fig. 3.1 shows a typical example of the dynamics in the patch, it shows how a more prudent type,

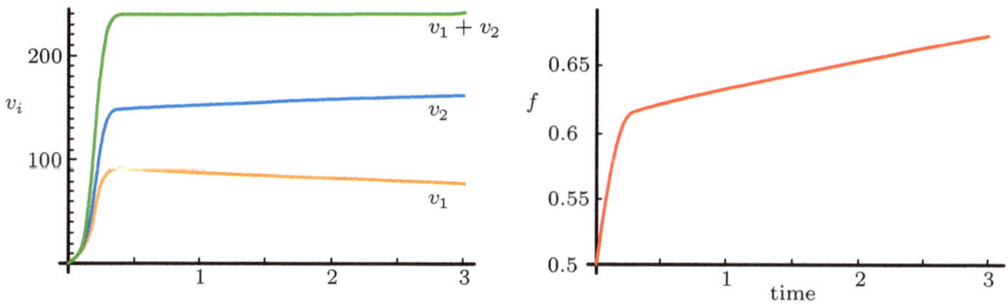

FIGURE 3.1. The local dynamics in a patch. In the top panel the densities of type 1 and type 2 versus time. Type 2 has a higher value of c and therefore this type is competitively superior. In the bottom panel the fraction of type 2 in the patch is depicted. Note how after an initial phase, which lasts as long as the local population has not reached its quasi equilibrium, this fraction changes fast. Following this phase a slow process of replacement takes place. Parameters: $c_1 = 0.1$, $c_2 = 0.11$, $U = 250$, $T = 3$, $d = 1$

that sequesters resources at a low rate, is outcompeted by a type that sequesters resource more readily.

Following the seeding of the patch, system (3.1) will determine the local dynamics. After the patch has incubated for an amount of time, say T, the season ends and the patch has to produce migrants. We assume that this happens through the conversion of all freely available resource that are then availably. These resources will be shared out pro-rata to types in the patch.

The amount of freely available resource at time T is given by $u(T) = U - s(T)$, where $s = v_1 + v_2$. The fraction of type 2 individuals is given by $f = v_2/s$, and consequently, the fraction of type 1 is than $1 - f = v_1/s$. Let the number of type k migrants in a patch that received i fundatrices of type 1, and j of type 2 be given by $m_k(i, j)$. We can then express these quantities as

$$m_2\left(v_1(0), v_2(0)\right) = f(T)(U - s(T)) ,$$
$$m_1\left(v_1(0), v_2(0)\right) = (1 - f(T))(U - s(T)) .$$

The number of migrants is proportional to the amount of free resource, so that more prudent types can produce more migrants. For a patch that is composed of a single type, for sufficiently long T, the number of free resources, and therefore the number of migrants is approximately d/c_i.

To fully specify the model we need to detail how the patches are seeded. We will denote the probability that a patch receives i fundatrices of type 1 and j of type 2 with $Q(i, j; N_1, N_2)$, where N_i is the average number of that type per patch. We will assume that Q, which details the way fundatrices are distributed over the patches, does not depend on the traits c_i. It does, of course, depend on the number of individuals that carry the trait through N_1 and N_2. As a consequence the total number of individuals in a patch depends only on the total number in the population:

$$P(n; N_1 + N_2) = \sum_{i=1}^{n} Q(n-i, i; N_1, N_2).$$

Furthermore, as the allocation of individuals to patches does not depend on their traits, a fundatrice is of type 1 with probability ϕ and of type 2 with probability $1-\phi$ where $\phi = N_2/N$ and $N = N_1 + N_2$. Therefore, for a patch that receives n fundatrices in total, the distribution over the types is binomial, so that we have:

$$Q(i, j; N_1, N_2) = \binom{i+j}{i}(1-\phi)^i \phi^j P(i+j; N). \tag{3.2}$$

The average density in the next season is found by harvesting all individuals from the patches, and exposing them to overwintering mortality:

$$N_1' = \mu \sum_{i=0}^{\infty} \sum_{j=0}^{\infty} Q(i, j; N_1, N_2) m_1(i, j) ,$$

$$N_2' = \mu \sum_{i=0}^{\infty} \sum_{j=0}^{\infty} Q(i, j; N_1, N_2) m_2(i, j),$$

where μ is the survival between seasons. This completely defines the dynamics of the haystack model.

This model assumes that the local population growth is a purely deterministic process. This, of course, will rarely be true as in particular the initial phases of the colonisation of the local patches will go through a phase in which stochastic effects can dominate. This will give an advantage to faster growing types (larger c_i). These issues are discussed in more detail in [11].

4. Fitness calculation

The above description is sufficient to simulate the change in frequency of the two types in the population by simulation. However, rather than studying the dynamics, we are interested in how evolution will shape the parameter c_i. We will therefore calculate the rate of invasion of type c_2 in a population dominated by c_1. Once we have established the pattern of invadability we can from that conclude if it is possible to have evolutionarily stable levels of prudence, i.e., parameter c_1 that cannot be invaded by any c_2.

We will apply the concept of invasion dynamics to find out the pattern of evolutionary change. We will therefore assume that there is a resident population of which all individuals are take up resources with rate c_1. We assume that in this population individuals with rate c_2 very infrequently appear, as it would occur through a process of mutation. Because these mutants occur very infrequently, the resident population will converge to its equilibrium value. The dynamics of type 1, in the absence of type 2, is given by:

$$N_1' = \mu \sum\nolimits_{i=0}^{\infty} Q(i,0;N_1,0)m_1(i,0) = \mu \sum\nolimits_{i=0}^{\infty} P(i;N_1)m_1(i,0).$$

At equilibrium we have

$$N_1^* = \mu \sum\nolimits_{i=0}^{\infty} P(i;N_1^*)m_1(i,0).$$

If in this population a new type appears, it will be very rare initially. To find out if such type can invade we will therefore calculate the rate of invasion when it is rare. If it is rare, it will have very little influence on the equilibrium density, which we therefore assume to be, to a good approximation, at N_1^*. The dynamics of type 2 when it is rare is then approximately given by:

$$\begin{aligned} N_2' &= \mu \sum_{i=0}^{\infty}\sum_{j=0}^{\infty} \left(Q(i,j;N_1^*,0) + N_2 \left. \frac{\mathrm{d}Q(i,j;N_1,N_2)}{\mathrm{d}N_2} \right|_{N_2=0,N_1=N_1^*} \right) m_2(i,j) \\ &= \frac{\mu N_2}{N_1^*} \sum_{i=0}^{\infty} (i+1)P(i+1,N_1^*)m_2(i,1), \end{aligned}$$

where we used $m_2(i,0)=0$. This shows that if type 2 is rare, one would not expect to find more than one individual of this type in a haystack. Following [12] we define the fitness of type c_2 in an environment dominated by c_1 as:

$$S_{c_1}(c_2) = \frac{N_2'}{N_2} = \frac{\mu}{N_1^*} \sum\nolimits_{i=0}^{\infty} (i+1)P(i+1,N_1^*)m_2(i,1).$$

Note that the dependence on the resident trait c_1 comes through both N_1^* and $m_2(i,j)$ both are potentially dependent on c_1. If $c_1 = c_2$ there is no selection. It is easy to show, using that if $c_1 = c_2$ then $m_2(i,1) = m_1(i+1,0)/(i+1)$, that $S_{c_1}(c_1) = 1$

This allows us to find out if a type carrying c_2 can invade a population dominated by c_1, and from this, deduce some properties of the evolutionary process. In Figs. 4.1 and 4.2 we have shown such plots for 2 different choices of P. In Fig. 4.1 the patches are seeded by a constant number of individuals. We see that in this case any trait can always invaded by types that sequester marginally more resources. We can deduce that in this case the evolutionary process leads to an ever increasing value of c_i, which only stops at the point where the c_i is so large that the population can not persist. In Fig. 4.2 the number of fundatrices is Poisson distributed. In this case for low values of c_1 types with a (somewhat) larger value than c_1 can invade, but for very high values of c_1 types with a lower value of c_1 can invade. If evolution would involve a sequence of invasions with small steps, this will eventually lead to the value of found at the crossing of the two invasion boundaries: this is a type that cannot be invaded by other types which have a marginally different value of. This can be seen in the figure because below the crossing point the marginal fitness is negative (but only just for these parameter values).

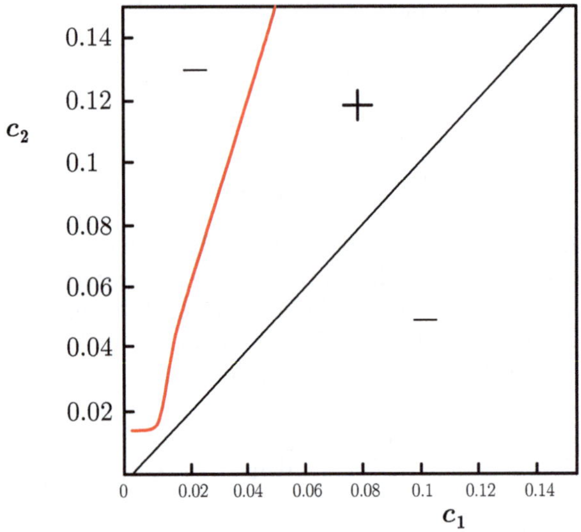

FIGURE 4.1. A pairwise invasibility diagram for the case where all occupied patches have received 3 fundatrices. Regions in which type 2 can invade a population of type 1, and for which combination of trait values the marginal fitness is positive (labelled +) or negative (labelled −). If $c_2 > c_1$ there is a large range of values for which invasion if possible. Because the area just above the diagonal is labelled +, types with a marginally larger value of c than the resident population can always invade. This will lead to a process of replacement leading to increasing values of c.

5. Weak selection and marginal fitness calculation

We can work out how the process of evolution works under small mutational steps through knowing the marginal fitness. To derive the marginal fitness we will revisit the local dynamics. Much of the above argument is based on a process where the differences between c_1 and c_2 are small. In these case, the pressure of selection is weak, and we will derive an approximation of the selection coefficient under weak selection. To do so, we make use of the fact that if selection is weak the replacement of one strain by another through competition is a slow process. We will use throughout $\epsilon = c_2 - c_1$ and $c = c_1$

The dynamics in the variables s, and f are given by:

$$\dot{s} = s\left[u(c+\epsilon f) - d\right], \tag{5.1}$$

$$\dot{f} = \epsilon f(1-f)u, \tag{5.2}$$

and remember that $u = U - s$. If ϵ is small clearly the change in f is slow compared to the changes in s. For sufficiently large T the dynamics of u will settle on a quasi

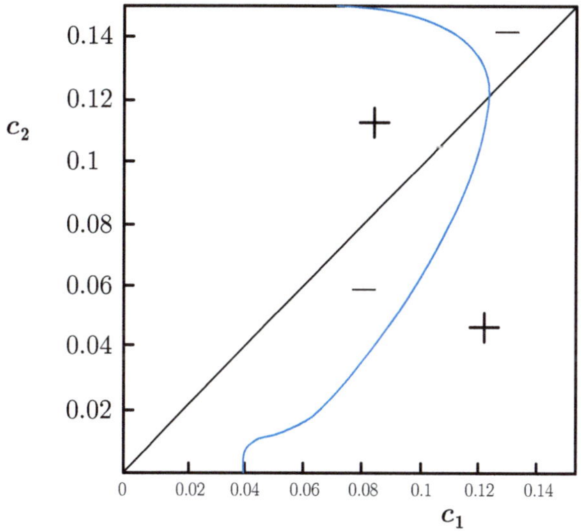

FIGURE 4.2. A pairwise invasibility diagram for the case where patches have received a number of fundatrices that is Poisson distributed. A process of repeated invasion of marginally different types will make that the evolutionary dynamics proceed to a value of c_1 for which the 2 curves cross. Here marginally different types cannot invade (although the region of stability is small). Parameters: $U = 250$, $T = 3$, $d = 1$, $\mu = 0.15$

steady state, given by

$$\tilde{u}(\epsilon) = \frac{d}{c + \epsilon f}.$$

For small ϵ this is approximately $\tilde{u}(0)\left(1 - \epsilon\frac{f(0)}{c}\right)$. The dynamics of s will also settle at a quasi steady state $\tilde{s} = U - \frac{d}{c+\epsilon f}$. Based on this argument it is possible to find a approximation for the number of dispersers produced under weak selection (see appendix)

$$m_2(i,j) \approx \tilde{u}(0)f(0) - \epsilon\frac{\tilde{u}(0)}{c}f(0)^2 + \epsilon\tilde{u}(0)f(0)(1 - f(0))g(0,T),$$

where $g(0,T) = \int_0^T u(t)\mathrm{d}t\Big|_{\epsilon=0}$ which for sufficiently large T is approximately

$$g(0,T) \approx \tilde{u}(0)T + \frac{1}{c}\ln\frac{\tilde{s}}{s(0)}.$$

The fitness is now given by

$$\frac{\mu}{N_1^*}\sum_{i=0}^{\infty}(i+1)P(i+1;N_1^*)\tilde{u}(0)\frac{1}{i+1}\left[1+\epsilon\left(-\frac{1}{c}\frac{1}{i+1}+\left(1-\frac{1}{i+1}\right)g(0,T)\right)\right]$$
$$=1+\epsilon\frac{\sum_{i=1}^{\infty}P(i;N_1^*)\left[-\frac{1}{ci}+\left(1-\frac{1}{i}\right)\left(\tilde{u}(0)T+\frac{1}{c}\ln\frac{\tilde{s}}{i}\right)\right]}{1-P(0;N_1^*)},$$

and the marginal increase in fitness is

$$\epsilon\sum_{i=1}^{\infty}\frac{P(i;N_1^*)}{1-P(0;N_1^*)}\left[-\frac{1}{ci}+\left(1-\frac{1}{i}\right)\left(\frac{dT}{c}+\frac{1}{c}\ln\frac{U-d/c}{i}\right)\right].$$

The marginal fitness can be used to find candidate end points of the evolutionary process. These are values of c for which the marginal fitness is zero. If these points in the phenotype space are stable against invasion, and if the evolutionary process leads towards them, they are called evolutionary stable states. Here, we are not so much interested in the value of the ESS, but much more in the interpretation of the marginal fitness equation. We will discuss two ways to interpret the marginal fitness

5.1. Inclusive fitness representation

To interpret the marginal fitness in terms of inclusive fitness, we will introduce the concept of relatedness. This is the normalised probability to pick two individuals of the same type from the same patch over and above the probability of picking two of the same type from the overall population. We show in the appendix that this is

$$R=\frac{\sum_{i=1}^{\infty}P(i;N_1^*)\frac{1}{i}}{1-P(0;N_1^*)}. \tag{5.3}$$

Using this, we can rewrite the marginal fitness as:

$$\epsilon\left[-\frac{R}{c}+(1-R)\frac{dT}{c}-\frac{1}{c}\sum_{i=1}^{\infty}\frac{P(i;N_1^*)}{1-P(0;N_1^*)}\left(1-\frac{1}{i}\right)\ln\frac{U-d/c}{i}\right].$$

The last term in the marginal fitness results from the difference in growth rate between individuals. In the initial stages of the exploitation of the patch the growth is approximately exponential, and a faster growing type will come to occupy a larger proportion of the patch. How long the period of approximate exponential growth lasts, depends on the logarithm of the number of fundatrices a patch receives.

One can interpret the terms in the sum as relatedness measures by defining

$$R_i'=\frac{1}{i},$$

that is, as the normalised probability to pick 2 identical individuals from a patch that was seeded by i individuals, relative to the probability in the population as a whole (see appendix). With this the last term in the marginal fitness can be written as

$$\sum_{i=1}^{\infty} \frac{P(i;N_1^*)}{1-P(0;N_1^*)}(1-R_i')\ln R_i'(U-d/c).$$

Strictly speaking we can thus interpret the marginal fitness in terms of relatedness measures. Although this is formally correct, this is practically of little value. The usefulness of the relatedness measure comes in part through the fact that one can assess the relatedness through sampling of neutral genes. By introducing the measures R_i' which depend on the number of fundatrices a patch receives, one can only assess such measures if one knows how many fundatrices a patch has received; information which would normally not be available after the incubation period.

It is, in fact, possible to link the sum $\sum_{i=1}^{\infty} \frac{P(i;N_1^*)}{1-P(0;N_1^*)}\left(1-\frac{1}{i}\right)\ln\frac{1}{i}$ to the average rate of finding a rare mutant in a patch (see appendix). This shows that it is, in principle, possible to uncover this information from a population, without having to know how many fundatrices funded a patch. However, where the relatedness is independent of the frequency of a particular gene in the population, this measure does depend on the frequency (hence the requirement of the gene to be rare) which makes the applicability limited.

Therefore, it is helpful to approximate the sum in the last term by Taylor expanding the logarithmic term around the average for $i = \frac{N_1^*}{1-P(0,N_1^*)}$, which if all patches that have at least one fundatrix go to quasi equilibrium is approximately $\frac{N_1^*}{1-P(0,N_1^*)} \approx \frac{\mu d}{c}$ so that we find for the last term in the fitness equation:

$$\sum_{i=1}^{\infty} \frac{P(i;N_1^*)}{1-P(0;N_1^*)}\left(1-\frac{1}{i}\right)\ln\frac{U-d/c}{i} \approx (1-R)\ln\left(\frac{cU-d}{\mu d}\right)$$

(see [13] for a similar argument and an application to the evolution of social behaviour in aphids). With this we can now write the marginal fitness as:

$$-R\frac{\epsilon}{c}+(1-R)\frac{\epsilon}{c}\left(dT+\ln\left(\frac{cU-d}{\mu d}\right)\right),$$

in which we recognize the effects of the change in the free resource which a change in c will cause (note $\frac{d\tilde{u}(0)}{dc} = -\frac{1}{c}$). This effect will contribute to the fitness through all related individuals. The term preceded by $(1-R)$ represents the effects of competition. A increase in c will increase the competitive ability.

Alternatively, we can partition the marginal fitness as

$$\underbrace{\frac{\epsilon}{c}\left(dT+\ln\left(\frac{cU-d}{\mu d}\right)\right)}_{-C}+\underbrace{R\frac{\epsilon}{c}\left(-1-dT-\ln\left(\frac{cU-d}{\mu d}\right)\right)}_{B}.$$

A decrease the amount of resource sequestered, i.e., a negative ϵ, is costly in that the first term is negative and will constitute in a negative direct effect, yet will result in a positive indirect effect and is therefore an act of altruism.

5.2. Group selection

For a group selection interpretation of the same model we will partition the fitness differently. A group, or multi-level, selection perspective considers the variation within and between patches. One can then interpret the advantage of an altruistic act as a consequence of the higher yield of groups which largely consist of altruists.

Now that we have an expression in terms of costs and benefits we can relate this to personal costs and benefits. We will denote the phenotype of individual j in patch i with x_{ij} which takes the value 0 if the individual has phenotype c_1, and 1 if it has phenotype c_2. We can assign the approximate personal fitness to individual j in patch i

$$W_{ij} = W_0 - Cx_{ij} + B\bar{x}_i,$$

where W_0 is the fitness if all individual carry trait c_1 (which will be 1, but for generality we keep this open) and $\bar{x}_i = \frac{1}{n_i}\sum_{j=1}^{n} x_{ij}$ is the mean phenotype among the individuals that i interacts with, which here is the same as the mean phenotype of the individuals in the patch. The number of fundatrices in patch i is n_i. It is straightforward, if tedious, to calculate the variation in fitness between groups $\mathrm{E}(\mathrm{var}_g(W)) = -(1-R)C\mathrm{var}(x)$, and the average variation within groups $\mathrm{var}(\overline{W}_i) = R(-C+B)\mathrm{var}(x)$, where $\mathrm{var}(x) = \phi(1-\phi)$ is the variation in the variable x.

We can now rewrite the marginal fitness as:

$$-C + RB = \underbrace{\frac{\mathrm{E}(\mathrm{var}_g(W))}{\mathrm{var}(x)}}_{\text{variation within patch}} + \underbrace{\frac{\mathrm{var}(\overline{W}_i)}{\mathrm{var}(x)}}_{\text{variation between patches}}.$$

6. Strong selection and fitness calculation

The above shows how the marginal fitness can be partitioned in a difference in variances within and between group, as it is frequently done in the context of multilevel (group) selection. However, to explain group selection verbally an argument along the lines of that of Maynard Smith [6] is sometimes given, stating that altruists are at a competitive disadvantage to the point that they disappear in all patches, but the ones in which they are the sole occupants. If the output in the solely occupied patches is sufficiently high, it is possible for altruists to invade a population of selfish individuals. This implies that selection is strong and is therefore not covered by the arguments above. To find out if this is the case in our model we consider the situation in which selection is strong (large ϵ) or the interaction is of sufficiently long duration so that local competition would always oust the inferior competitor.

If $\epsilon > 0$ than the type with c_2 will outcompete all c_1 in mixed patches. The fitness then is

$$\frac{\mu}{N_1^*}\sum_{i=0}^{\infty}(i+1)P(i+1, N_1^*)m_2(i,1).$$

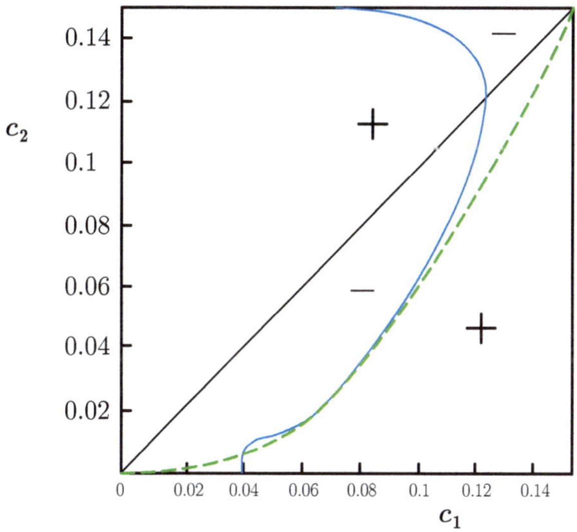

FIGURE 6.1. A pairwise invasibility diagram for the case where patches have received a number of fundatrices that is Poisson distributed to which the approximate invasion boundary is added for which it is assumed that replacement is immediate and complete and that the local dynamics go to equilibrium within the lifetime of a local patch (dashed line). Note that there is good agreement if $c_1 \gg c_2$ and c_1 not too small. Parameters: $U = 250$, $T = 3$, $d = 1$, $\mu = 0.15$

If c_2 the local dynamics are sufficiently fast for type 2 to outcompete type 1, and for the dynamics of type 2 to settle at the equilibrium $u(\epsilon, T) \approx \frac{d}{c_2}$ we find that the fitness is

$$\frac{\mu}{N_1^*} \sum_{i=0}^{\infty} (i+1) P(i+1, N_1^*) \frac{d}{c_2} = \frac{d}{c_2} \frac{\mu}{N_1^*} \sum_{i=0}^{\infty} i P(i, N_1^*) = \frac{d\mu}{c_2}.$$

If $\epsilon < 0$ than the type with c_2 will be outcompeted all c_1 in mixed patches. Type 2 will be able to produce dispersers only in patches which had a type 2 fundatrix and no type 1 fundratices will. The fitness then is

$$\frac{\mu}{N_1^*} P(1, N_1^*) m_2(0, 1).$$

If we assume the local dynamics approximately go to equilibrium this is

$$\frac{d\mu}{c_2} \frac{P(1, N_1^*)}{N_1^*}.$$

Note that because the equilibrium level N_1^* depends on c_1, the fitness of type 2, does depend on the trait that type 1 carries because the number of dispersers that type 1 produces depends on how it exploits its local resources.

This argument formalises the group selection argument: a trait that is competitively inferior will be ousted from any group in which it would not have sole occupancy. However, if groups which exclusively consist of such a trait produce a sufficient excess of migrants so that it compensates for the loss due to competition, such a trait can invade in a population. This analysis also shows that this argument is based on a number of implicit assumptions. If the interaction within the groups is time limited, which is a realistic requirement for this mechanism to work, the difference between the traits needs to be large. Therefore the group selection argument assumes implicitly that strong selection is at work.

If selection is strong, this assumption gives a reasonably good indication of when an altruist can invade in an selfish population. In Figure 6.1 we have plotted the invasion boundaries as numerically calculated with our earlier derived fitness function, and it can be seen that if the difference between c_1 and c_2 is sufficiently large that the agreement can be good (provided c_1 is not too small). This shows that if the difference between the two types is sufficiently large the replacement argument can be valid. If the difference is small, the argument is clearly not valid. Extrapolating the immediate replacement argument to marginal differences between two types leads to qualitatively incorrect decision (the crossing point between the curves moves away to a point where no population can be sustained.)

7. Discussion

The analysis given above demonstrates that the inclusive fitness argument and the group selection argument can both be valid, albeit under different conditions. Inclusive fitness arguments which rely on the fact that the fitness can be partitioned in additive components normally require weak selection. Moreover, if the distribution of genes depends on the trait, one can only validly infer the statistical association of two traits from the association of neutral genes if one assumes weak selection.

The partitioning of fitness in between and within group variance generally works if one actually can identify the groups to which it applies. It is generally not possible to independently measure or assess these fitness components in a real world population.

For the group selection argument based on local replacement to work the traits under study need to be sufficiently different. This requires strong selection, as it would result from a finite, and sufficiently different set of different phenotypes, or from sufficiently large mutation steps. What the group selection argument can predict is whether two types can replace each other or whether they can coexist, but the argument is ill suited to predict the long term course of evolution under small mutations steps.

Whilst this analysis confirms that most of the views in the debate on kin and group selection can be shown to operate in the haystack model, it also demonstrates why this debate is so persistent. In most of the mathematical arguments that are presented an arbitrarily chosen functions are used. Mostly these define what the fitnesses of different types are, without being specific how the fitness will vary under a change of trait values. Therefore, it is not immediately obvious whether these statements refer to a weak or a strong selection scenario. I hope that the above analysis demonstrates how these ecological details are of importance in interpreting the results from mathematical models as used in evolutionary ecology.

Appendix A. Relatedness calculation

We first observe that the probability to pick an single individual from a patch is simply ϕ. This follows directly from the assumption that the types are distributed according to a binomial distribution in patches of equal size.

Let us call the probability to pick, with replacement, two individuals of type 1, from the same patch

$$p_1(\phi) = \frac{\sum_{i=1}^{\infty}\sum_{j=0}^{\infty} Q(i,j;N(1-\phi),\phi)\left(\frac{i}{i+j}\right)^2}{1-Q(0,0;N(1-\phi),\phi)},$$

and likewise

$$p_2(\phi) = \frac{\sum_{i=0}^{\infty}\sum_{j=1}^{\infty} Q(i,j;N(1-\phi),\phi)\left(\frac{j}{i+j}\right)^2}{1-Q(0,0;N(1-\phi),\phi)},$$

where ϕ is the fraction of individuals of type 2, and N is the total population size. Note that $p_1(\phi) = 1-2(1-\phi)+p_2(\phi)$

The relatedness is measure of how likely it is that two genotypically individuals are found in the same haystack, relative to the probability of picking the same individuals in the population at large (this is $(1-\phi)^2+\phi^2$). This measure is normalised so that if haystacks only contain a single type the relatedness is unity. The relatedness is defined as [14]

$$R = \frac{p_1(\phi)-(1-\phi)^2+p_2(\phi)-\phi^2}{1-(1-\phi)^2-\phi^2} = \frac{p_2(\phi)-\phi^2}{\phi(1-\phi)}.$$

If we now evaluate this $p_2(\phi)$ to find

$$\begin{aligned}
\frac{\sum_{i=0}^{\infty}\sum_{j=1}^{\infty} Q(i,j;N(1-\phi),\phi)\left(\frac{j}{i+j}\right)^2}{1-P(0;N)} &= \frac{\sum_{n=1}^{\infty}\sum_{j=0}^{n} P(n;N)\frac{n!}{j!(n-j)!}\left(\frac{j}{n}\right)^2}{1-P(0;N)} \\
&= \frac{\sum_{n=1}^{\infty} P(n;N)\left(\frac{\phi(1-\phi)}{n}+\phi^2\right)}{1-P(0;N)} \\
&= \phi(1-\phi)\frac{\sum_{n=1}^{\infty} P(n;N)\frac{1}{n}}{1-P(0;N)}+\phi^2,
\end{aligned}$$

from which (5.3) follows.

Similarly, we have for R'_n :

$$R'_n = \frac{\frac{\sum_{i=0}^{n} Q(i,n-i,N(1-\phi),N\phi)((\frac{i}{n})^2+(\frac{i-n}{n})^2)}{P(n,N)} - \phi^2 - (1-\phi)^2}{1-\phi^2-(1-\phi)^2}$$

$$= \frac{\left(\sum_{i=0}^{n} \phi^i (1-\phi)^{n-i} \frac{n!}{i!(n-i)!} (2\left(\frac{i}{n}\right)^2 - 2\frac{i}{n} + 1)\right) - \phi^2 - (1-\phi)^2}{2\phi(1-\phi)}$$

$$= \frac{\left(\frac{\phi(1-\phi)}{n} + \phi^2\right) - \phi^2}{\phi(1-\phi)} = \frac{1}{n}.$$

Appendix B. Interpretation of logarithmic term

Here we will demonstrate that the term $\sum_{i=1}^{\infty} \frac{P(i;N_1^*)}{1-P(0;N_1^*)} \left(1-\frac{1}{i}\right) \ln \frac{1}{i}$ can be interpreted in terms of a sampling procedure.

The sampling procedure we apply is as follows: identify a rare mutant, which is not selected for in the process. Within a patch, keep drawing individuals, with replacement, until this mutant is encountered. Subtract 1 from the number of individuals, and, if the result is larger than 0, take the average of the reciprocal of this number by sampling over different patches.

In a patch containing i mutants among n individuals, the probability of encountering the mutant for the first time after k draws is given by the hypergeometric distribution $\frac{i}{n}\left(1-\frac{i}{n}\right)^k$. The expectation of reciprocal of positive values of $k-1$, provided $i > 0$, is

$$\sum_{k=2}^{\infty} \frac{i}{n} \left(1 - \frac{i}{n}\right)^k \frac{1}{k-1} = -\frac{i}{n}\frac{n-i}{n} \log\left(\frac{i}{n}\right).$$

If $i = 0$ then

$$\lim_{i/n \to 0} \sum_{k=2}^{\infty} \frac{i}{n} \left(1 - \frac{i}{n}\right)^k \frac{1}{k-1} = 0.$$

When sampled over occupied patches, and if ϕ is the mutant's fraction of the population the expectation is

$$-\sum_{n=1}^{\infty} \frac{P(n,N)}{1-P(0,N)} \sum_{i=0}^{n} \frac{n!}{i!(n-i)!} \phi^i (1-\phi)^{n-i} \frac{i}{n}\frac{n-i}{n} \ln\left(\frac{i}{n}\right).$$

For small ϕ this is approximately

$$-\sum_{n=1}^{\infty} P(n,N) \left(1 - \frac{1}{n}\right) \ln\left(\frac{1}{n}\right).$$

Appendix C. Derivation of marginal fitness under weak selection

The find the change over the incubation time of the fraction of type 2, f, we will solve (5.2). To do so we first rearrange:

$$\left(\frac{1}{f} + \frac{1}{1-f}\right)\mathrm{d}f = \epsilon u(\epsilon, t)\mathrm{d}t,$$

which we can solve by integration from $t = 0$ to $t = T$:

$$\ln \frac{f(T)}{f(0)} - \ln \frac{1 - f(T)}{1 - f(0)} = \epsilon \int_0^T u(\epsilon, t)\mathrm{d}t$$

and thus

$$f(T) = \frac{f(0)}{f(0) + (1 - f(0))e^{-\epsilon \int_0^T u(\epsilon,t)\mathrm{d}t}}.$$

To make the dependence of u on ϵ explicit we have written $u(\epsilon, T)$. If ϵ is small the we can now find the approximate value of $f(T)$ as:

$$f(T) \approx f(0) + \epsilon f(0)(1 - f(0)) \int_0^T u(0, t)\mathrm{d}t.$$

To solve the integral we will turn our attention to s. The dynamics of s for $\epsilon = 0$ are given by the logistic growth model: from a positive initial condition the total number of individuals will increase and saturate at $\tilde{s}(0)$. The solution to the logistic equation is

$$s(t) = \frac{s(0)\tilde{s}}{s(0) + (\tilde{s} - s(0))e^{-c\tilde{s}t}}. \tag{C.1}$$

This also prescribes the dynamics of $u(t) = U - s(t)$.

For $g(0, T) = \int_0^T u(0, t)\mathrm{d}t$ we have

$$g(0, T) = \tilde{u}(0)T + \frac{1}{c} \ln \frac{s(T)}{s(0)}.$$

For sufficiently large T this is approximately

$$g(0, T) \approx \tilde{u}(0)T + \frac{1}{c} \ln \frac{\tilde{s}}{s(0)}.$$

If we assume that the resource levels settle on the quasi equilibrium $u(0, T) = \tilde{u}$ we find that if there is a marginal difference in the traits number of type 2 dispersers that will be produced for large T is equal to

$$\begin{aligned} m_2(i, j) &= u(\epsilon, T)f(T) \\ &\approx \tilde{u}(0)f(0) - \epsilon\frac{\tilde{u}(0)}{c}f(0)^2 + \epsilon\tilde{u}(0)f(0)(1 - f(0))g(0, T). \end{aligned}$$

Appendix D. Calculation of within and between group variance in fitness

Assume a local group is seeded by n_i individuals with k with trait value c_2, and $n-k$ with trait value c_1. Such a group has as mean value $\bar{x}_i = \frac{k}{n_i}$, and the variance in x in the local group is

$$\operatorname{var}_g(x) = \frac{1}{n_i}\sum_{j=1}^{n_i}\left(x_{ij} - \frac{k}{n_i}\right)^2 = \frac{k}{n_i}\left(1 - \frac{k}{n_i}\right).$$

The expected value of this variance is

$$\mathrm{E}[\operatorname{var}_g(x)] = \sum_{n_i=1}^{\infty}\sum_{k=0}^{n_i} P(n_i, N)\binom{n_i}{k}\phi^k(1-\phi)^{n_i-k}\frac{i}{n}\left(1-\frac{k}{n_i}\right)\phi(1-\phi)(1-R).$$

The expected value of the group mean is

$$\mathrm{E}(\bar{x}_i) = \sum_{n_i=1}^{\infty}\frac{P(n_i,N)}{1-P(0,N)}\sum_{k=0}^{n}\binom{n_i}{k}\phi^k(1-\phi)^{n_i-k}\frac{k}{n_i} = \phi,$$

and the variance

$$\begin{aligned}\operatorname{var}(\bar{x}_i) &= \sum_{n_i=1}^{\infty}\sum_{k=0}^{n_i}\frac{P(n_i,N)}{1-P(0,N)}\binom{n_i}{k}\phi^k(1-\phi)^{n_i-k}\left(\frac{k}{n_i}-\phi\right)^2\\ &= \sum_{n=1}^{\infty}\frac{P(n,N)}{1-P(0,N)}\phi(1-\phi)\frac{1}{n} = \phi(1-\phi)R.\end{aligned}$$

The overall variance in x is given by $\operatorname{var}(x) = \phi(1-\phi)$.

A group selection interpretation partitions the fitness in the variation in fitness within and between patches. Now that we have identified the individual fitness we can calculate the average mean fitness within a patch j as

$$\overline{W}_i = \frac{1}{n_i}\sum_{j=1}^{n} W_{ij} = W_0 + (-C+B)\bar{x}_i.$$

The average within patch variance is given by

$$\begin{aligned}\mathrm{E}(\operatorname{var}_g(W)) &= \mathrm{E}\left(\frac{1}{n_i}\sum\nolimits_{j=1}^{n_i}(W_{ij}-\overline{W}_i)^2\right) = -C\mathrm{E}\left(\frac{1}{n_i}\sum\nolimits_{j=1}^{n_i}(x_{ij}-\bar{x}_i)^2\right)\\ &= -C\mathrm{E}[\operatorname{var}_g(x)] = -C\phi(1-\phi)(1-R).\end{aligned}$$

The variance between the mean fitness of patches is

$$\begin{aligned}\operatorname{var}(\overline{W}_i) &= \mathrm{E}[(\overline{W}_i - W_0 - (-C+B)\phi)^2]\\ &= (-C+B)\mathrm{E}[(\bar{x}_i-\phi)^2]\\ &= (-C+B)\operatorname{var}(\bar{x}_i)\\ &= (-C+B)\phi(1-\phi)R.\end{aligned}$$

References

[1] C. Darwin, *The Descent of Man, and Selection in Relation to Sex.* John Murray, London, 1871.

[2] C. Darwin, *On the Origin of Species by Means of Natural Selection or the Preservation of Favoured Races in the Struggle for Life.* John Murray, London, 1859.

[3] R.A. Fisher, *The Genetical Theory of Natural Selection.* Clarendon Press, Oxford, 1930.

[4] J.B.S. Haldane, *Population genetics.* New Biol. **18** (1955), 34–51.

[5] W.D. Hamilton, *Genetical evolution of social behaviour I.* J. Theor. Biol. **7** (1964), 1.

[6] J. Maynard Smith, *Group selection and kin selection.* Nature **201** (1964), 1145.

[7] A. Grafen, *A geometric view of relatedness.* Oxford Surveys in Evolutionary Biology **2** (1985), 28–89.

[8] V.C. Wynne-Edwards, *Animal Dispersion in Relation to Social Behaviour.* Oliver and Boyd, Edinburgh, 1962.

[9] D.S. Wilson, *Theory of group selection.* Proc. Natl. Acad. Sci. USA **72** (1975), 143–146.

[10] V.A.A. Jansen and G.S.E.E. Mulder, *Evolving biodiversity.* Ecol. Lett. **2** (1999), 379–386.

[11] V.A.A. Jansen and R. Vitalis, *The evolution of dispersal in a Levins' type metapopulation model.* Evolution **61** (2007), 2386–2397.

[12] J.A.J. Metz, R.M. Nisbet, and S.A.H. Geritz, *How should we define "fitness" for general ecological scenarios?* Trends Ecol. Evol. **7** (1992), 198–202.

[13] J. Bryden and V.A.A. Jansen, *The impact of clonal mixing on the evolution of social behaviour in aphids.* P. Roy. Soc. B: Biological Sciences **277** (2010), 1651–1657.

[14] D.C. Queller and K.F. Goodnight, *Estimation of genetic relatedness using allozyme data.* Evolution **43** (1989), 258–275.

Vincent A.A. Jansen
School of Biological Sciences
Royal Holloway, University of London
Egham, Surrey TW20 0EX, United-Kingdom
e-mail: vincent.jansen@rhul.ac.uk

Population Formulation of Adaptative Meso-evolution: Theory and Numerics

Sepideh Mirrahimi, Benoît Perthame, Emeric Bouin
and Pierre Millien

Abstract. The population formalism of "adaptive evolution" has been developed in the last twenty years along ideas presented in other chapters in this volume. This mathematical formalism addresses the question of explaining how selection of a favorable phenotypical trait in a population occurs. In the language of Metz's Chapter, it refers to meso-evolution. It uses models based, usually, on integro-differential equations for the population structured by a phenotypical trait. A self-contained mathematical formulation of adaptive evolution also contains the description of mutations and leads to partial differential equations. Then the complete evolution picture follows from the model ingredients mostly driven by the changing adaptive landscape.

It is possible to introduce scaling parameters and perform asymptotic analysis. Then highly concentrated population densities (well-separated Dirac masses) arise that can undergo branching patterns. This phenomenon is interpreted as the speciation process.

The process in which concentrated solutions occur and a continuous set of traits cannot be present is subtle and numerical methods can induce artifacts if not correctly shaped. Simulations on Monte-Carlo methods can be compared to deterministic numerical methods as finite differences.

Mathematics Subject Classification (2000). 35K57, 47G20, 49L25, 65C05, 65M06, 92D15.

Keywords. Adaptive evolution, Lotka-Volterra equation, Darwinian evolution, Hamilton-Jacobi equation, Viscosity solutions, Dirac concentrations, Monte-Carlo simulations, Finite differences.

1. Introduction

Since the 1980's the word "adaptive evolution" has been coined to describe the mathematical formalisms addressing the selection of a favorable trait in a population structured by a continuous phenotypical trait. In the language of [1] it refers to meso-evolution. Closely related to the concept of "Evolutionary game theory"

[2, 3, 4], the models ingredient are the three principles underlying Darwin's explanation of Evolution:

- multiplication of the population,
- selection by competition for resources,
- variability (mutations).

Simple models based on these ingredients explain how the fittest traits can emerge and populations characterized by several well-separated traits (also called strategies) can possibly coexist. The theory and numerical simulations show the appearance of clusters and speciation that can be explained simply: the limited resources lead to competition and individuals with close traits use similar resources, therefore competition between them is higher. The question of understanding how, in such a population, a mutant can invade or not a population has been initiated in [5, 6, 7] and a recent survey can be found in [8], see also [1]. In a self-contained population model, the mutations are part of the dynamics and take into account that the newborn may inherit a slightly different trait than its parent.

The formalism for describing selection, in an asexual population, uses integro-differential equations for the population density $n(x,t)$ where x denotes the phenotypical trait and several models have been derived or postulated for mutations, leading to parabolic partial differential equations (PDEs) [9, 10, 11]. In this Chapter, we aim at explaining how speciation occurs in such PDE models. This corresponds to highly concentrated population densities, which means that $n(x,t)$ is close to well-separated Dirac masses. Because of their regularizing effects, parabolic PDEs cannot sustain such singular solutions and this phenomenon can only happen asymptotically. With this respect, two typical asymptotic regimes are possible. The first one consists in introducing a small parameter for mutations frequency or size and considers the limiting behavior when this parameter vanishes [12, 13, 14, 15]. The second asymptotic is to consider long times and this leads to singular steady state solutions, very similar to the pure selection case [16, 17]. We present these models in Sections 2 and 3 on two different types of competition kernels that we have chosen for their simplicity.

The appearance of these singular solutions is related to an instability mechanism of Turing type. Numerical methods may produce artificially this Turing mechanism in particular because artificial boundary conditions are needed. We discuss this fact in Section 4 based on finite differences or Monte-Carlo simulations.

2. A model with a single nutrient

2.1. The chemostat

Following [8, 12], the simplest example to build up a self-contained mathematical model for adaptive evolution is the *chemostat*. Micro-organisms characterized by a parameter $x \in \mathbb{R}$ (it can be thought of as the logarithm of their size) live in a bath containing a nutrient which is continuously renewed with a rate $d > 0$. The nutrient concentration is denoted by $S(t) \geq 0$ (for substrate) and the fresh nutrient $S_{\text{in}} > 0$, the population density of the micro-organism is denoted by $n(x,t)$ and the uptake rate for individuals of trait x is $\eta(x) > 0$.

In such a simple situation, the standard equations for the chemostat is written

$$\begin{cases} \frac{\mathrm{d}}{\mathrm{d}t} S(t) = d\,(S_{\text{in}} - S(t)) - S(t) \int_{-\infty}^{\infty} \eta(x) n(x,t) \mathrm{d}x, \\ \frac{\partial}{\partial t} n(x,t) = -dn(x,t) + (1-\mu) S(t) \eta(x) n(x,t) \\ \qquad\qquad + \mu S(t) \int_{-\infty}^{\infty} M(y,x) \eta(y) n(y,t) \mathrm{d}y. \end{cases}$$

The first two principles mentioned earlier from Darwin theory are directly included in the model: the population growth comes from the equation on $n(x,t)$ and the competition comes from the limited amount of nutrients. We assume that initially $S(0) \leq S_{\text{in}}$, then all along the dynamics we have $S(t) \leq S_{\text{in}}$ because $S(t)$ decreases if it attains S_{in}. The term $(1-\mu)\eta(x)n(x,t)$ represents the birth rate without mutations. The parameter $0 < \mu < 1$ represents the proportion of birth undergoing mutations.

Mutations are represented by the probability $M(y,x)$ that a newborn has the trait x when its parent has the trait y. We therefore assume $M(y,x) \geq 0$, $\int \int_{-\infty}^{\infty} M(y,x) \mathrm{d}x = 1$.

We may simplify the model in various ways to make it more amenable to analysis. One can suppose that the nutrients reach quickly an equilibrium compared to the evolution time scale for the population. Then one can replace the differential equation on $S(t)$ by the relation

$$S(t) = \frac{dS_{\text{in}}}{d + \int_{-\infty}^{\infty} \eta(x) n(x,t) \mathrm{d}x}.$$

One can also replace the mutation term by a mere diffusion leading to

$$\frac{\partial}{\partial t} n(x,t) = -dn(x,t) + S(t)\eta(x)n(x,t) + \lambda \Delta n(x,t).$$

Note however that both representations of mutations by integral terms or by a Laplace term $\lambda\Delta$ can be derived from stochastic individual based models (IBM) depending on the scaling of microscopic mutations, [18, 19, 20]. See also [21].

We can write a general form of the resulting model, that we will keep for the end of this section

$$\begin{cases} \frac{\partial}{\partial t} n(x,t) = n(x,t) R\,(x, I(t)) + \lambda \Delta n(x,t), \quad x \in \mathbb{R},\ t > 0, \\ I(t) = \int_{-\infty}^{\infty} \eta(x) n(x,t) \mathrm{d}x. \end{cases} \tag{2.1}$$

With these notations, the neat growth rate $R(x, I)$ contains both birth and death terms. In the case at hand, it is given by

$$R(x, I) = -d + \frac{dS_{\text{in}}}{d+I}\eta(x).$$

It is natural to handle more general models and then we need some general hypothesis. We assume that R is smooth enough and there are $I_{\text{M}} > I_{\text{m}} > 0$ such that

$$\begin{cases} \sup_{x \in R} R_I(x, I) < 0, \qquad \forall I \geq 0, \\ \max_{x \in \mathbb{R}} R(x, I_{\text{M}}) = 0, \\ \min_{x \in \mathbb{R}} R(x, I_{\text{m}}) = 0. \end{cases} \tag{2.2}$$

We also assume that there are positive constants η_{m}, η_{M} such that

$$0 < \eta_{\text{m}} \leq \eta(x) \leq \eta_{\text{M}} < \infty, \quad \text{with} \quad \eta \in W^{2,\infty}(\mathbb{R}). \tag{2.3}$$

2.2. Rescaling

As mentioned earlier, such parabolic models cannot exhibit high concentrations as long as the diffusion coefficient $\mu > 0$ is fixed. This is the reason why we rescale the problem and set $\lambda = \varepsilon^2$. Having in mind that the mutation rate is small we consider the limit $\varepsilon \to 0$. Such a limit only leads to the same equation with $\lambda = 0$, the selection model. This is because the effect of rare mutations on the population can be observed only on a very long time. This leads us naturally to change time and replace t by t/ε so as to consider the evolution on a long time rather than a generation time scale. Then equation (2.1) is changed to

$$\begin{cases} \varepsilon \dfrac{\partial}{\partial t} n_\varepsilon(x,t) = n_\varepsilon(x,t) R\left(x, I_\varepsilon(t)\right) + \varepsilon^2 \Delta n_\varepsilon(x,t), \quad x \in \mathbb{R},\ t > 0, \\ I_\varepsilon(t) = \displaystyle\int_{-\infty}^{\infty} \eta(x) n_\varepsilon(x,t) \mathrm{d}x. \end{cases} \tag{2.4}$$

But we can point out that other scales are also interesting [10].

We are now ready for a possible interpretation of the speciation phenomena

Theorem 2.1 ([14, 15]). *We assume* (2.2)–(2.3), *that R is monotonic in x and the initial data is "well prepared" (see below). Then, there are two constants $\rho_m > 0$, $\rho_M > 0$ such that*

$$\rho_{\text{m}} \leq \int_{-\infty}^{\infty} n_\varepsilon(x,t) \mathrm{d}x \leq \rho_{\text{M}} \tag{2.5}$$

and $I_\varepsilon(t) \to \bar{I}(t)$ almost everywhere and in the weak sense of measures

$$n_\varepsilon(x,t) \rightharpoonup \bar{\rho}(t) \delta\big(x - \bar{x}(t)\big).$$

The above assumptions, and in particular monotonicity of R in x, can be replaced by strong concavity on R with quadratic behavior at infinity [22].

This theorem is a mathematical version of the famous *competitive exclusion principle* in ecology. With a single nutrient, a single species will be selected. With N nutrients, we expect in general that N species will coexist.

It is not easy to characterize the fittest trait $\bar{x}(t)$ and the total population size $\bar{\rho}(t)$. In the situations covered by Theorem 2.1, it is proved (see [14, 22]) that

$$R\big(\bar{x}(t), \bar{I}(t)\big) = 0, \qquad \bar{I}(t) = \bar{\rho}(t)\eta\big(\bar{x}(t)\big).$$

Such points appear naturally in the language of evolutionary game theory and are called "singular points'. Of course this identity only relates $\bar{x}(t)$ and $\bar{I}(t)$. It is possible to go further and establish an analogue of the so-called *canonical equation* [23]

$$\dot{\bar{x}}(t) = \big(-D^2 u(\bar{x}(t), t)\big)^{-1} \cdot \nabla_x R\big(\bar{x}(t), \bar{I}(t)\big),$$

where $u(x,t)$ is introduced below. Such a differential equation was formally introduced in [12] and it can be established rigorously in a multidimensional framework, see [22].

2.3. The constrained Hamilton-Jacobi equation

The proof of Theorem 2.1 relies on a WKB approach, as in front propagation [24, 25, 26]. In the context of adaptive dynamics the method was introduced in [12] and yields a new type of Hamilton-Jacobi equation because an algebraic constraint appears. It is based on the real phase defined by the Hopf-Cole transform

$$u_\varepsilon = \varepsilon \ln(n_\varepsilon).$$

This requires that the initial data itself is "well prepared', that is "exponentially" concentrated as $u^0_\varepsilon = \varepsilon \ln(n^0_\varepsilon)$ with u^0_ε a function that behaves nicely as $\varepsilon \to 0$ (even though this can be somehow relaxed, see [15]).

The equation on u_ε is written

$$\frac{\partial}{\partial t} u_\varepsilon(x,t) = R\big(x, I_\varepsilon(t)\big) + \varepsilon \Delta u_\varepsilon(x,t) + \big|\nabla u_\varepsilon(x,t)\big|^2.$$

One can prove that u_ε is uniformly Lipschitzian (this requires that u^0_ε is so) and that I_ε is uniformly with bounded variations. This allows us to pass to the limit $\varepsilon \to 0$ and obtain the *constrained Hamilton-Jacobi* equation

$$\begin{cases} \dfrac{\partial}{\partial t} u(x,t) = R\big(x, I(t)\big) + |\nabla u(x,t)|^2. \\ \max\limits_{x \in \mathbb{R}} u(x,t) = 0, \qquad \forall t > 0. \end{cases} \tag{2.6}$$

The algebraic constraint $\max_{x\in\mathbb{R}} u(x,t) = 0$ comes from the uniform a priori bound on the total mass stated in (2.5) together with the definition of u_ε by the Hopf-Cole transform.

Being a parabolic limit, the solution $u(x,t)$ should be understood as a viscosity solution to (2.6), see [27].

As mentioned earlier, the originality of this problem stems from the two unknowns $u(x,t)$ et $I(t)$ which should be solved together. The latter is a Lagrange multiplier associated with the algebraic constraint. This makes the main difference with the standard eikonal equation arising in geometrical optics. A uniqueness result is proved in [14], however under restrictive assumptions. The method of Hopf-Cole transform is very general and, in the present context, it has been extended to systems in [28] (for fronts see [25]).

3. Competition models

In a chemostat, the competition between species is global because it arises through the substrate described by $S(t)$. All individuals are equally competing for the resource. This is not always the case and, in many situations, it is more realistic to assume that there is higher competition between individuals with closer traits. This is the reason why other models have been proposed that implement a trait dependent competition. A class of such models (see [16, 29, 30, 31, 32]) are given by the population dynamics of Lotka-Volterra type

$$\frac{\partial n(x,t)}{\partial t} - \lambda \frac{\partial^2 n(x,t)}{\partial x^2} = n(x,t)\big(R(x) - (K * n)(x,t)\big), \qquad t \geq 0,\ x \in \mathbb{R}. \quad (3.1)$$

The model is completed by an initial data $n(x, t = 0) = n^0(x)$ which we take highly concentrated for the numerical simulations presented below in Section 4.

The interpretation of the quantities arising in this model are

- $n(x,t)$ still denotes the population density at position x and time t,
- $R(x) > 0$ is the intrinsic growth rate of individuals with trait x (if isolated without competition)
- $K \in L^\infty(\mathbb{R})$ is called the competition kernel. It is a probability density: $K \geq 0$, $\int_{-\infty}^{\infty} K(z)dz = 1$. The convolution $(K*n)(x) = \int_{-\infty}^{\infty} K(x-y)n(y,t)dy$ represents the competition for resource,
- λ is the mutation rate that is supposed to be a constant.

When derived from stochastic IBM, as in [9, 19, 20] such models are called *mean field* equations [33, 34]. They arise not only in evolution theory but also in ecology for non-local resources (and x denotes the location then) [35, 36, 37, 38].

The large variety of regimes that can appear in such models can be seen in special cases. Below, we use simple examples to describe two of them, regularly distributed traits, or concentration as a Dirac mass. The main interest of the model (3.1) is mostly from the branching patterns that correspond to multiple concentration points which can either die out or branch again and create new structures (see [39]).

3.1. The Gaussian case without mutations

Firstly we consider the case

$$\lambda = 0, \qquad R(x) = \frac{1}{\sqrt{2\pi\sigma_1}} e^{-\frac{|x|^2}{2\sigma_1}}, \qquad K(z) = \frac{1}{\sqrt{2\pi\sigma_2}} e^{-\frac{|x|^2}{2\sigma_2}}. \quad (3.2)$$

This corresponds to widely used standard forms of the input parameters because of their statistical meaning.

As usual for pure selection models, $\lambda = 0$, there are Dirac mass stationary solutions $N(x) = \bar{\rho}\delta(x - \bar{x})$ with $R(\bar{x}) = \bar{\rho}K(0)$. But this can be obtained in a long time asymptotic only when

$$R(x) < \bar{\rho}K(x - \bar{x}) , \qquad \forall x \neq \bar{x},$$

or, replacing $\bar{\rho}$ from the first condition

$$\frac{R(x)}{R(\bar{x})} < \frac{K(x - \bar{x})}{K(0)} , \qquad \forall x \neq \bar{x}.$$

One can deduce from this calculation the

Proposition 3.1. *For $\sigma_1 > \sigma_2$ there is a smooth steady state to* (3.1) *given by*

$$N(x) = \frac{1}{\sqrt{2\pi\sigma}} e^{-\frac{|x|^2}{2\sigma}}, \qquad \sigma = \sigma_1 - \sigma_2,$$

and Dirac masses are not stable steady states.

For $\sigma_1 < \sigma_2$ the Dirac mass $\bar{\rho}\delta(x)$ is a stable steady state (*and only the Dirac mass at* 0 *is stable*).

The authors in [17] prove that the corresponding stable states are also the long time limits of the dynamics described by equation (3.1). They use a relative entropy method built on the corresponding steady state. The construction of this entropy is rather easy when the positive steady state exists. It is much more difficult in the case where the Dirac masses have to be handled.

3.2. The Nonlocal-Fisher equation

We now consider the case

$$R \equiv 1. \tag{3.3}$$

Then, the equation (3.1) is called the *Nonlocal-Fisher* (NLF) equation. It also arises in mathematical ecology, as an extension of the Fisher/KPP equation. As mentioned earlier, the nonlocal aspect induced by the convolution represents long range access to resources, see [32, 36, 38] and the references therein.

The positive steady state is simply given by $N \equiv 1$ but a result from [30] states that it can be Turing unstable (i.e., only a bounded set of linearly unstable modes occur). In order to explain this, we may use the Fourier transform of the competition kernel K defined as

$$\widehat{K}(\xi) = \int_{-\infty}^{\infty} K(x) e^{-ix\xi} dx.$$

Then one has

Proposition 3.2 ([30]). *Assume there is a ξ_0 such that*

$$\widehat{K}(\xi_0) < 0, \tag{3.4}$$

then for λ small enough the steady state $N \equiv 1$ is linearly unstable.

The result of this statement corresponds qualitatively to the case $\sigma_1 < \sigma_2$ in Proposition 3.1 (with mutations neglected).

The Fourier transform also characterizes a nonlinear stability result; this is the case in the

Theorem 3.3 ([32]). *Take $R \equiv 1$ and assume*

$$\widehat{K}(\xi) > 0 \qquad \forall \xi \in \mathbb{R}. \tag{3.5}$$

Then $n \equiv 0$ and $n \equiv 1$ are the only two nonnegative and bounded steady states of (3.1).

Furthermore, there are traveling waves connecting the states $n = 0$ and $n = 1$.

The result of this theorem corresponds to the situation $\sigma_1 > \sigma_2$ in Proposition 3.1.

In the Turing unstable case it is possible to rescale the problem as we did it in Section 2.2 and it is observed numerically that, in general, the asymptotic limit leads to Dirac concentrations characterized again by a constrained Hamilton-Jacobi equation [31].

4. Numerical methods and branching patterns

In general it is very difficult, in the direct competition model (3.1), to distinguish between the two behaviors: convergence towards a continuous state or speciation. Numerical methods are useful to get an intuition but they can create artifacts and we explain this now.

We present two numerical approaches that allow to simulate solutions to equation (3.1). The first is a standard finite difference scheme, the second one is a Monte-Carlo simulations related to IBM that solves the same equation.

For the sake of simplicity we concentrate on the Nonlocal Fisher equation as in Section 3.2 with a Gaussian competition kernel

$$R \equiv 1, \qquad K(x) = \frac{1}{\sqrt{2\pi\sigma}} e^{-\frac{|x|^2}{2\sigma}}. \tag{4.1}$$

Because the Fourier transform of K is positive (a Gaussian), we do not expect appearance of concentrations (speciation).

At this stage we insist that the Monte Carlo algorithms are only seen here as an approximation to (3.1). From this point of view, the closer it is from the PDE, the better it is because one looks only for possible computational cost reduction. Monte-Carlo methods are also used as a modeling tool and allow to include further stochastic effects. One of them is "demographic stochasticity" which makes that too small populations can die out by statistical effects [40, 41]. These effects are not included in the models under consideration here and give quantitatively different answers (in terms of evolution speed, branching patterns). It is shown in [39] that the notion of "survival threshold" in the equations as (3.1) is able to reproduce these effects in great details.

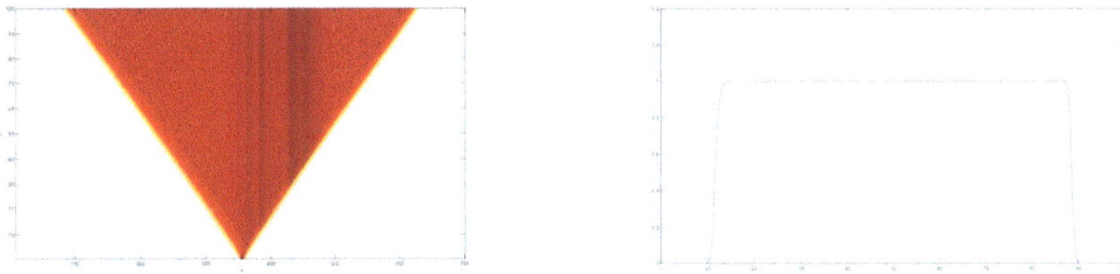

FIGURE 4.1. Left: Numerical population density dynamics obtained for model (3.1)–(4.1) when the initial population is concentrated in the center of the computational domain. Horizontally is x and vertically is t, in gray zone $n \equiv 1$ and the white zone corresponds to $n \equiv 0$. Right: The population density $n(x, T)$ at final time. The deterministic finite difference scheme (4.2)–(4.4) has been used with parameters in (4.5). We observe convergence toward the constant solution in accordance with Theorem 3.3.

4.1. Finite differences

We consider the solution on interval $\left[-\frac{L}{2}, \frac{L}{2}\right]$. We use a uniform grid with N points on the segment, with $\Delta x = \frac{L}{N}$ the space step. We denote by $n_i^k \geq 0$ the numerical solution at grid point $x_i = i\Delta x$, $1 \leq i \leq N$, and time $t^k = k\Delta t$ where Δt is the time step

$$n(x_i, k\Delta t) \approx n_i^k.$$

We use a time splitting algorithm between the growth term and the diffusion that is we solve alternatively the two equations

$$\frac{\mathrm{d}}{\mathrm{d}t} n(x,t) = n(x,t)\left[1 - (K * n)(t)\right],$$

and

$$\frac{\partial n(x,t)}{\partial t} - \lambda \frac{\partial^2 n(x,t)}{\partial x^2} = 0.$$

1. First compute, with a semi-implicit method, the solution to the discrete reaction term

$$\frac{\mathrm{d}}{\mathrm{d}t} n_i(t) = n_i(t)\left[1 - K_d * n_i^k\right].$$

The exact solution is

$$n_i^{k+\frac{1}{2}} = n_i^k \exp\left(\frac{\Delta t}{\lambda}\left(1 - K_d * n_i^k\right)\right), \qquad 1 \leq i \leq N. \tag{4.2}$$

The discrete convolution is computed according to

$$\begin{cases} K_d * n_i^k = \Delta x \cdot \displaystyle\sum_{j=-N}^{N} K_d(j\Delta x) n_{i-j}^k, \\ n_{i-j}^k = 0 \ \text{ for } i - j \notin [1, N]. \end{cases} \tag{4.3}$$

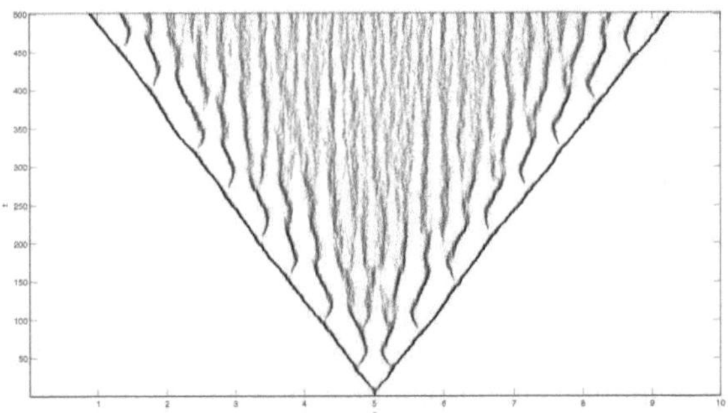

FIGURE 4.2. Numerical solution with the Monte-Carlo algorithm in section 4.2. Horizontally is the trait x and vertically is time t. Initially the population is concentrated in one Dirac mass at the center of the domain. We observe that the population distribution converges weakly towards the constant solution as expected (see also Fig. 4.1).

Indeed, as a consequence of the domain truncation, only those terms satisfying $1 \leq i - j \leq N$ are well defined and the extension by zero amounts to extend n by 0 outside $\left[-\frac{L}{2}, \frac{L}{2}\right]$. This is some kind of Dirichlet boundary condition.

2. As for the Laplace term, we use a three points explicit scheme

$$n_i^{k+1} = n_i^{k+\frac{1}{2}} + \frac{\lambda \Delta t}{2\Delta x^2}\left(n_{i+1}^{k+\frac{1}{2}} + n_{i-1}^{k+\frac{1}{2}} - 2n_i^{k+\frac{1}{2}}\right), \qquad 1 \leq i \leq N. \tag{4.4}$$

Because we choose λ small, the explicit scheme is not penalizing in terms of computational time. We use Neumann boundary condition, $n_0^{k+1} = n_1^{k+1}$ and $n_N^{k+1} = n_{N-1}^{k+1}$, but as far as the wave does not reach the boundary, the Dirichlet boundary condition $n_1^{k+1} = n_N^{k+1} = 0$ gives equivalent results.

The stability of the scheme is ensured by the CFL condition $\frac{\lambda \Delta t}{2\Delta x^2} \leq 1$, which is verified for

$$\begin{aligned} \lambda &= 0.004, \qquad & \sigma &= 0.04, \qquad & \Delta t &= 0.025, \\ \Delta x &= 0.1, \qquad & L &= 100, \qquad & N &= 1000. \end{aligned} \tag{4.5}$$

We have implemented this method. We choose the initial data concentrated in the center of the domain. The numerical results are depicted in Fig. 4.1. We can observe that the population propagates as a traveling wave. For L large enough, for $0 \leq t \leq T$ the front does not reach the numerical boundary and there is almost no mass on the boundary of the interval $[-\frac{L}{2}, \frac{L}{2}]$. This is in accordance to the theory in [32] and the statement in Theorem 3.3.

4.2. The stochastic individual-based method

We also compare the finite volume simulation with a Monte Carlo algorithm. Then, the solution is approximated by a sum of Dirac masses

$$n(t) \approx \omega \sum_{j=1}^{N(t)} \delta\big(x - y_j(t)\big).$$

Here the weight ω is taken constant. The simulation starts with a number $N(0)$ of "individuals located" distributed on an interval of length L. Then $N(0)$ and ω are related by the approximation $n(0) \approx \omega \sum_{j=1}^{N(0)} \delta\big(x - y_j(0)\big)$ in the weak sense of measures.

Several Monte Carlo algorithms are possible. See for instance [35, 38] for another algorithm motivated by models from ecology.

Here we use the method proposed in [33, 34]. The number of individuals is denoted by $N(k)$ at iteration k. The algorithm uses also a time splitting but not with the same operators as in Section 4.1. We solve alternatively the two equations

$$\frac{d}{dt} n(x,t) = -n(x,t)(K * n)(t),$$

and

$$\frac{\partial n(x,t)}{\partial t} - \lambda \frac{\partial^2 n(x,t)}{\partial x^2} = n(x,t).$$

Finally, in the rationale of small mutations and long times, as in section 2.2, we choose $\Delta t = 1$. Then the algorithm [33, 34] reads as follows.

1. The competition term is now computed as (this makes a difference with [33, 34])

$$C(x) = \frac{\omega}{\sqrt{2\pi\sigma}} \sum_{j=1}^{N(k)} \exp\left(-\frac{|x - y_j|^2}{2\sigma}\right). \tag{4.6}$$

   Because the value of $C(x)$ is small, it defines the probability that an individual located at x dies. For a given j, we compute this probability and set $N(k+1) = N(k) - 1$ if this individual dies.
2. If the individual survives, it reproduces. The newborn undergoes a mutation from its parent trait to a new trait given by a Gaussian distribution with variance $\lambda' = 2\lambda$. Then $N(k+1) = N(k) + 1$.

   We notice that for n the solution of

$$\partial_t n = \lambda' \Delta n, \qquad n(x, t^k) = n^k(x),$$

   we have $n(t^{k+1}) = n^k * \frac{1}{\sqrt{4\pi\lambda'}} e^{\frac{-x^2}{4\lambda'}}$. Hence the choice $\lambda' = 2\lambda$ in the second step of the Monte Carlo method. We act a Gaussian mutation to the new-born only but with twice stronger intensity.

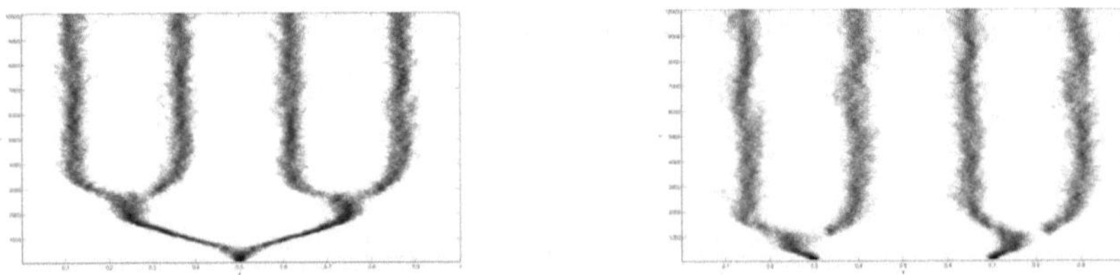

FIGURE 4.3. Dynamics of the concentration points with the Monte-Carlo algorithm in section 4.3 based on periodizing the convolution. Horizontally is the trait x and vertically is time t. Initially the population is concentrated in one Dirac mass on the left and two Dirac masses on the right.

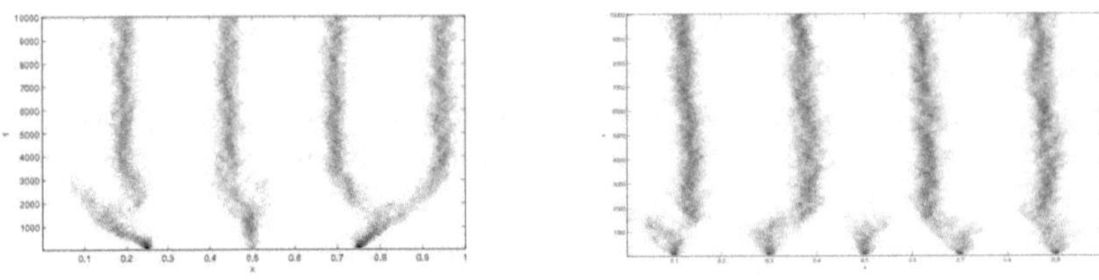

FIGURE 4.4. Dynamics of the concentration points. Same as above but with different initial data. A new phenomena occurs with extinction of branches.

We have used the following parameters values which take into account the small time step in the deterministic algorithm

$$\lambda' = 10^{-6}, \quad \sigma = 0.04, \quad L = 10, \quad N = 3000, \quad \frac{\omega}{\sqrt{2\pi\sigma}} = \frac{1}{18000}.$$

These values are such that the mutations are very weak compared to intraspecific competition, again in accordance with the parameters used in the finite difference method. The numerical results are depicted in Fig. 4.2. We can observe that the population propagates as a traveling wave as in Fig. 4.1 and according to the theoretical prediction in Theorem 3.3.

4.3. The convolution formula

Surprisingly, in [33, 34] the authors observed that simulations based on this Monte-Carlo method may yield concentration patterns too (clusters). The main difference is that, rather than with equation (4.6), the convolution kernel is computed assuming the y_j are on the circle

$$C(x) = \frac{\omega}{\sqrt{2\pi\sigma}} \sum_{j=1}^{N(k)} \exp\left(-\frac{d(x, y_j)^2}{2\sigma}\right), \tag{4.7}$$

where d is the shortest distance on the circle.

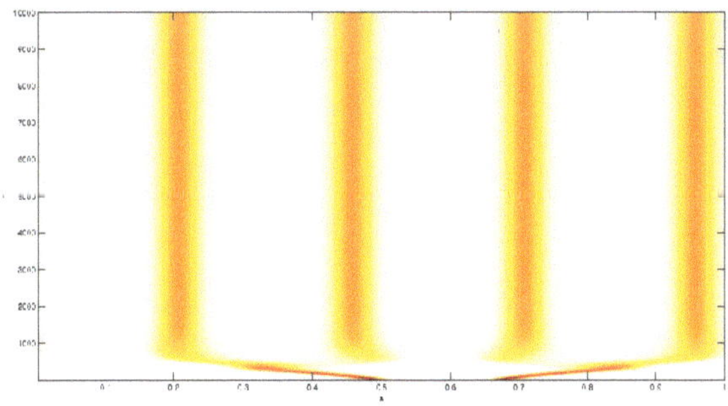

FIGURE 4.5. Numerical population density dynamics obtained by deterministic simulations for model (3.1)–(4.1) with periodic boundary conditions. We have used the following parameter values: $\lambda = 0.001$, $\sigma = 0.04$, $\Delta t = 0.0001$, $\Delta x = 0.001$, $L = 1$, $N = 1000$.

This can be interpreted as periodic boundary conditions rather than extension by zero or as a periodic convolution kernel

$$K_s(x) \propto \exp\left(-\frac{(x\,[L])^2}{2\sigma}\right), \qquad x[L] = x \mod L, \;\; x \in \mathbb{R}.$$

In opposition with the Gaussian kernel because it has some Fourier coefficients with a negative real part. In this case the Fourier condition (3.4) is not fulfilled. Therefore according to the linear analysis in [30], and Proposition 3.2, the constant state is unstable for problem (3.1)–(4.1) and we expect to observe pattern formation.

We have run both the Monte Carlo and finite difference approximations with this periodic kernel. The numerical results are in accordance with those obtained in different contexts in [30, 31, 33, 34]. They can be found in Fig. 4.3 and Fig. 4.4 for Monte Carlo simulations and Fig. 4.5 for finite differences.

5. Conclusion

Mathematical models explaining how speciation occurs in biological population have been developed since the 1980s. They involve a population dynamics under local competition and with mutations. A self-contained formalism can be established. It allows to represent the speciation phenomena as the convergence of the solution to a sum of Dirac masses, either in the large time limit or the small mutation rate limit. However, competition models not always yield speciation and a population with a continuous set of traits can occur. It is difficult to predict between these two alternatives.

Numerical methods are therefore useful tools to observe the model prediction. We presented two numerical methods: finite differences and the individual based approach. These methods give compatible numerical results either in the case when a uniform trait distribution is produced by the model or when patterns are obtained.

References

[1] J.A.J. Metz, *Thoughts on the geometry of meso-evolution: Collecting mathematical elements for a post-modern synthesis. In* F.A.C.C. Chalub and J.F. Rodrigues (eds.), *The Mathematics of Darwin's Legacy*, 193–231, Birkhäuser, Basel, 2011, *This issue.*

[2] J. Maynard Smith, *Theory of games and evolution of animal conflicts.* J. Theor. Biol. **47** (1974), 209–221.

[3] J. Hofbauer and K. Sigmund, *Evolutionary Games and Population Dynamics.* Cambridge Univ. Press, Cambridge, UK, 1998.

[4] J. Hofbauer and K. Sigmund, *Evolutionary game dynamics.* Bull. Amer. Math. Soc. (N.S.) **40** (2003), 479–519.

[5] J. Hofbauer and K. Sigmund, *Adaptive dynamics and evolutionary stability.* Appl. Math. Lett. **3** (1990), 75–79.

[6] J.A.J. Metz, S.A.H. Geritz, G. Meszéna, F.J.A. Jacobs, and J.S. van Heerwaarden, *Adaptive dynamics, a geometrical study of the consequences of nearly faithful reproduction. In Stochastic and spatial structures of dynamical systems* (*Amsterdam,* 1995), Konink. Nederl. Akad. Wetensch. Verh. Afd. Natuurk. Eerste Reeks, 45, 183–231, North-Holland, Amsterdam, 1996.

[7] S.A.H. Geritz, É. Kisdi, G. Meszena, and J.A.J. Metz, *Evolutionarily singular strategies and the adaptive growth and branching of the evolutionary tree.* Evol. Ecol. **12** (1998), 35–57.

[8] O. Diekmann, *A beginner's guide to adaptive dynamics. In Mathematical Modelling of Population Dynamics*, volume 63 of *Banach Center Publ.*, 47–86, Polish Acad. Sci., Warsaw, 2004.

[9] A. Sasaki and S. Ellner, *The evolutionarily stable phenotype distribution in a random environment.* Evolution **49** (1995), 337–350.

[10] A. Calsina and S. Cuadrado, *Small mutation rate and evolutionarily stable strategies in infinite dimensional adaptive dynamics.* J. Math. Biol. **48** (2004), 135–159.

[11] A. Calsina and C. Perello, *Equations for biological evolution.* Proc. R. Soc. Edinb. Sect. A-Math. **125** (1995), 939–958.

[12] O. Diekmann, P.E. Jabin, S. Mischler, and B. Perthame, *The dynamics of adaptation: An illuminating example and a Hamilton-Jacobi approach.* Theor. Popul. Biol. **67** (2005), 257–271.

[13] G. Barles and B. Perthame, *Concentrations and constrained Hamilton-Jacobi equations arising in adaptive dynamics. In* D. Danielli (ed.), *Recent Developments in Nonlinear Partial Differential Equations*, volume 439 of *Contemporary Mathematics Series*, 57–68, 2007.

[14] B. Perthame and G. Barles, *Dirac concentrations in Lotka-Volterra parabolic PDEs.* Indiana Univ. Math. J. **57** (2008), 3275–3301.

[15] G. Barles, S. Mirrahimi, and B. Perthame, *Concentration in Lotka-Volterra parabolic or integral equations: a general convergence result.* Methods Appl. Anal. **16** (2009), 321–340.

[16] L. Desvillettes, P.E. Jabin, S. Mischler, and G. Raoul, *On selection dynamics for continuous structured populations.* Commun. Math. Sci. **6** (2008), 729–747.

[17] P.-E. Jabin and G. Raoul, *On selection dynamics for competitive interactions.* Preprint CMLA-ENS Cachan 17 (2009).

[18] N. Champagnat, *A microscopic interpretation for adaptive dynamics trait substitution sequence models.* Stoch. Process. Their Appl. **116** (2006), 1127–1160.

[19] N. Champagnat, R. Ferrière, and S. Méléard, *Unifying evolutionary dynamics: From individual stochastic processes to macroscopic models.* Theor. Popul. Biol. **69** (2006), 297–321.

[20] N. Champagnat, R. Ferrière, and S. Méléard, *Individual-based probabilistic models of adaptive evolution and various scaling approximations. In* R.C. Dalang, M. Dozzi, and F. Russo (eds.), *Seminar On Stochastic Analysis, Random Fields And Applications V*, volume 59 of *Progress In Probability*, 75–113, 2008, 5th Seminar on Stochastic Analysis, Random Fields and Applications, Ascona, Switzerland, May 30–Jun 03, 2005.

[21] S. Méléard, *Random modeling of adaptive dynamics and evolutionary branching. In* F.A.C.C. Chalub and J. Rodrigues (eds.), *The Mathematics of Darwin's Legacy*, 175–192, Birkhäuser, Basel, 2011, *This issue.*

[22] A. Lorz, S. Mirrahimi, and B. Perthame, *Dirac concentration in a multidimensional nonlocal parabolic equation. In preperation.*

[23] U. Dieckmann and R. Law, *The dynamical theory of coevolution: A derivation from stochastic ecological processes.* J. Math. Biol. **34** (1996), 579–612.

[24] L.C. Evans and P.E. Souganidis, *A PDE approach to geometric optics for certain semilinear parabolic equations.* Indiana Univ. Math. J. **38** (1989), 141–172.

[25] G. Barles, L.C. Evans, and P.E. Souganidis, *Wavefront propagation for reaction-diffusion systems of PDE.* Duke Math. J. **61** (1990), 835–858.

[26] P.E. Souganidis, *Front propagation: theory and applications. In Viscosity solutions and applications* (*Montecatini Terme,* 1995), volume 1660 of *Lecture Notes in Math.*, 186–242, Springer, Berlin, 1997.

[27] M.G. Crandall, H. Ishii, and P.L. Lions, *User's guide to viscosity solutions of second order partial differential equations.* Bull. Amer. Math. Soc. **27** (1992), 1–67.

[28] J.A. Carrillo, S. Cuadrado, and B. Perthame, *Adaptive dynamics via Hamilton-Jacobi approach and entropy methods for a juvenile-adult model.* Math. Biosci. **205** (2007), 137–161.

[29] G. Meszéna, M. Gyllenberg, F.J. Jacobs, and J.A.J. Metz, *Link between population dynamics and dynamics of Darwinian evolution.* Phys. Rev. Lett. **95** (2005), 078105.

[30] S. Genieys, V. Volpert, and P. Auger, *Adaptive dynamics: modelling darwin's divergence principle.* C. R. Biol. **329** (2006), 876–879.

[31] B. Perthame and S. Génieys, *Concentration in the nonlocal Fisher equation: the Hamilton-Jacobi limit.* Math. Model. Nat. Phenom. **2** (2007), 135–151.

[32] H. Berestycki, G. Nadin, B. Perthame, and L. Ryzhik, *The non-local Fisher-KPP equation: travelling waves and steady states.* Nonlinearity **22** (2009), 2813–2844.

[33] E. Brigatti, V. Schwammle, and M.A. Neto, *Individual-based model with global competition interaction: fluctuation effects in pattern formation.* Phys. Rev. E **77** (2008).

[34] V. Schwämmle and E. Brigatti, *Speciational view of macroevolution: Are micro and macroevolution decoupled?* Europhys. Lett. **75** (2006), 342–348.

[35] B. Bolker and S.W. Pacala, *Using moment equations to understand stochastically driven spatial pattern formation in ecological systems.* Theor. Popul. Biol. **52** (1997), 179–197.

[36] S.A. Gourley, *Travelling front solutions of a nonlocal Fisher equation.* J. Math. Biol. **41** (2000), 272–284.

[37] Z.C. Wang, W.T. Li, and S.G. Ruan, *Travelling wave fronts in reaction-diffusion systems with spatio-temporal delays.* J. Differ. Equ. **222** (2006), 185–232.

[38] N. Fournier and S. Méléard, *A microscopic probabilistic description of a locally regulated population and macroscopic approximations.* Ann. Appl. Probab. **14** (2004), 1880–1919.

[39] M. Gauduchon and B. Perthame, *Survival thresholds and mortality rates in adaptive dynamics: conciliating deterministic and stochastic simulations.* Mathematical Medicine and Biology **27** (2010), 195–210.

[40] D. Claessen, J. Andersson, L. Persson, and A.M. de Roos, *Delayed evolutionary branching in small populations.* Evol. Ecol. Res. **9** (2007), 51–69.

[41] S.R. Proulx and T. Day, *What can invasion analyses tell us about evolution under stochasticity in finite populations?* Selection **2** (2002), 2–15.

Sepideh Mirrahimi
UPMC Univ Paris 06, UMR 7598
Laboratoire Jacques-Louis Lions
F-75005, Paris
e-mail: `mirrahimi@ann.jussieu.fr`

Benoît Perthame
Institut Universitaire de France & UPMC Univ Paris 06, UMR 7598
Laboratoire Jacques-Louis Lions
F-75005, Paris
e-mail: `benoit.perthame@upmc.fr`

Emeric Bouin and Pierre Millien
UPMC Univ Paris 06, UMR 7598
Laboratoire Jacques-Louis Lions
F-75005, Paris
e-mail: `emeric.bouin@ens-lyon.fr`
`pierre.millien@ens-cachan.fr`

Random Modeling of Adaptive Dynamics and Evolutionary Branching

Sylvie Méléard

Abstract. We are interested in modeling the Darwinian dynamics of a polymorphic asexual population, as driven by the interplay of phenotypic variation and natural selection through ecological interactions. Our modeling is based on a stochastic individual-based model that details the dynamics of heritable traits characterizing each individual. We consider the specific scales of the biological framework of adaptive dynamics: rare mutations and large population. We prove that under a good combination of these two scales, the population process is approximated in an evolution long time scale by a Markov pure jump process describing successive equilibria of the population. Then we consider this polymorphic evolution process in the limit of small mutations. From a fine study in the neighborhood of evolutionary singularities, we obtain a full mathematical justification of a heuristic criterion for the phenomenon of evolutionary branching.

Mathematics Subject Classification (2000). 92D25, 60J80, 37N25, 92D15, 60J75.

Keywords. Mutation-selection individual-based model, fitness of invasion, adaptive dynamics, polymorphic evolution sequence, evolutionary branching.

1. Darwinian or Adaptive Evolution

We are interested in modeling the Darwinian dynamics of a polymorphic asexual population, as driven by the interplay of phenotypic variation and natural selection through ecological interactions. We assume that the individual ability to survive or to reproduce is characterized by a quantitative parameter, called (phenotypic) trait, as the body size, the age at maturity, or the rate of food intake. Evolution, acting on the trait distribution of the population, is the consequence of three basic mechanisms: *heredity*, which transmits traits to new offspring, *mutation* driving a variation in the trait values in the population, and *selection* between these different trait values, which is due to the competition between individuals for limited resources or area. Adaptive dynamics models aim at studying the interplay between these different mechanisms [1, 2, 3]. Our goal in this paper is to show how an individual-based stochastic model can help for the understanding of the long time evolutive behavior of the population. This survey mainly summarizes works detailed in [4, 5, 6, 7].

Simulations can be obtained from an initial monomorphic population, following the individual dynamics (see Figure 2.1). In the left, the population support is concentrated around a single trait value evolving continuously in order to maximize the growth rate, whereas in the right, a phenotypic separation in two separate sub-populations appears (evolutionary branching). These sub-populations are still in interaction but are centered around distinct traits at a distance increasing with time. This phenomenon is called *Evolutionary Branching*. It is related to the phenomenon of sympatric speciation in the sexual case [8], and to the creation of ecological niches in the asexual setting. In this paper, we will explain the dynamics of the evolution process and highlight the evolutionary branching phenomenon from the mathematical point of view.

We follow the basic description of adaptive dynamics based on the biologically motivated assumptions of rare mutations and large population. Under these assumptions, we prove that the individual-based process can be approximated by a Markov pure jump process on the set of point measures on the trait space. The transitions of this process are given by the long time behavior of competitive Lotka-Volterra systems. They describe the succession of mutant invasions followed by a fast competition phase between the mutant population and the resident one. In the mutation time scale, and for large populations, the successful traits in the competition are given by the nontrivial equilibria of Lotka-Volterra systems which model the dynamics of the sizes of each sub-population corresponding to each resident or mutant trait. We thus generalize the situation introduced by Metz *et al.* [9] and mathematically developed by Champagnat [6], when the parameters of the model prevent the coexistence of two traits. In that case, the microscopic model converges to a monomorphic (one trait support) pure jump process, called Trait Substitution Sequence (TSS). This limit involves a timescale separation between the mutations and the population dynamics driving the competition between traits. Here, we will relax the assumption of non-coexistence and obtain a polymorphic evolution sequence (PES), allowing coexistence of several traits in the population at this evolutionary scale (cf. [7]). Thus ensuring that mutations are small we give a full mathematical justification of the criterion for evolutionary branching proposed in [9].

Let us stress the delicate combination of the limits. Here we are concerned by the combination of the limits of large populations and rare mutations, followed by a limit of small mutations. An alternative approach would be firstly to study the limit of large population alone, giving in the limit an integro-differential or partial differential equation for the density of traits [5]; and next to study a limit of small mutations on this equation with a proper time scaling that would lead to some dynamics on the set of finite sums of Dirac masses on the trait space. The second part of this program has already been partly studied by Diekmann *et al.* [10] in a specific model, but is related to difficult problems on Hamilton-Jacobi equations with constraints [11]. In this case, evolutionary branching is numerically observed, but not yet fully justified. Another approach would be to combine the three limits we consider directly at the level of the microscopic model, allowing one to study the evolutionary process on several time scales [12]. This requires a finer analysis of the invasion and competition phases after the appearance of a new mutant. Note that all these approaches are based on the same idea of separation between the time scales of mutation and competition, whereas the model of Yu [13] does not satisfy this assumption. This may explain why his results are different from ours (in particular, he does not observe evolutionary branching in a model where the fitness satisfies our branching criterion). This shows the delicate influence of parameters scaling and of the specific ecological model on the phenomenon of evolutionary branching.

2. The microscopic model

2.1. The transitions of the microscopic process – simulations

We describe the dynamics of the population by a birth and death process with mutation and selection. This individual-based approach, taking into account the parameters of each individual and describing all birth and death events has been heuristically introduced by biologists, cf. [14] for a seed dispersion process, [15, 16] for size models. It has been mathematically defined in the spatial case by Fournier-Méléard [4] and developed in the ecological framework by Champagnat-Ferrière-Méléard [5].

Each individual is characterized by a phenotypic trait $x \in X$ where X is a bounded and closed subset of $\mathbb{R}$. (We have chosen the real case by simplicity but X could also be a compact of $\mathbb{R}^d$.) The size of the population is scaled by a parameter K which can also be interpreted as a resource parameter. Bigger is K, smaller is the biomass of each individual, in the sense that its influence on the other individuals is smaller. We assume that each individual has a weight $\frac{1}{K}$, and for each integer K, we define the population process

$$\nu_t^K = \frac{1}{K}\sum_{i=1}^{N_t^K} \delta_{X_t^i}, \tag{2.1}$$

where N_t^K is the number of individuals living at time t, δ_x denotes the Dirac mass at point x and $X_t^1, \dots, X_t^{N_t^K}$ are the traits of the individuals. Therefore, for each bounded and measurable function f,

$$\int f(x)\nu_t^K(\mathrm{d}x) = \frac{1}{K}\sum_{i=1}^{N_t^K} f(X_t^i).$$

Let us now describe the transitions of the process. Each individual with trait x gives birth at rate $b(x)$ to a unique individual. With probability $1 - u_K\, p(x)$, the reproduction is clonal and the offspring wears the trait x. The function p takes values in $[0,1]$ and $u_K \in [0,1]$ is the order of the mutation rate with respect to K. It will further play a main role. With probability $u_K\, p(x)$, the offspring is a mutant with trait $x+h$ distributed following $m(x,h)\mathrm{d}h$. We assume that $m(x,h) \geq 0$ and $m(x,h) = 0$ as soon as $x+h \notin X$ and $\int m(x,h)\mathrm{d}h = 1$. Each individual with trait x dies from natural death at rate $d(x)$ (depending on its trait x), and from the competition with an individual with trait y at rate $\alpha(x,y)/K$. Thus, the total individual death rate is

$$d(x) + \frac{1}{K}\sum_{i=1}^{N_t^K} \alpha(x,x_i) = d(x) + \int_X \alpha(x,y)\nu_t^K(\mathrm{d}y).$$

For a measure ν, we will denote $\alpha * \nu(x) = \int \alpha(x,y)\nu(\mathrm{d}y)$.

The assumptions on the coefficients are the following.

Assumptions (A):

A1) Initial size of order K.

A2) b, d, α, p et m are smooth functions (at least continuous, and more if necessary).

A3) b and d are bounded functions.

A4) The growth rate of the population of type x is positive (the population is supercritical if there is no competition): $\forall x \in X$, $b(x) - d(x) > 0$.

A5) The interaction rate between two individuals is upper and lower bounded by positive constants: $0 < \underline{\alpha} \leq \alpha(x,y) \leq \bar{\alpha} < +\infty$.

A classical example is the one developed in [17] then [8] which models the evolution of bird's beak size. We assume that $X = [-2,2]$, that $d(x) \equiv 0$ and that the mutation probability is p. The mutation kernel describing the mutant distribution issued from an individual with trait x is given by the Gaussian law centered on x and with variance σ^2, conditioned to $x+h \in X$. We assume that the birth rate and the (symmetric) competition kernel are given by

$$b(x) = \exp\left(-x^2/2\sigma_b^2\right), \qquad \text{and} \qquad \alpha(x,y) = \exp\left(-(x-y)^2/2\sigma_\alpha^2\right).$$

Let us remark that without competition, the growth rate $b(x)$ attains its maximum for $x=0$. The population should adapt thanks to favorable mutations in order to attain this trait $x=0$. Figure 2.1 presents simulations of this process for different ecological parameters. The initial population is a monomorphic population and the K individuals have the same trait -1. The simulation algorithm follows the jumps of the birth and death process. In every case, the upper figure shows the trait distribution and the lower one the population size.

Figure 2.1, left and right differ by the respective positions of σ_b and σ_α. In the left, the population maximizes its growth rate and stabilizes around 0. In the right, a branching occurs in a long time scale. As the range competition is small enough (σ_α is smaller in the right figure than in the left), the decrease of the competitive pressure compensates the loss of reproductive efficiency near 0 and allows the appearance of new branches.

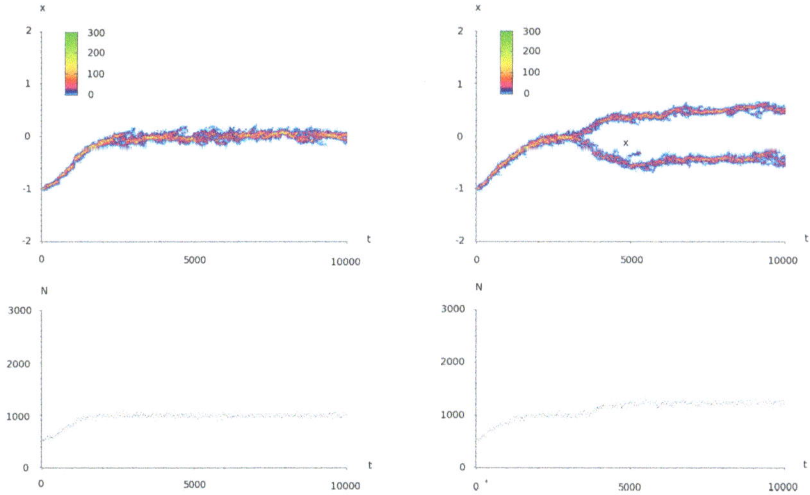

FIGURE 2.1. Left: $p = 0.1$, $K = 1000$, $\sigma = 0.01$, $\sigma_b = 0.9$, $\sigma_\alpha = 1.0$. Right: $p = 0.1$, $K = 1000$, $\sigma = 0.01$, $\sigma_b = 0.9$, $\sigma_\alpha = 0.7$.

We will rigorously analyze the phenomenon of evolutionary branching from the individual-based model (see details in [7]). We are going to consider three biological assumptions: large populations (K tends to infinity), rare mutations (u_K tends to zero), and small mutation amplitude (h replaced by εh with ε tending to 0). These scales and the biological heuristics of this approach were introduced in [9]. The combination of the first two scales (large population and rare mutation) will give convergence of the individual-based process to the so-called polymorphic evolutionary sequence which we observe in Figure 2.2. The mutation probability has been drastically reduced and we observe similar differences in the asymptotic behaviors as in Figure 2.1. The microscopic process behaves as a jump process without or with evolutionary branching points. The heuristic explanation of the convergence when $K \to \infty$ and $u_K \to 0$ is the following one. The assumption of rare mutations implies a separation between ecological (or population dynamics) and evolutionary (or mutation) time scales: the selection process has sufficient time between two mutations to eliminate disadvantaged traits. The large population assumption allows one to assume a deterministic population dynamics between mutations, so that the outcome of the competition can be predicted. Then evolution proceeds by a succession of phases of mutant invasion and very short phases of competition between traits, and only few traits remain after competition between each mutation. The only randomness remaining in the system comes from the mutation times and the mutant traits appearing in the population.

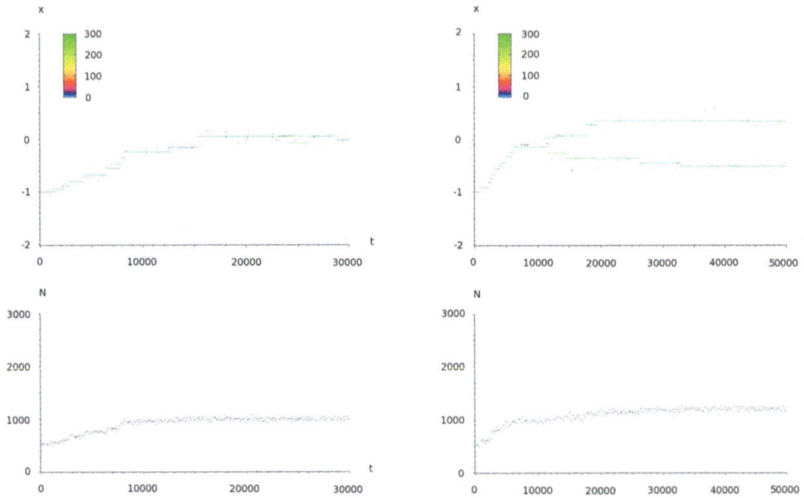

FIGURE 2.2. Left: $p = 0.0001$, $K = 1000$, $\sigma = 0.08$, $\sigma_b = 0.9$, $\sigma_\alpha = 1.0$ Right: $p = 0.0001$, $K = 1000$, $\sigma = 0.08$, $\sigma_b = 0.9$, $\sigma_\alpha = 0.7$.

2.2. The microscopic process

For each parameter K, the population process is a Markov process with jumps taking values in the point measure space with weights $\frac{1}{K}$. The initial state is denoted by ν_0^K. In what follows, we will assume that

$$\sup_K \mathbb{E}\left(\left(\int_X \nu_0^K(\mathrm{d}x)\right)^3\right) < \infty. \tag{2.2}$$

Theorem 2.1 ([4, 5]). *Assume* (2.2). *Then for any K, There exists a Markov process $(\nu_t^K, t \geq 0)$ whose dynamics is the one described in Section* 2.1 *and given by the following infinitesimal generator. For $\nu = \frac{1}{K}\sum_{i=1}^N \delta_{x_i}$,*

$$\begin{aligned} L^K\phi(\nu) &= \sum_{i=1}^N b(x_i)(1-u_K p(x_i))\left(\phi\left(\nu + \frac{1}{K}\delta_{x_i}\right) - \phi(\nu)\right) \\ &\quad + \sum_{i=1}^N b(x_i)u_K p(x_i)\int\left(\phi\left(\nu+\frac{1}{K}\delta_{x_i+h}\right)-\phi(\nu)\right)m(x_i,h)\mathrm{d}h \\ &\quad + \sum_{i=1}^N\left(d(x_i)+\frac{1}{K}\sum_{j=1}^N \alpha(x_i,x_j)\right)\left(\phi\left(\nu-\frac{1}{K}\delta_{x_i}\right)-\phi(\nu)\right) \end{aligned} \tag{2.3}$$

Moreover, for any $T > 0$, we get $\sup_K \mathbb{E}\left(\sup_{t\in[0,T]}\left(\int_X \nu_t^K(\mathrm{d}x)\right)^3\right) < \infty$.

We deduce from (2.3) that

$$\phi(\nu_t^K) - \phi(\nu_0^K) - \int_0^t L^K\phi\left(\nu_s^K\right)\mathrm{d}s \tag{2.4}$$

is a martingale as soon as each term is integrable. It implies that for each bounded and measurable function f on X, the process $t \to \int_X f(x)\nu_t^K(\mathrm{d}x)$ is a semi-martingale. More precisely, we have

$$\int_X f(x)\nu_t^K(\mathrm{d}x) = \frac{1}{K}\sum_{i=1}^{N_t^K} f(X_t^i) = \frac{1}{K}\sum_{i=1}^{N_t^K} f(X_0^i) + M_t^{K,f} + A_t^{K,f},$$

where $M^{K,f}$ is a martingale and $A^{K,f}$ a finite variation process. Taking $\phi(\nu) = \int_X f(x)\nu(\mathrm{d}x)$ in (2.4), we obtain

$$\begin{aligned} A_t^{K,f} = \int_0^t \int_X \Big\{ &\big(b(x)(1-u_K p(x)) - (d(x) + \alpha * \nu_s^K(x))\big) f(x) \\ &+ u_K p(x) b(x) \left(\int_{\mathbb{R}} f(x+h) m(x,h)\mathrm{d}h\right)\Big\}\nu_s^K(\mathrm{d}x)\mathrm{d}s. \end{aligned} \tag{2.5}$$

Moreover, by Itô's Formula and (2.4) for $\phi(\nu) = \left(\int_X f(x)\nu(\mathrm{d}x)\right)^2$, we get

$$\begin{aligned} \mathbb{E}\left(\left(M_t^{K,f}\right)^2\right) = \frac{1}{K}\Big\{\int_0^t \int_X \Big\{ &(b(x)(1-u_K p(x)) + d(x) + \alpha * \nu_s^K(x)) f^2(x) \\ &+ u_K p(x) b(x) \left(\int_{\mathbb{R}} f^2(x+h) m(x,h)\mathrm{d}h\right)\Big\}\nu_s^K(\mathrm{d}x)\mathrm{d}s\Big\}. \end{aligned} \tag{2.6}$$

The scaling of the parameters plays a main role in the study of the asymptotic behavior of $(\nu_t^K, t \geq 0)$ as K tends infinity (cf. [5]).

3. Ecological equilibria

Let us firstly recall the convergence theorem proved in [4] and [5].

Theorem 3.1. *Assume Hypotheses* (A) *are fulfilled, that $K \to \infty$ and that $u_K \to u > 0$ and that the sequence of random measures ν_0^K converges in law to the deterministic measure ξ_0. Then for any $T > 0$, the process $(\nu_t^K, t \geq 0)$ converges in law to the unique deterministic continuous measure-valued function ξ, weak solution of the following integro-differential equation: for any bounded and measurable* *function* $f : X \to \mathbb{R}$,

$$\int_X f(x)\xi_t(\mathrm{d}x) = \int_X f(x)\xi_0(\mathrm{d}x) + \int_0^t \int_X \Big\{ f(x)\left[(1-up(x))b(x) - d(x) - \alpha * \xi_s(x)\right] + up(x)b(x)\left(\int_{\mathbb{R}} f(x+h)m(x,h)\mathrm{d}h\right)\Big\}\xi_s(\mathrm{d}x)\mathrm{d}s. \tag{3.1}$$

Let us now develop some particular cases. Let us firstly remark that if ξ_0 is absolutely continuous with respect to Lebesgue's measure, then it's also true for any ξ_t: $\xi_t(\mathrm{d}x) = \xi_t(x)\mathrm{d}x, t > 0$ (cf. [4]). The probability density $\xi_t(x)$ is thus weak solution of

$$\partial_t \xi(x) = \left((1-up(x))b(x) - d(x) - \alpha * \xi_t(x)\right)\xi_t(x) + \int m(y,x)up(y)b(y)\xi_t(y)\mathrm{d}y.$$

The asymptotic behavior of ξ_t, as $t \to \infty$ is a difficult open question and is explored in recent works of L. Desvillettes, P.-E. Jabin, S. Mischler, G. Raoul, see [18] or works in progress.

Let us now study the case where the mutation rate is null ($u_K = 0,\ \forall K$). We assume that the initial population is monomorphic and composed of individuals with trait x: $\nu_0^K = n_0^K \delta_x$, with $n_0^K \to n_0$. Then the population remains monomorphic, and Theorem 3.1 implies that $\nu_t^K = n_t^K \delta_x$ converges to $n_x(t)\delta_x$, and $(n_x(t), t \geq 0)$ is solution of the logistic equation

$$\frac{\mathrm{d}}{\mathrm{d}t} n_x(t) = n_x(t)(b(x) - d(x) - \alpha(x,x)n_x(t)), \text{ with } n_0(x) = n_0. \tag{3.2}$$

Since $\forall x \in X, b(x) - d(x) > 0$ and $\alpha(x,x) > 0$ (Assumptions (A1)–(A4)), Equation (3.2) has a unique stable stationary state, called charge capacity,

$$\bar{n}(x) = \frac{b(x) - d(x)}{\alpha(x,x)}. \tag{3.3}$$

Let us now assume that the initial population is dimorphic: $\nu_0^K = n_x^K(0)\delta_x + n_y^K(0)\delta_y$, with $n_x^K(0) \to n_x(0)$ and $n_y^K(0) \to n_y(0)$. Without mutation, it remains dimorphic for any time t, with traits x et y. In addition, $n_x^K(t)\delta_x + n_y^K(t)\delta_y$ converges to $n_x(t)\delta_x + n_y(t)\delta_y$, where $(n_x(t), n_y(t))$ is solution of the Lotka-Volterra system with initial condition $(n_x(0), n_y(0))$ and satisfying

$$\begin{aligned} \dot{n}_x &= n_x\left(b(x) - d(x) - \alpha(x,x)n_x - \alpha(x,y)n_y\right); \\ \dot{n}_y &= n_y\left(b(y) - d(y) - \alpha(y,x)n_x - \alpha(y,y)n_y\right). \end{aligned} \tag{3.4}$$

There are in this case 4 equilibria: $(0,0)$, $(\bar{n}(x), 0)$, $(0, \bar{n}(y))$, and a nontrivial equilibrium $(\bar{n}^1_{xy}, \bar{n}^2_{xy}) \in (\mathbb{R}_+^*)^2$.

More generally, if the initial population is composed of individuals with d distinct traits $x_1, \dots, x_d \in X$, then the population dynamics

$$(n^K_{x_1}(t), \dots, n^K_{x_d}(t))$$

converges to the solution of the competitive Lotka-Volterra system $LV(d, \mathbf{x})$, (with $\mathbf{x} = (x_1, \dots, x_d)$), defined for $n(t) = (n_1(t), \dots, n_d(t))$ by

$$\dot{n}_t = F^{\mathbf{x}}(n(t)), \tag{3.5}$$

where

$$F_i^{\mathbf{x}}(n) = n_i G_i^{\mathbf{x}}(n), \quad \text{and} \quad G_i^{\mathbf{x}}(n) = b(x_i) - d(x_i) - \sum_{j=1}^{d} \alpha(x_i, x_j) n_j.$$

Definition 3.2. We say that $x_1, \dots, x_d \in X$ coexist if the system $LV(d, \mathbf{x})$ has a unique non trivial stable equilibrium $\bar{n}(\mathbf{x}) \in (\mathbb{R}^*_+)^d$, in the sense that any eigenvalue of the Jacobian matrix $DF^{\mathbf{x}}(\bar{n}(\mathbf{x})) = ((-\alpha(x_i, x_j)\bar{n}_i(\mathbf{x})))_{1 \le i,j \le d}$ of $LV(d, \mathbf{x})$ at point $\bar{n}(\mathbf{x})$ has a negative real part.

In what follows, we will describe the invasion of a mutant trait y in a resident population with d coexisting traits $x_1, \dots, x_d$. Immediately after its birth, the population's size issued from the mutant individual is almost zero and we can neglect it. We may define the *invasion fitness of the mutant trait y in the resident population with traits $x_1, \dots, x_d$*. This function approximates the mutant population growth rate at the beginning of its appearance in the resident population at equilibrium. It describes the ability of the mutant trait y to invade the resident population $\mathbf{x} = (x_1, \dots, x_d)$ and is given by

$$f(y; x_1, \dots, x_d) = b(y) - d(y) - \sum_{j=1}^{d} \alpha(y, x_j)\bar{n}_j(\mathbf{x}), \tag{3.6}$$

where $\bar{n}(\mathbf{x})$ is the non trivial equilibrium of $LV(d, \mathbf{x})$.

Let us give some examples:

Example. If the resident population is monomorphic with trait x, the invasion fitness of y is

$$f(y; x) = b(y) - d(y) - \alpha(y, x)\bar{n}(x). \tag{3.7}$$

Let us remark that $f(x; x) = 0$, and that f is actually not symmetric.

We can characterize the stability of the equilibria of $LV(2, (x, y))$ thanks to the sign of the fitness function:

1. $(\bar{n}(x), 0)$ is unstable if $f(y; x) > 0$ and stable if $f(y; x) < 0$.
2. If $f(y; x) > 0$ and $f(x; y) < 0$, $(0, \bar{n}(y))$ is stable (fixation of the trait y).
3. If $f(y; x) > 0$ and $f(x; y) > 0$, there is a nontrivial and stable equilibrium (coexistence of traits x and y).

Example. If the resident population is dimorphic with trait x and y, the invasion fitness of a mutant z is

$$f(z;x,y) = b(z) - d(z) - \alpha(z,x)\bar{n}^1_{xy} - \alpha(z,y)\bar{n}^2_{xy}. \tag{3.8}$$

Let us remark that $f(x;x,y) = f(y;x,y) = 0$. Here again, the positivity of the fitness function describes the possible fixation of the mutant z.

4. Limit of rare mutations – convergence to the polymorphic evolutionary sequence

Here we will consider the scales of the adaptive dynamics: large population and small biomass, rare mutations and long mutation time scale.

The jump process that we will obtain has been heuristically introduced in [9], and rigorously studied in [6], as Trait Substitution Sequence, in the case where the ecological coefficients impede the coexistence of two traits. We consider here a more general case, when coexistence of two traits is authorized.

We assume that there is a time scale separation between the ecological fast time scale in which the population comes back to equilibrium after competition, and the mutation time scale which is much longer. If mutations are rare, the selection has time to eliminate the deleterious traits or to fix advantageous traits before a new mutant arrives. Large population assumption allows us to approximate the birth and death dynamics between mutations by a deterministic Lotka-Volterra system which is more tractable. That will allow us to predict the issue of the competition after a mutant arrival.

Initial condition (CI): The initial population is composed of individuals with coexisting traits $x_1, \ldots, x_d \in X$ and $\nu_0^K = \sum_{i=1}^d n_i^K \delta_{x_i}$. Moreover when $K \to \infty$, $n_i^K \to \bar{n}_i(\mathbf{x})$, $\bar{n}(\mathbf{x}) = (\bar{n}_1(\mathbf{x}), \ldots, \bar{n}_d(\mathbf{x}))$ denoting the nontrivial equilibrium of $LV(d,\mathbf{x})$.

Since u_K is the individual mutation rate, (going to 0 when $K \to \infty$), the population mutation rate is thus Ku_K, and if t is the ecological time scale, $\frac{t}{Ku_K}$ will represent the mutation time scale.

We study the asymptotic behavior of $(\nu^K_{\frac{t}{Ku_K}}, t \geq 0)$. We need to assume

Assumption (B): For coexisting traits $x_1, \ldots, x_d$ and for almost all mutant y with positive fitness $f(y;x_1,\ldots,x_d) > 0$, the system $LV(d+1,(x_1,\ldots,x_d,y))$ issued from $(\bar{n}^1(x),\ldots,\bar{n}^d(x),0)$ converges to a unique equilibrium n^*, which is a stable equilibrium.

Theorem 4.1. *Assume that* (A), (CI) *and* (B) *are fulfilled and assume moreover the following time scale separation: for all* $C > 0$,

$$\ln K \ll \frac{1}{Ku_K} \ll e^{CK}. \tag{4.1}$$

(where $f \ll g$ means $\frac{f}{g} \overset{K\to\infty}{\longrightarrow} 0$). Then the process $(\nu^K_{\frac{t}{Ku_K}}, t \geq 0)$ converges in the sense of finite marginals to a pure jump process $(\Lambda_t, t \geq 0)$, with values in

$$\mathcal{M}_0 = \left\{ \sum_{i=1}^{d} \bar{n}_i(\mathbf{x})\delta_{x_i}; (x_1, \dots, x_d) \text{ coexist and } \bar{n}(\mathbf{x}) \text{ equilibrium of } LV(d, \mathbf{x}) \right\},$$

and with transitions from

$$\sum_{i=1}^{d} \bar{n}_i(\mathbf{x})\delta_{x_i} \text{ to } \sum_{i=1}^{d} n_i^*(x_1, \dots, x_d, x_{j+h})\delta_{x_i} + n_{d+1}^*\delta_{x_{j+h}}$$

at rate

$$p(x_j)b(x_j)\bar{n}_j(x)\frac{[f(x_{j+h}; \mathbf{x})]_+}{b(x_j + h)} m(x_j, h)\mathrm{d}h.$$

The process $(\Lambda_t, t \geq 0)$ will be called polymorphic evolution sequence (PES).

Let us remark that if the initial population is monomorphic with trait x, then it remains monomorphic until the first coexistence time, that is the first time τ with $f(X_\tau; X_{\tau-}) > 0$ and $f(X_{\tau-}; X_\tau) > 0$. If the coefficients impede coexistence, then the process Λ writes $\Lambda_t = \bar{n}_{X_t}\delta_{X_t}$. The process $(X_t, t \geq 0)$ (with $X_0 = x_0$) describing the support of $(\Lambda_t, t \geq 0)$ is a pure jump Markov process with infinitesimal generator given by

$$A\varphi(x) = \int (\varphi(x+h) - \varphi(x))p(x)b(x)\,\bar{n}(x)\,\frac{[f(x+h; x)]_+}{b(x+h)} m(x, h)\mathrm{d}h. \tag{4.2}$$

The process $(X_t, t \geq 0)$ is called the Trait Substitution Sequence (TSS).

Let us now give ideas of the proof of Theorem 4.1. For a detailed proof, we refer to [7]. One describes the steps of the invasion of a mutant and the stabilization of the population which follows, with or without fixation of the mutant trait. Let us fix $\eta > 0$. Assume that at $t = 0$, the traits $x_1, \dots, x_d$ coexist and that Assumption (CI) is satisfied. For t and K large enough, the density process $(\langle \nu_t^K, \mathbf{1}_{x_1}\rangle, \dots, \langle \nu_t^K, \mathbf{1}_{x_d}\rangle)$ belongs to a η-neighborhood of $\bar{n}(\mathbf{x})$ with large probability (cf. Theorem 3.1). We need the process to stay in this neighborhood before the first mutation occurs. We thus use a large deviations result for exit time of domains, stated in Freidlin-Wentzell [19]. The time taken by the density process to leave the η-neighborhood of $\bar{n}(\mathbf{x})$ is larger than $\exp(CK)$, for some $C > 0$, with high probability. Hence, the first mutant will appear with large probability before the population exits the η-neighborhood of $\bar{n}(\mathbf{x})$, if one assumes that $\frac{1}{Ku_K} \ll \exp(CK)$.

Let us now study the invasion dynamics of the mutant with trait y. We will divide this event in three steps, as described in the Figure 4.1 and developed in [6].

First step: Between 0 and t_1, the number of individuals with mutant trait y is small and the resident population's size is close to $\bar{n}(\mathbf{x})$. Thus the mutant dynamics is close to the one of a birth and death process with rates $b(y)$ and $d(y)+\sum_{i=1}^{d} \alpha(y, x_i)\bar{n}_i(x)$. Its growth rate approximately equals the invasion fitness $f(y; x_1, \dots, x_d) = b(y)-d(y)-\sum_{i=1}^{d} \alpha(y, x_i)\bar{n}_i(\mathbf{x})$. If $f(y; x_1, \dots, x_d) > 0$,the birth

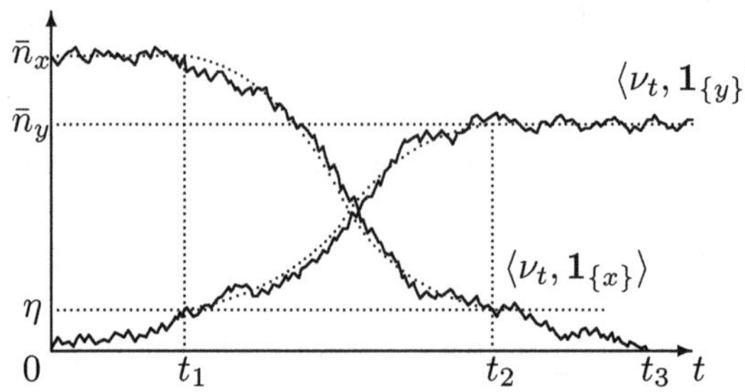

FIGURE 4.1. he three steps of the invasion of a mutant trait y in a monomorphic population with trait x.

and death process is supercritical, and therefore, for large K,

$$\mathbb{P}(\text{the mutant population's size attains } \eta) \simeq \mathbb{P}(\text{the branching process attains } \eta K)$$
$$\simeq \frac{[f(y; x_1, \ldots, x_d)]_+}{b(y)} \quad \text{(survival probability)}.$$

Between t_1 and t_2, it's the competition step. When K increases, the density process $(\langle \nu_t^K, \mathbf{1}_{x_1} \rangle, \ldots, \langle \nu_t^K, \mathbf{1}_{x_d} \rangle, \langle \nu_t^K, \mathbf{1}_y \rangle)$ tends to the solution of the system $LV(d+1, x_1, \ldots, x_d, y)$. Thus the population process will attain with large probability a η-neighborhood of n^* (defined in Assumptions (B)) in time t_2, for small η. In the figure above, $n^* = (0, \bar{n}_y)$.

The third step describes the stabilization of the population. The density processes for traits x_j such that $n_j^* = 0$ can be approximated by subcritical birth and death processes. Thus they will attain 0 in finite time and the population will stabilize around the traits x_j such that $n_j^* > 0$.

If the initial population is of order K, then the time taken for these three steps is of order $\ln K$, which is the order of the expectation of the extinction time for a birth and death process. Hence, if $\ln K \ll \frac{1}{K u_K}$, with a large probability these three phases of competition-stabilization will happen before a next mutation occurs and we can reiterate the reasoning after every mutation event.

Thanks to this analysis, we obtain the PES $(\Lambda_t, t \geq 0)$ which describes the successive stationary states and only keeps in its support the favorable mutations. It takes its values in $\mathcal{M}_0 = \left\{ \sum_{i=1}^d \bar{n}_i(\mathbf{x}) \delta_{x_i};\ x_1, \ldots, x_d \text{ coexist} \right\}$. Let us assume that at some time t, $\Lambda_t = \sum_{i=1}^d \bar{n}_i(\mathbf{x}) \delta_{x_i}$. If the process belongs to a η-neighborhood of $\bar{n}(\mathbf{x})$, the mutation rate from an individual with trait x_i is close to $b(x_i) u_K p(x_i) K \bar{n}_i(\mathbf{x})$. Hence, in the time scale $\frac{t}{K u_K}$, it is approximately $b(x_i) p(x_i) \bar{n}_i(\mathbf{x})$. When a mutation occurs, the mutant trait $x_i + h$ is chosen following $m(x_i, h) dh$. The invasion probability is thus approximately the survival probability of the mutant with trait y in the resident population, given by $\frac{[f(y; x_1, \ldots, x_d)]_+}{b(x_i + h)}$.

The process will jump to the equilibrium of $LV(d+1, x_1, \ldots, x_d, x_i + h)$, given by $\sum_{i=1}^d n_i^*(x_1, \ldots, x_d, x_{j+h}) \delta_{x_i} + n_{d+1}^* \delta_{x_{j+h}}$.

5. Canonical equation of the adaptive dynamics

We will now add an assumption of small mutations. In this section, we assume by simplicity that the trait space X is convex and symmetric and that the measure $m(x,h)\mathrm{d}h$ is symmetric with bounded three-order moments. We introduce the parameter $\varepsilon > 0$ which will scale the mutation's size. Let H_ε be the function defined by $H_\varepsilon(h) = \varepsilon h$. The distribution of mutant traits from an individual with trait x is given by $m_\varepsilon(x, \mathrm{d}h) = (m(x,h)\mathrm{d}h) \circ H_\varepsilon^{-1}$. Replacing m by m_ε in (4.2), we obtain the TSS X^ε. To observe a non trivial limit, we need to change the time scale t in t/ε^2. Let us now make ε tend to 0.

Theorem 5.1. *The process $(X^\varepsilon_{t/\varepsilon^2}, t \geq 0)$ converges in law as $\varepsilon \to 0$, to the deterministic monomorphic process $t \to \bar{n}(x(t))\delta_{x(t)}$, where $x(\cdot)$ is the unique solution of the ordinary differential equation*

$$\frac{\mathrm{d}x}{\mathrm{d}t} = \frac{1}{2} p(x)\bar{n}(x)\partial_1 f(x;x) \int_{\mathbb{R}} h^2 m(x,h)\mathrm{d}h. \tag{5.1}$$

This equation has been heuristically introduced by Dieckmann and Law [20] and is called *canonical equation of the adaptive dynamics*. When the mutant distribution m is not symmetric, it involves the whole measure m, instead of its variance.

Idea of the proof. It is based on a compactness-uniqueness argument. The process $(X^\varepsilon_{t/\varepsilon^2}, t \geq 0)$ has the generator

$$L^\varepsilon \varphi(x) = \frac{1}{\varepsilon^2} \int_{\mathbb{R}} (\varphi(x+\varepsilon h) - \varphi(x)) p(x) b(x) \bar{n}(x) \frac{[f(x+\varepsilon h; x)]_+}{b(x+\varepsilon h)} m(x,h)\mathrm{d}h. \tag{5.2}$$

Its uniqueness is obtained by a standard theorem (boundedness of the coefficients). As $f(x;x) = 0$, and by an expansion of $\varphi(x+\varepsilon h)$ and $f(x+\varepsilon h; x)$ at order 2 around $\varepsilon = 0$, one can show that

$$L^\varepsilon \varphi(x) \xrightarrow{\varepsilon \to 0} \frac{1}{2} p(x)\bar{n}(x)\partial_1 f(x;x)\varphi'(x) \int_{\mathbb{R}} m(x,h) h^2 \mathrm{d}h. \tag{5.3}$$

The process $(X^\varepsilon_{t/\varepsilon^2}, t \geq 0)$ is a semimartingale. Uniform tightness (in ε) of the laws of $(X^\varepsilon_{\cdot/\varepsilon^2})$ comes from uniform estimates for the martingale part and for the total variation part of its decomposition. The characterization of the limiting martingale problem is deduced from (5.3).

We assume that for all $x \in X$, $\int_{\mathbb{R}} h^2 m(x,h)\mathrm{d}h \neq 0$ (biologically, it means that effective mutations can occur from each ancestor trait). The equilibria are the points x^* such that $\partial_1 f(x^*; x^*) = 0$. These points will play a fundamental role in the characterization of the evolutionary branching as explained below.

Definition 5.2. An evolutionary singularity (ES) is a point $x^* \in X$ satisfying

$$\partial_1 f(x^*; x^*) = 0. \tag{5.4}$$

Let us remark that since for any $x \in X$, $f(x;x) = 0$, we have for small h,

$$f(x+h;x-h) = h(\partial_1 f - \partial_2 f)(x;x) + O(h^2),$$
$$f(x-h;x+h) = h(\partial_2 f - \partial_1 f)(x;x) + O(h^2).$$

Two traits x et y can coexist if and only if $f(x;y) > 0$ and $f(y;x) > 0$. Thus coexistence can only happen in neighborhoods of ES. One can show that the solution x of (5.1) cannot attain an ES in finite time. Moreover, if $I(T) = \{x(t), t \in [0,T]\}$, then for x and y in $I(T)$ and close enough, the sign of $(y-x)f(y;x)$ is constant and x and y cannot coexist.

If $\Lambda_0^\varepsilon = \bar{n}(x)\delta_x$, the support of $\Lambda^\varepsilon_{t/\varepsilon^2}$ before the first coexistence time

$$\tau^\varepsilon = \inf\left\{t > 0, f\left(\Lambda^\varepsilon_{t/\varepsilon^2}; \Lambda^\varepsilon_{t-/\varepsilon^2}\right) > 0 \; ; \; f\left(\Lambda^\varepsilon_{t-/\varepsilon^2}; \Lambda^\varepsilon_{t/\varepsilon^2}\right) > 0\right\}, \qquad (5.5)$$

has the dynamics of the process $(X^\varepsilon_{t/\varepsilon^2}, t \geq 0)$. Therefore, before coexistence, it takes values in a η-neighborhood of $I(T)$ with probability 1. It turns out that an evolutionary branching can only happen on a longer time scale than T/ε^2, for any $T > 0$. We get

Theorem 5.3. *Assume Hypotheses* (A) *and* (B) *are fulfilled and that* $\Lambda_0^\varepsilon = \bar{n}_x\delta_x$. *Then for all* $T > 0$, $\lim_{\varepsilon\to 0}\mathbb{P}(\tau^\varepsilon > T) = 1$ *and the process* $\left(\Lambda^\varepsilon_{t/\varepsilon^2}, t \geq 0\right)$ *converges in law, as* ε *tends to* 0, *to the deterministic process* $\left(\bar{n}\left(x(t)\right)\delta_{x(t)}, t \geq 0\right)$, *where the function* x *is the unique solution of* (5.1). *Moreover, with high probability and* η *small enough, the function* $t \to \Lambda^\varepsilon_{t/\varepsilon^2}$ *is monotonous before the first entrance in the* η*-neighborhood of an ES* x^*, *which is smaller than the coexistence time.*

The proof of Theorem 5.3 is based on the properties of the canonical equation that we have stated above, and on the convergence of the PES killed at the first coexistence time, to the solution of the canonical equation (cf. Theorem 4.1).

The PES can only become polymorphic by attaining an ES x^*. Nevertheless some of these singularities are repulsive for the canonical equation. The point x^* is proved to be attractive if

$$\partial_{22}f(x^*;x^*) \geq \partial_{11}f(x^*,x^*). \qquad (5.6)$$

6. Evolutionary branching

Following [7], let us now properly define the evolutionary branching. Let us consider an attractive ES x^* and let us fix $\eta > 0$.

Definition 6.1. Fix $\varepsilon > 0$. We say that there is an η-branching at x^* for the PES $\left(\Lambda^\varepsilon_{t/\varepsilon^2}, ct \geq 0\right)$ if

1. $\exists\, t_1$ such that $\mathrm{Supp}\left(\Lambda^\varepsilon_{t_1/\varepsilon^2}\right)$ is a singleton belonging to $[x^*-\eta; x^*+\eta]$.

2. $\exists\, t_2 > t_1$ such that $\mathrm{Card}\left(\mathrm{Supp}\left(\Lambda^{\varepsilon}_{t_2/\varepsilon^2}\right)\right) = 2$, and the points of this support are distant of more than $\frac{\eta}{2}$.
3. Between t_1 and t_2, the support of the PES is a subset of $[x^* - \eta; x^* + \eta]$ composed of at most two points, and has an increasing diameter.

Let us now state our main theorem, which proves a conjecture proposed by Metz *et al.* [9]. Let $\mathbb{P}^{\varepsilon}$ be the law of the process $\left(\Lambda^{\varepsilon}_{t/\varepsilon^2}, t \geq 0\right)$.

Theorem 6.2. [7] *Assume that* (A) *and* (B) *are fulfilled and that* x^* *is an evolutionary singularity satisfying*

$$\partial_{22} f(x^*; x^*) > \partial_{11} f(x^*, x^*); \qquad \text{and} \qquad \partial_{22} f(x^*; x^*) + \partial_{11} f(x^*, x^*) \neq 0.$$

Then for any $\eta > 0$, *there exists* ε_0 *such that for all* $\varepsilon < \varepsilon_0$,

1. *If* $\partial_{11} f(x^*, x^*) > 0$, $\mathbb{P}^{\varepsilon}(\eta\text{-branching}) = 1$.
2. *If* $\partial_{11} f(x^*, x^*) < 0$, $\mathbb{P}^{\varepsilon}(\eta\text{-branching}) = 0$.

We thus have an explicit condition that we can check on the ecological coefficients and which allows us to predict, with large probability a phenotypic separation.

In the example of Section 2 (cf. simulations), the unique evolutionary singularity is $x^* = 0$ and it satisfies the conditions of Theorem 6.2. One checks that

$$\partial_{11} f(0;0) = \frac{1}{\sigma_\alpha^2} - \frac{1}{\sigma_b^2}; \qquad \text{and} \qquad \partial_{22} f(0;0) = \frac{1}{\sigma_\alpha^2} + \frac{1}{\sigma_b^2}.$$

Moreover, $\partial_{11} f(0;0) > 0 \Longleftrightarrow \sigma_\alpha < \sigma_b$.

Idea of the proof. Step 1. The first step consists in giving an expansion of the fitness for 2 traits and 3 traits, at the neighborhood of x^*. Noting $a = \partial_{11} f(x^*, x^*)$; $c = \partial_{22} f(x^*, x^*)$, one shows that

$$\begin{aligned} f(y;x) &= \frac{1}{2}(y-x)(a(y-x^*) - c(x-x^*)) + o\left(|y-x|\left(|x-x^*| + |y-x^*|\right)\right); \\ f(z;x,y) &= \frac{a}{2}(z-x)(z-y) + o\left(|z-x||z-y|\right). \end{aligned} \tag{6.1}$$

Step 2. The condition $a + c > 0$ ensures the existence of coexisting traits x and y in any neighborhood $\mathcal{U}$ of x^*. The coexistence area is delimited by the line $\{x = y\}$, and by curves γ with tangent $a(y - x^*) = c(x - x^*)$ at point (x^*, x^*), and γ^s defined as γ by exchanging the variables x and y. This region is described in Figure 2 in case $c > a > 0$. The fitness expansion shows that it changes of sign at the boundaries of this region.

For each coexisting couple (x, y), and mutant trait z, the possible equilibria in the Lotka-Volterra system have been given by [21]. By a fine study of each equilibrium, one can show that for every ES x^* satisfying the assumptions of Theorem 6.2 and such that $a \neq 0$, there exists a neighborhood $\mathcal{U}_0$ of x^* such that any triplet x, y, z belonging to $\mathcal{U}_0$ cannot coexist.

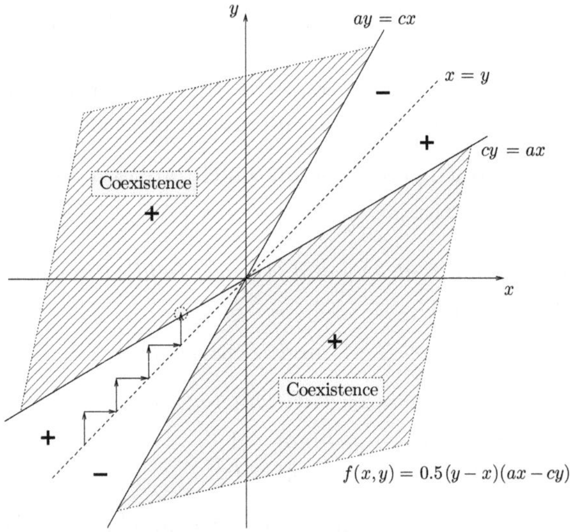

FIGURE 6.1. Coexistence region, case $c > a > 0$

Step 3. Let us firstly assume that $a < 0$ and that two traits x and y coexist in a neighborhood $\mathcal{U}_0$ of x^*. Then, (6.1) shows that $f(z; x, y)$ is locally concave and is zero in x and y. It is thus positive on $[x, y]$ and the mutant traits z which could fix in the population belong to this interval. After competition between the three traits, the population becomes monomorphic again or dimorphic, but in this case, the distance between the two traits of the PES' support decreased. Therefore, almost surely, no branching can happen.

Let us now assume that $a > 0$. The fitness function $f(z; x, y)$ is then locally convex in a neighborhood of x^* and is zero at x and y. It is thus positive on the complement of $[x, y]$, which will allow to mutant traits out of $[x, y]$ to fix in the population.

Let us fix $\eta > 0$ small enough. If $c > a > 0$, and for ε small enough, the first coexistence time τ^ε is finite almost surely and $\mathrm{Supp}(\Lambda^\varepsilon_{t/\varepsilon^2}) \subset (x^* - \eta; x^* + \eta)$. Indeed, $c > a$ ensures that x^* isn't repulsive and the PES (whose jumps are drawn by arrows in Figure 2), only jumps to mutant traits with positive invasion fitness and will attain the coexistence region with probability 1. After τ^ε, the distance between two points of the PES' support is increasing. We show more precisely that the process cannot become monomorphic again. Moreover, a study of fitness' sign shows that if $x < y$, the support $\{x, y\}$ of the PES can jump in $\{z, y\}$ with $z < x$ or in $\{x, z\}$ with $z > y$. In addition, almost-surely, the distance between two points of the PES' support after τ^ε becomes larger than $\frac{\eta}{2}$ in finite time, before the PES exits $(x^* - \eta; x^* + \eta)$. The probability to obtain a η-branching in this case is equal to 1. □

References

[1] J. Hofbauer and K. Sigmund, *The Theory of Evolution and Dynamical Systems. Mathematical Aspects of Selection.* London Mathematical Society Student Texts, 7. Cambridge University Press, U,K., 1988.

[2] P. Marrow, R. Law, and C. Cannings, *The coevolution of predator prey interactions – ESSs and Red Queen dynamics.* Proc. R. Soc. Lond. Ser. B-Biol. Sci. **250** (1992), 133–141.

[3] J.A.J. Metz, R.M. Nisbet, and S.A.H. Geritz, *How should we define "fitness" for general ecological scenarios?* Trends Ecol. Evol. **7** (1992), 198–202.

[4] N. Fournier and S. Méléard, *A microscopic probabilistic description of a locally regulated population and macroscopic approximations.* Ann. Appl. Probab. **14** (2004), 1880–1919.

[5] N. Champagnat, R. Ferrière, and S. Méléard, *From individual stochastic processes to macroscopic models in adaptive evolution.* Stoch. Models **24** (2008), 2–44, 9th Symposium on Probability and Stochastic Process, Guanajuato, Mexico, Nov. 20–24, 2006.

[6] N. Champagnat, *A microscopic interpretation for adaptive dynamics trait substitution sequence models.* Stoch. Process. Their Appl. **116** (2006), 1127–1160.

[7] N. Champagnat and S. Méléard, *Polymorphic evolution sequence and evolutionary branching (to appear)*, Probab. Theory Relat. Fields.

[8] U. Dieckmann and M. Doebeli, *On the origin of species by sympatric speciation.* Nature **400** (1999), 354–357.

[9] J.A.J. Metz, S.A.H. Geritz, G. Meszéna, F.J.A. Jacobs, and J.S. van Heerwaarden, *Adaptive dynamics, a geometrical study of the consequences of nearly faithful reproduction. In Stochastic and spatial structures of dynamical systems (Amsterdam, 1995)*, Konink. Nederl. Akad. Wetensch. Verh. Afd. Natuurk. Eerste Reeks, 45, 183–231, North-Holland, Amsterdam, 1996.

[10] O. Diekmann, P.E. Jabin, S. Mischler, and B. Perthame, *The dynamics of adaptation: An illuminating example and a Hamilton-Jacobi approach.* Theor. Popul. Biol. **67** (2005), 257–271.

[11] G. Barles and B. Perthame, *Concentrations and constrained Hamilton-Jacobi equations arising in adaptive dynamics. In* D. Danielli (ed.), *Recent Developments in Nonlinear Partial Differential Equations*, volume 439 of *Contemporary Mathematics Series*, 57–68, 2007.

[12] A. Bovier and N. Champagnat, *Time scales in adaptive dynamics: directional selection, fast and slow branching. (in preparation).*

[13] F. Yu, *Stationary distributions of a model of sympatric speciation.* Ann. Appl. Probab. **17** (2007), 840–874.

[14] B. Bolker and S.W. Pacala, *Using moment equations to understand stochastically driven spatial pattern formation in ecological systems.* Theor. Popul. Biol. **52** (1997), 179–197.

[15] É. Kisdi, *Evolutionary branching under asymmetric competition.* J. Theor. Biol. **197** (1999), 149–162.

[16] U. Dieckmann and R. Law, *Relaxation projections and the method of moments. In The Geometry of Ecological Interactions: Symplifying Spatial Complexity*, 412–455, Cambridge University Press, UK, 2000.

[17] J.A. Roughgarden, *Theory of Population Genetics and Evolutionary Ecology: an Introduction.* Macmillan, New York, 1979.

[18] L. Desvillettes, P.E. Jabin, S. Mischler, and G. Raoul, *On selection dynamics for continuous structured populations.* Commun. Math. Sci. **6** (2008), 729–747.

[19] M.I. Freidlin and A.D. Wentzel, *Random Perturbations of Dynamical Systems.* Springer-Verlag, Germany, 1984.

[20] U. Dieckmann and R. Law, *The dynamical theory of coevolution: A derivation from stochastic ecological processes.* J. Math. Biol. **34** (1996), 579–612.

[21] M.L. Zeeman, *Hopf bifurcations in competitive three-dimensional Lotka-Volterra systems.* Dynam. Stability Systems **8** (1993), 189–217.

Sylvie Méléard
CMAP, Ecole Polytechnique, CNRS
route de Saclay
F-91128 Palaiseau Cedex, France
e-mail: sylvie.meleard@polytechnique.edu

Thoughts on the Geometry of Meso-evolution: Collecting Mathematical Elements for a Postmodern Synthesis

J.A.J. (Hans) Metz

Abstract. Adaptive dynamics (AD) is a recently developed framework geared towards making the transition from micro-evolution to long-term evolution based on a time scale separation approximation. This assumption allows defining the fitness of a mutant as the rate constant of initial exponential growth of the mutant population in the environment created by the resident community dynamics. This definition makes that all resident types have fitness zero. If in addition it is assumed that mutational steps are small, evolution can be visualized as an uphill walk in a fitness landscape that keeps changing as a result of the evolution it engenders. The chapter summarises the main tools for analysing special eco-evolutionary models based on these simplifications. In addition it describes a number of general predictions that directly derive from the AD perspective a such, without making any further assumptions.

AD arguments are largely local in character. Hence they can only deal with what might be called meso-evolution. For longer timescales (macro-evolution) it becomes necessary to look at general developmental and morphological (sensu lato) arguments that bear on the larger scale geometry of fitness landscapes. From this enlarged perspective the low dimensional fitness landscapes studied in AD can be seen as the surfaces at the top of ridges in a much higher dimensional landscape over potential morphologies, with the abyss around the ridges created by the lack of a proper development, or functioning. The location of the ridges and abysses is grossly the same for large sets of possible environmental conditions. Biological parlance expresses this constancy by referring to the corresponding selective processes as internal. High dimension and ridgyness turn out to conspire in a number of ways:

1. Developmental systems leading to mutation distributions that are in some way aligned with the ridges evolve much faster than systems where such is not the case.
2. The stabilizing selection in the off-ridge directions has a great robustness of the developmental system as an indirect evolutionary consequence. Yet, the high dimension of genotype space makes that this robustness can never as such lead to the conservation of features. Hence, the fact that evolution largely proceeds through the quantitative variation in the

size and shape of homologous parts should be due to stabilizing internal selection.

3. So-called allopatric speciation supposedly occurs by separated populations wandering around along the high fitness maze, so that after a while their mixed offspring, having intermediate properties, ends up in the abyss. As random fitness landscapes almost never engender allopatric speciation the question arises whether, and if so for what reason, evolved genotype to phenotype maps may be more speciation prone.
4. Large mutational steps far more often than not make an individual land in the fitness abyss, and only the small steps keep it on the top. This may provide a theoretical justification for the assumption made in AD.

Mathematics Subject Classification (2000). 92D15.

Keywords. Adaptive dynamics, fitness landscapes, evo-devo, meso-evolution, macro-evolution, internal selection, speciation.

1. Introduction

1.1. On micro-, macro- and meso-evolution

The present chapter is complementary to the ones by Warren Ewens [1] and by Reinhard Bürger [2]. It contains almost no equations or genetics (or rather, both are implicitly there, but stay well hidden), fitnesses are not assumed to be constant, but keep changing (in genetical terms, selection is always seen as frequency dependent), the evolutionary path is considered as being shaped by the repeated substitution of novel mutations (as opposed to gene frequency change), and adaptive landscapes depict the fitness of potential mutant types over a space spanned by traits (instead of the mean fitness of a population over a space of gene frequencies). The reason for this difference in emphasis is that my first training is not as a mathematician or population geneticist but as a naturalist, i.e., field oriented taxonomist-ecologist, which shapes the questions that have my interest. As a consequence, within evolutionary biology I am interested primarily in meso-evolution, defined here as evolutionary changes in the values of traits of representative individuals and concomitant patterns of taxonomic diversification. This in contrast to micro-evolution, a term reserved for the changes in gene frequencies on a population dynamical time scale (the topic of Warren Ewens' and Reinhard Bürger's chapters [1, 2].), and macro-evolution, a term that then can be reserved for large scale changes like anatomical innovations, where one cannot even speak in terms of a fixed set of traits. Thus meso-evolution acts on a time scale above the micro-evolutionary scale of gene substitutions but below the scale on which the intricacies of the developmental process start to have a large influence.

Meso-evolution is far more than micro-evolution writ large, and a similar statement holds for macro- versus meso-evolution. Each of these levels has its own emergent phenomena, and its own explanatory frameworks, which should in the end be based at least in part on idealised summaries of large scale regularities in the outcome of lower level mechanisms, in a similar manner as pressure and temperature can be treated as macroscopic causes, although they themselves are but statistics of the underlying process of molecular motion. And where the results of thermodynamics are predicated on man-made or naturally evolved macroscopic structures confining these motions, so do trait changes result from the micro-evolutionary process of mutant substitutions taking place against the backdrop of a genetic architecture and developmental system as deliverers of the required mutational variation, internal selection caused by the necessity that the machinery of a body stays in concert (think of Cuvier's famous dictum that given one part he could deduce the rest of an organism), and ecological selection due to the interactions of individuals with their conspecifics, resources, predators, parasites and diseases. In this chapter I focus on these encompassing mechanisms rather than on the motion of gene frequencies.

In order to get a clean story I assume time scale separations all over. Not that I believe that such time scale separations hold good even most of the time. In a rigorous sense they only hold good very rarely. However, it looks as if arguments based on them lead to fair outcomes in more than a fair fraction of the cases. Moreover, it is only by such time scale arguments that I can easily make the transition from population genetics to the views common among morphologists and taxonomists. As I am aiming at contributing to a postmodern synthesis (cf. Subsection 1.2), so be it. I have chosen for being wrong in the details, although I believe often close, over being strictly correct but unable to address the larger picture.

1.2. The so-called modern synthesis

Figure 1.1 gives a schematic representation of the "integrated biology" view of life. The small inner loop (fat arrows) can be seen as abstracting a life cycle. The genome produces shapes that change over an individual's life. Shapes should be interpreted here in a generalised sense, e.g., including the distribution of all sorts of chemicals within a body. What these forms do is called function. What they can do depends on the environment. What they do, together with other aspects of the environment like availability of resources or density of predators, determines how many descendants they on average contribute to future generations. What they do and how many they are necessarily affects the environment. Their number of descendants if the environment were not to change, measured as the asymptotic average rate of exponential growth, is called fitness (see Subsection 2.2). Only when its fitness is positive a mutant type has a positive probability to invade, and a mutant can only take over if the fitness of its parent type becomes negative when the mutant further grows in numbers. This filtering determines which types will be present in the future. Finally, reproduction needs to be almost faithful in order to get evolution; if it were fully faithful no mutants would occur, if it were too unfaithful, the effects of selection would be swamped by a mutational oozing out over trait space.

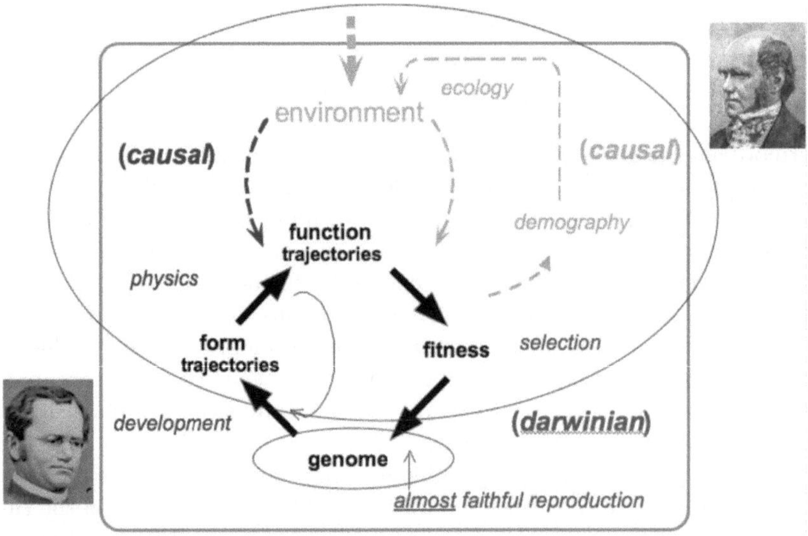

FIGURE 1.1. The "integrated biology" view. People, clockwise from lower left: Gregor Johan Mendel, Charles Darwin.

For lack of better, Darwin thought of inheritance as a blending of the parental types, which would make evolution impossible (cf. the chapter by Warren Ewens [1]). Luckily Mendel saved the theory of evolution by natural selection by discovering the faithfully reproducing genes. Seen from first principles the genes are the primary units of evolution, while the usual biological individuals are no more than uneasy coalitions of genes. Luckily, under some simplifying conditions, among which a separation between the time scales of population dynamics and evolution, the Mendelian mechanism fairly often, but certainly not always, allows calculating evolutionary trajectories and outcomes as if it were the individuals that reproduce faithfully (cf. Subsections 2.2, 2.3 and 2.5).

The synthesis between the Darwinian and Mendelian view was made by the three great theoretical population geneticists: Ronald Fisher, J.B.S. Haldane and Sewall Wright. In North America this material reached the less mathematically inclined biologist community through the experimental work of Theodosius Dobzhanski. This let taxonomists like Ernst Mayr and paleontologists like George Gaylord Simpson embrace the so-called population view of evolution. Although espoused by them as the "modern synthesis" (see, e.g., [3, 4, 5]), the result was more like a mutual admiration society than a synthesis. The discrepancy between terminology and reality was caused largely by the remaining but blissfully ignored lack of sufficient understanding of both development and population dynamics.

When Mayr grew older, he elevated himself to the role of philosopher and historian of biology, and in this role pushed the simplified picture of integrated

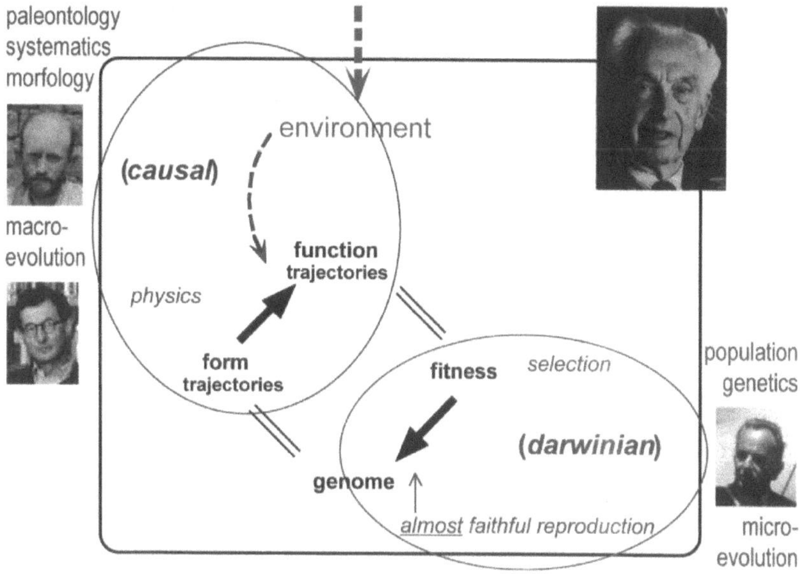

FIGURE 1.2. The simplified view effectively espoused in the "modern synthesis" of the nineteen-fourties. People, clockwise from lower left: Ernst Mayr (young), George Gaylord Simpson, Ernst Mayr (old), Theodozius Dobzhanski.

biology that you see in Figure 1.2. It was through this ploy that the modern synthesis indeed could be perceived as a synthesis[1]. In simple textbook evolutionary scenarios this simplified worldview indeed works remarkably well. However, it unravels when one starts to focus on complicated trait spaces and ecologies as they occur in real life.

1.3. Reinstating the missing components

In short, the so-called modern synthesis has not actually achieved a connection between micro-evolutionary mechanisms and meso- let alone macro-evolutionary patterns. All that has been shown is compatibility in principle, as you may see when you put the two ovals from Figure 1.2 in Figure 1.1.

Presently two developments that attempt to close the gaps are in full swing. Evo-Devo started around 1980 with the discovery of the various genes underlying developmental switches, but for many of its ideas goes back to the German *Entwicklungsmechanik* from the first part of the 20th century and to the morphology of an even earlier century. Adaptive dynamics (AD) started in the early nineteen-nineties as the simplest dynamic extension of the evolutionary statics covered by ESS theory (see Subsection 2.3), which itself started with the work of Hamilton

[1]See [6] for a further discussion, in particular of the political reasons for this purposeful oversimplification.

[7] and Maynard Smith and Price [8], but already had precursors in some of the work of Fisher [9][2].

Evo-Devo focuses mainly on intra-individual processes, macro-evolution, and the post-hoc explanation of realised patterns. Its goal is to fill the left lower quadrant of Figure 1.1. AD focuses mainly on the shaping of the selective arena by ecological processes, on meso-evolution, and on prediction oriented theory. Its goal is to fill the right upper quadrant of Figure 1.1.

In this chapter I will sketch some mathematical aspects of what directions a postmodern synthesis based on these newer developments might take.

2. A short introduction to adaptive dynamics and its ramifications

2.1. Codifying the ecological principles of meso-evolution

Adaptive dynamics (AD) was developed as an aid for making the transition from micro- to meso-evolution. Meso-evolution proceeds by the selective filtering by the ecology of a continual stream of mutants. AD concentrates on the ecological side of this process, as there are much clearer a priori mathematical structures to be found at that end. The basic theory assumes clonal reproduction, and only a subset of the results extend to the Mendelian case, for monomorphic populations directly (cf. Subsections 2.3 and 2.5) and for polymorphic populations after appropriate modification (cf. Subsection 2.8).

One of the immediate consequences of stressing the ecological side of the equation is a strong awareness that fitnesses are not given quantities, but depend both on the traits of an individual and on the environment in which it lives. The ecological feedback loop makes that in the monomorphic and clonal cases necessarily the fitnesses of all types present on an ecological time scale are zero (see Subsection 2.2). Only the fitnesses of potential mutants can be positive or negative. The signs and sizes of these mutant fitnesses determine the direction and speed of evolutionary progress. Evolution corresponds to permanent uphill movement in a fitness landscape that keeps changing so as to keep the resident types exactly at zero. See Figure 2.1.

The main general insight from the mathematical analyses of this picture has been the discovery of a potential mechanism for adaptive speciation that appears with a certain ubiquity in ecological models (see Subsection 2.8). Apart from that, the theory has produced a good number of very effective tools for analysing special families of eco-evolutionary models (Subsections 2.4, 2.5 and 2.7).

[2]There are also precursors at the dynamic end. In particular Ilan Eshel already in the nineteen-eighties developed a mutation limitation based theory for long-term evolution, with a focus on genetics instead of on ecology and traits. For a summary see [10] and [11]. Adaptive dynamics theories not based on a time scale separation but combining population dynamical ODEs with phenomenologically justified ODEs for their parameters were developed by Tom Vincent and Joel Brown and independently by Peter Abrams, see, e.g., [12, 13].

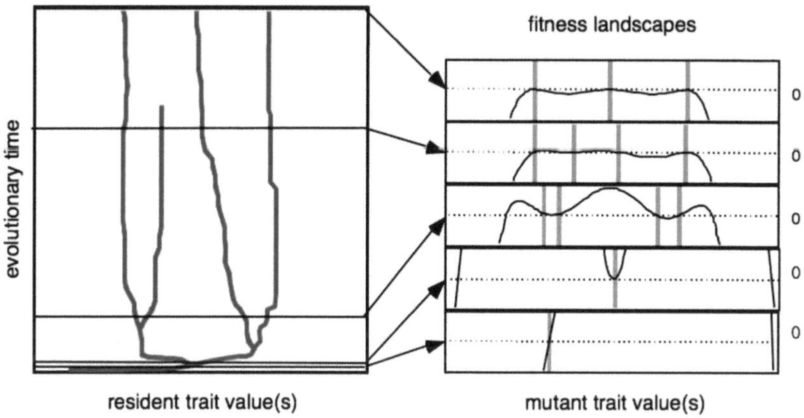

FIGURE 2.1. Left: Evolutionary path simulated on the basis of a population dynamical model, assuming clonal reproduction. Only the traits that are dominantly present in the population are shown. The second ascending branch finishes since the subpopulation under consideration went extinct. Right: The fitness landscapes for five population compositions as these occurred at the indicated times. The vertical bars indicate the types that at that moment were present in the population. At the second selected time the population resided at a branching point (see Subsection 2.8). At the final time the remaining three subpopulations reside at an evolutionarily stable ESC (see Subsection 2.3).

Below I will keep exploiting the landscape analogy and refer to zero as sea level, etc. Moreover, to keep the story simple I shall each time initially proceed on the assumption that individuals reproduce clonally.

2.2. Fitness

The ecological perspective. The concept of fitness as a quantitative measure of competitive prowess is a modern invention. Darwin never used the term in this meaning, and neither did population genetics' founding triumvirate (with the exception of [14], otherwise they use terms like selective advantage, see [15]). In population genetics, fitness is generally used for the probability to survive to reproduction. However, this only works for the relatively simple ecological scenarios considered there, where the different life phases are both neatly separated and synchronised. In ecology one has to account for a messier world where populations have age, size, spatial or other structures, and where demographic properties vary with the weather over an individual's life and over the generations.

Let the environment be defined as anything outside an individual that influences its population dynamical behaviour, which by definition consists of impinging on the environment, giving birth, and dying (see, e.g., [15, 16, 17, 18]). It is always possible in principle to find a Markovian representation of that behaviour, in terms of a state space, transition probabilities that depend on the course of the environment, and outputs that are either deterministic or occur in a Poisson cluster process with rate and cluster (clutch) size dependent on an individual's state and the condition of the environment at the time. Given the course of the environment, individuals independently move through their state spaces, the population state is a measure over this space, and the expectation of this measure, which is again a measure, moves according to a positive linear evolutionary system. The theory of positive linear systems then tells that generally the expected size of a population in an ergodic environment will in the long run on average grow or decline exponentially (for details see [15, 19]). This growth rate ρ is what ecologists call fitness. It necessarily is a function of two variables, the type of the individuals Y, parametrised by their traits, and the environment E, to be written as $\rho(Y|E)$. The theory of branching processes moreover tells that when a population is started with a single individual it will, barring some technical conditions, either eventually go extinct or grow exponentially, with the probability of the latter being positive if and only if its fitness is so (see [20, 21, 22, 23]).

In the theory of long-term adaptive evolution one is mainly interested in populations in which the number of individuals exposed to similar environments are sufficiently large that the internal workings of these populations can be modelled in a deterministic manner, with possibly on top an external stochastic driver. In nature populations are necessarily bounded. (Thanks to the above definitions, this can be ascribed to changes in the environment brought about by the growth of those populations.) Hence the population state space is a closed bounded subset of the cone of positive measures over the state space of the individuals, and the state space of a community is the product of the state spaces of the comprising species, plus the state spaces of the dynamics of any inanimate resources. With an infinitesimal amount of noise the states of such communities will approach an "extinction preserving chain attractor" (this is a modification that accounts for the fact that extinct populations cannot be resurrected of the concept of Conley-Ruelle or chain attractor; see [24, 25]). With larger amounts of noise the community will in general end up in a stochastic attractor, that is, a stationary distribution of community states. I will throughout assume that this attractor generates an ergodic environment (the exceptions that I have seen constructed all appeared to need biologically pretty exceptional conditions). Let the environment generated by a coalition of clones $C = (X_1, \ldots, X_n)$ be written as $E_{\mathrm{attr}}(C)$. A combination of the preceding arguments then leads to the introduction of the *invasion fitness* $\rho(Y|E_{\mathrm{attr}}(C))$ of a new type Y in a C-community.

For ease of exposition I proceed as if $E_{\mathrm{attr}}(C)$ is unique. Most of my statements extend to the general case with only small modifications.

The extension of the previous framework to Mendelian populations turns out to be easier than perhaps expected (although implementing it in concrete cases tends to be horrible). For the community dynamics all one has to do is distinguish individuals according to their genotypes, and incorporate their mating opportunities with different genotypes into the description of the environment (cf. [26];

this in the case of casual matings, with more extended pair formation it becomes necessary to extend the state space of individuals to keep track of their marriage status). Alleles, of course, reproduce clonally and as such have fitnesses. It is also possible to define a mock fitness of phenotypes by introducing a parallel clonal model with individuals passing through their lives like their Mendelian counterparts and having a reproduction equal to the average of the contributions through the micro- and macro-gametic routes (for humans semen and ova) of those counterparts. With such a definition some essential, but certainly not all, fitness-based deductions for the clonal case go through for Mendelian inheritance. In particular, for genetically homogeneous populations the fitness of a so-called resident, that is, a type that is present in a community dynamical attractor, equals zero (since genetically homogeneous populations breed true and resident populations by definition do not in the long run grow or decline). Moreover, the invasion of a new mutant in a homogeneous population is correctly predicted, as that mutant initially only occurs in heterozygotes that breed true by backcrossing with the homogeneous resident.

The Evo-Devo perspective. The Evo-Devo view of fitness is much closer to the population genetical one. In Evo-Devo people generally only consider whether a developmental program works sufficiently well, that is, produces a functioning organism. The better an organism functions the higher its fitness, with this functioning more often than not being largely independent of the specific environment in which it lives. Moreover, in general a mutant either develops along lines compatible with the environments under consideration or it dies early in its life. Translating these considerations into population dynamics does not necessarily lead to the usual population genetical equations, but it often leads to model formulations supporting a so-called optimisation principle (see the chapter by Gyllenberg et al. [27]). For more refined eco-evolutionary models optimisation principles may be rare, but they occur more readily when one aims for a model that captures only grosser characteristics while glossing over the ecological fine detail. Hence optimality arguments can hold sway in Evo-Devo, where in evolutionary ecology they have been largely replaced by ESS and AD arguments (see Subsections 2.3 to 2.7). Thus, the picture espoused by Evo-Devo researchers tends to be like the ones in Figure 2.2.

2.3. Evolutionarily statics: ESS theory

Evolution stays put whenever the community produces an environment such that mutants have negative fitness whenever they differ from any of the residents. In the special case where there is but one resident type, we speak of an Evolutionarily Steady Strategy. (The old name Evolutionarily Stable Strategy, introduced by Maynard Smith and Price [8], is a bit of a misnomer, since, as first discovered by Eshel [28] and illustrated in Figure 2.4, ESSes need not be evolutionarily attractive.) In the general case when there may be more than one resident type I will speak of an Evolutionarily Steady Coalition. ESCs are the equilibria of evolution.

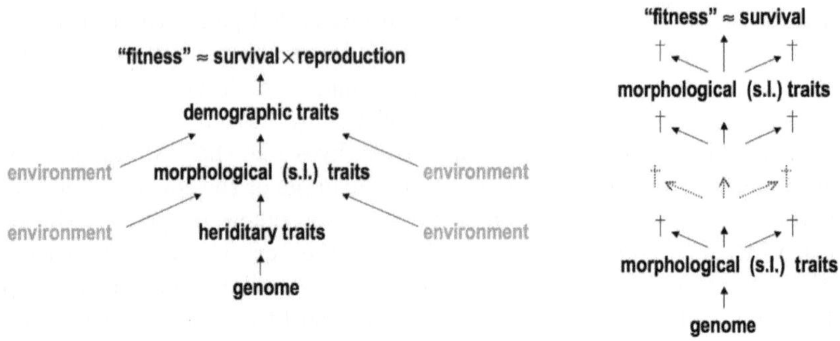

FIGURE 2.2. The Evo-Devo perspective on fitness, with left an ecologically enlightened and right a more narrow perspective. Note that what here is called "fitness" is at best proportional to $\exp(\rho)$.

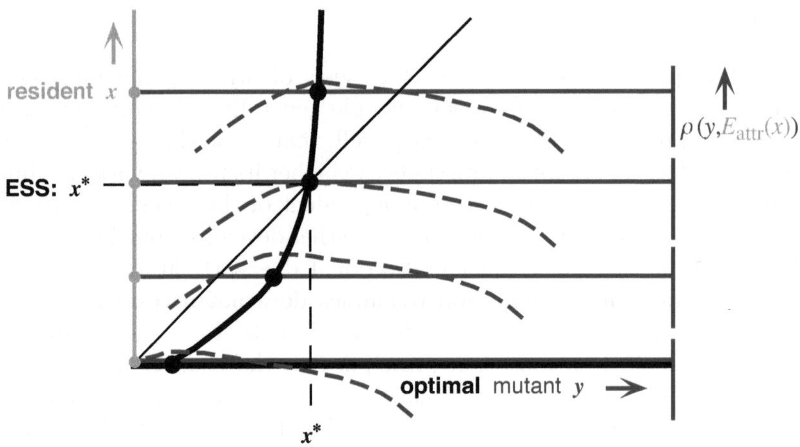

FIGURE 2.3. Scheme for calculating ESSes: For each of the possible resident populations, characterised by a scalar trait, the invasion fitness of all potential mutants is calculated (interrupted curves). The mutant axis is drawn on the same scale as the resident axis. From these fitness curves the optimal strategy for the corresponding resident environment is calculated (fat curve). The ESS is the optimal reply to itself, to be calculated by intersecting the fat curve with the 45° line.

One way of calculating ESSes is depicted in Figure 2.3. For each environment as generated by a possible resident the maximum of the invasion fitness landscape $\rho(Y|E_{\text{attr}}(X))$ is calculated. Next one intersects the resulting manifold $Y = Y_{\text{opt}}(X)$ with the linear manifold $Y = X$ to get the ESS $X^* = Y^*$. As necessarily any monomorphic resident has fitness zero, all potential mutants $Y \neq Y^*$ have negative fitness.

The situation for ESCs is a bit more complicated, as there may be so-called genetic constraints. So it may happen, for example, that a trait is under control of a single locus only and at the ESC the heterozygote has a higher fitness than the two homozygotes. The good message is that in the so-called Ideal Free (IF) case, as in the clonal case, all phenotypes comprising an ESC have fitness zero, at least when there is only a single birth state and the ESC engenders a community dynamical equilibrium[3]; this IF case is defined by the requirement that there are no genetic constraints whatsoever, that is, mutants can occur that produce any feasible type as heterozygotes in the genetic backgrounds supplied by the resident population. The bad message is that at the present state of knowledge about the genotype to phenotype map there is no way of predicting when genetic constraints may throw a spanner in the works and neither is there an inkling of this becoming feasible in the future.

2.4. Adaptive dynamics I: on traits, PIPs, MIPs and TEPs

Paleontologists and taxonomists are interested in the change of traits on an evolutionary time scale. What are traits to taxonomists are parameters to ecologists. So in AD one is after a dynamics in the parameter space of a community dynamics. The first trick for arriving at such a simple picture is to assume a time scale separation, such that favourable mutants come along singly after a community has relaxed to an attractor. The second trick is to assume clonal reproduction, on the assumption that this way one can find out where the ecology would drive evolution if the latter were not hampered by the constraints of Mendelian genetics[4].

To get at a purely trait oriented picture, first any reference to the environment should be removed from the expression for invasion fitness:

$$s(Y|C) := \rho\left(Y|E_{\text{attr}}(C)\right).$$

(Often this is written as $s_C(Y)$ to emphasize the interpretation as a family of fitness landscapes.)

In this subsection, I concentrate on scalar traits. I start with the case where there is only a single clonally reproducing resident, $C = x$. The first step in the analysis is plotting a contour plot of $s(y|x)$. Usually this is simplified to plotting only the zero contours, as those are the ones that matter by far the most. The result is customarily called Pairwise Invasibility Plot (PIP). See Figure 2.4.

---

[3]The proof goes by contradiction, see, e.g., [29]. It would be most useful if the result could be extended to situations with multiple birth states or non-equilibrium attractors.

[4]The set of ideas described in Subsections 2.4 and 2.7 was first worked out in some detail in [30] with [31] as counterpart geared to a biological rather than a mathematical audience. [32] as well as [30] independently argued that adaptive dynamics should be seen as a limit processes of sequences of stochastic models in which one lets the system size go to infinity and the probability of a mutation per birth event go to zero in such a manner that their product stays bounded, while appropriately rescaling time; a rigorous convergence proof, though thus far only for a particular special case, can be found in [33].

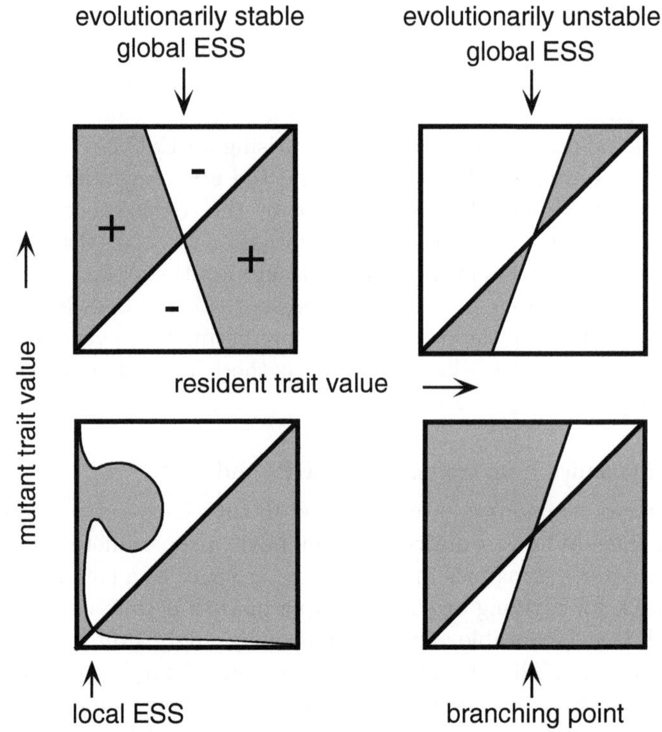

FIGURE 2.4. Pairwise Invasibility Plots: sign, as indicated in the upper left panel, of the fitness of potential mutants as a function of the mutant and the resident traits. The four panels show some alternative possible configurations, indicative of correspondingly different evolutionary phenomena. The abbreviation ESS stands for Evolutionarily Steady Strategy. The upper right panel explains my use of Steady instead of the still more common Stable as interpretation for the middle letter in ESS.

Note that the diagonal is always a zero contour as residents have fitness zero. The points where some other contour crosses the diagonal are referred to as evolutionarily singular strategies (ess-es). The ESSes are a subset of the ess-es.

Now assume that mutational steps are but small and that in the beginning there is only one resident trait value $x(0)$. Plot this value on the abscissa of the PIP, say the one in the top left panel in Figure 2.4. After some random waiting time mutation creates a new trait value y. This trait value can invade only when it has positive fitness, i.e., is in one of the plus areas of the PIP. It can be proved that an invading type replaces its progenitor if the latter is not too close to an ess or a bifurcation point of the community dynamics, and the mutational step

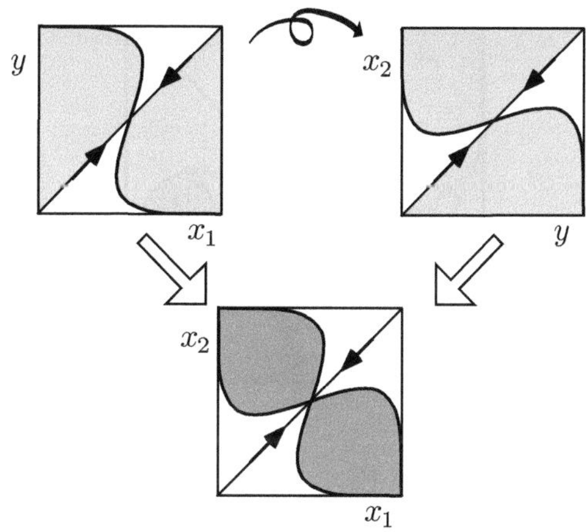

FIGURE 2.5. The construction of a Mutual Invasibility Plot, depicting the set in (trait space)^2 harbouring protected dimorphisms. Not all polymorphisms occurring in AD are protected, but unprotected polymorphisms have the habit of never lying close to a diagonal (Stefan Geritz, pers. com.).

was not too large ([34, 35] and [36, Appendix B][5]). If such a replacement has occurred we call the new trait value $x(1)$. In the PIP under consideration, if $x(0)$ lies to the left of the ESS then $x(1)$ lies to the right, and vice versa. By repeating this process it can be seen that in this case the evolutionary path converges to a close neighbourhood of the ESS. When the path has reached that neighbourhood it may become possible that the mutant and its progenitor persist together on a population dynamical time scale, that is, until a next mutant comes along that ousts one or both of the former residents.

To see how such coexisting pairs of strategies fare it is necessary to consider the set of so-called protected dimorphisms, i.e., pairs of strategies that can mutually invade, to be denoted as (x_1, x_2). The construction of this set is depicted in the Figure 2.5.

The evolutionary movement of the pair (x_1, x_2) is governed by $s(y|x_1, x_2)$. Under the assumption of small mutational steps a good deal of information can be extracted from the adaptive isoclines, calculated by setting the selection gradients

$$g_i(x_1, x_2) := \frac{\partial s}{\partial y}(y|x_1, x_2)\bigg|_{y=x_i}$$

equal to zero. As depicted in Figure 2.6, x_1 will move to the right when g_1 is positive and to the left when it is negative, and x_2 will move up when g_2 is positive and

[5]These proofs only consider relatively simple community dynamical scenarios. Extensions to more general structured populations would be very welcome!

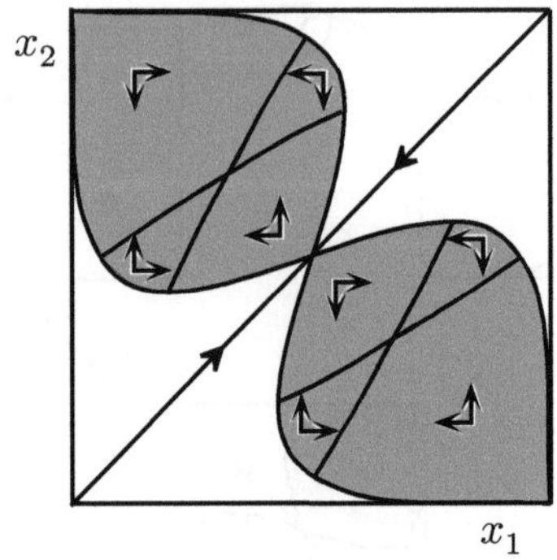

FIGURE 2.6. Trait Evolution Plot, i.e., MIP together with arrows that indicate the direction of the small evolutionary steps that result from the invasion by mutants that differ but little from their progenitor, and adaptive isoclines.

down when it is negative. The likes of Figure 2.6 are customarily referred to as Trait Evolution Plot (TEP).

Subsection 2.7 gives a classification of the possible dynamics near an ess. From that classification it can be seen that the ESS in the left upper PIP in Figure 2.5 also attracts in the dimorphic regime.

2.5. Adaptive dynamics II: the canonical equation

For vectorial traits the geometric constructions exhibited for the scalar case go through in a general sense, but not necessarily with the same consequences. In particular the neat dependence of the dynamical outcomes on no more than the sign of the invasion fitness hinges on the ordering properties of the real line.

The main workhorse in the vectorial case is the so-called Canonical Equation (CE) of AD, a differential equation that captures how the trait vector changes over evolutionary time on the assumption that mutational steps are sufficiently small. See Figure 2.7[6]. The CE moreover adds a quantitative slant to the analysis, by

[6] The form of the CE given in Figure 2.7 is for the case where the mutation distribution has mean zero and is symmetric around that mean. The CE was derived for ODE population models in [32] at a physicist level of rigour. A mathematically rigorous proof of the implied convergence followed seven years later [37]. These papers also give expressions applicable for more general mutation distributions. The derivation from the basic ingredients shows that the convergence to the CE is not uniform, becoming ever slower near ess-es. The extension to general structured population models was derived by [38], again at a physicist level of rigour; a mathematically rigorous proof for the purely age-dependent case can be found in [39]. The extension to the Mendelian case is made in [40] and [41]. The essential element in the latter extension is that for smooth genotype to phenotype maps, in the absence of any parental effects on gene expression, these maps are locally additive, so that the heterozygote between two different but similar homozygotes has a phenotype that is the average of the parental ones (Andrea Pugliese, pers. com. and [42]). The resident is invaded by heterozygotes, while after take-over there are only homozygotes left. This adds an additional factor 2. In [38] the term N_e in the version of the CE given in Figure 2.6

taking account not only of whether a mutant can invade, but also of the probability that it does so. Its equilibrium points are the ess-es mentioned previously.

There are still few results about the quantitative match of the CE to the "real thing". The main potential problem is that in reality many mutant substitutions will occur in parallel. Luckily, for small mutational steps this tends to affect the environment only in the higher-order terms that in the derivation of the CE disappear from sight [43]. A second problem is that in the clonal case the effects of the invading mutants do not add up since a good mutant may be supplanted by an even better one coming from the same parent type. Hence the CE may be supposed to do a better job in the Mendelian case where the substitutions occur in parallel on different loci, which to the required order of approximation should interact additively. Remains that any standing genetic variation enlarges the variance of the offspring number of an allele, which should roughly act to proportionally diminish the effective population size. Unfortunately, this variance is a random variable which appears to show little temporal stability (Jörgen Ripa, pers. com.).

A final point is that the more mechanistic detail that has to be brought in for the derivation of the CE for the Mendelian case brings to the fore that the CE is but the first term in a moment expansion, on top of which comes, except under very special assumptions, a similar equation for the change of the mutational covariance matrix, which in turn depends on third mutational moments, etc.

2.6. Links between adaptive dynamics and Evo-Devo

From an AD perspective the link with Evo-Devo is first of all through the mutational covariance matrices. At this point in time Evo-Devo unfortunately has yet little to offer in this area, although there are some promising developments (e.g., [44, 45]). Therefore, at present often the most AD researchers can do is work out how the outcomes of a specific eco-evolutionary model depend on the possible forms of the mutational covariance matrix. The answers from AD thus become Evo-Devo questions: is the mutational covariance matrix for these traits expected to fall within this or that class?

Just to show the importance of the missing Evo-Devo input in AD: mutational covariance matrices have an, often dominating, influence on the time scales of evolution (Figure 2.8, left), and the basins of attraction of ess-es (Figure 2.8, right), even to the extent that they often determine whether an ess attracts or not (cf. [46, 47, 48]).

On a more philosophical level it bears noting that the selection gradient points only in a single direction, while the components of the trait vector orthogonal to that gradient hitchhike with the selectively determined motion thanks to a developmental coupling as expressed in the mutational covariance matrix. The higher the dimension of the trait space the larger the contribution of development as a determinant of the direction of evolutionary motion. The dimensions of the

---

is still written as a product of the population size and some life history parameters. It is only recently that Vincent Jansen and I discovered that this product is actually equal to N_e whatever the ecological scenario.

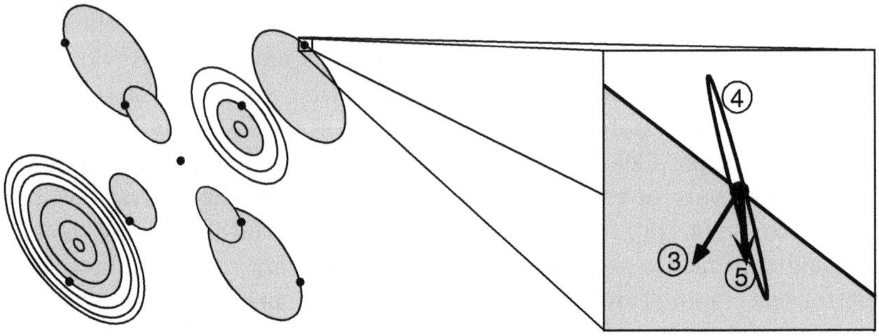

$$\frac{dX}{dt} = 2 \underbrace{N_e}_{\textcircled{1}} \underbrace{\overbrace{\varepsilon C}^{\textcircled{4}} \underbrace{\left[\left. \frac{\partial s(Y|X)}{\partial Y} \right|_{Y=X} \right]^{\mathrm{T}}}_{\textcircled{3}}}_{\textcircled{2}\ \text{on } \varepsilon C;\ \textcircled{5}\ \text{over } C[\ldots]^{\mathrm{T}}}$$

FIGURE 2.7. Upper left: Fitness landscapes for a selected number of residents (black dots; the dot in the center is an ESS (see Subsection 2.3). For two of the landscapes the contour lines are shown, for the other ones only the part is indicated where the fitness landscape is above sea level. Upper right: Enlargement of the coastal area around the resident for one of the fitness landscapes. Arrow ③ is the selection gradient. Ellipse ④ symbolises the probability distribution of mutational steps. Since the mutational steps in different directions are not equally probable an evolutionary movement results according to arrow ⑤. Below: The canonical equation of adaptive dynamics. The speed of evolutionary movement of the trait vector X equals twice the product of ① the effective population size (as defined in population genetics), ② a term summarising the nature of the mutational process, consisting of the mutation probability per birth event and the mutational covariance matrix as locally effective summary of the distribution of the mutational steps, and ③ the selection gradient.

trait spaces that are routinely considered thus makes for the contrast in attitudes of, for example, behavioural ecologists and morphologists, with the former stressing selection and the latter the developmental options for change. Formulated a bit facetiously, advocates of the absolute supremacy of selection basically think one-dimensionally, while harking on the evolutionary primacy of genes or development is like buying a car for its steering qualities without caring for its motor.

A final point is that I everywhere assume that the trait space for the largest part has the geometry of a manifold. However, in reality that geometry may be

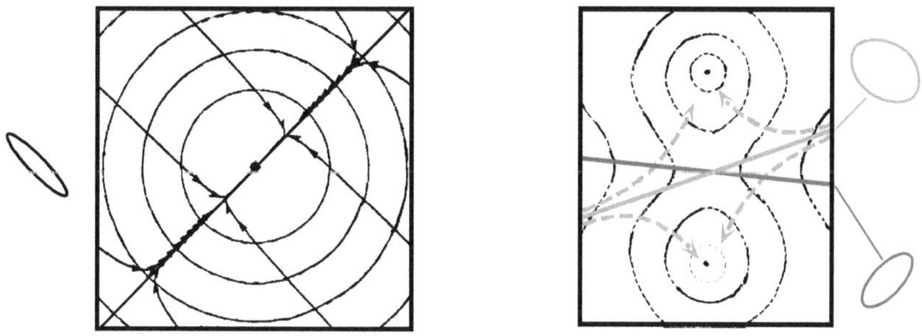

FIGURE 2.8. Two fitness landscapes that are supposed to keep their shape and only to sink when the adaptive trajectory moves uphill (as is the case if and only if the population regulation is through an additional state-independent death rate). Distributions of mutational steps are symbolised by ovals. Left: The shape of the mutation distribution induces a time scale separation between the movement along the diagonal and anti-diagonal direction. Right: The difference in mutation distributions causes a difference in the domains of attraction of the two ESS-es.

much more complicated, as it basically should reflect everything that can be generated by the developmental system. In particular, the trait space may consist of components with different dimension, as when a so-called key innovation adds a new trait that evolution then can seize upon to achieve some rapid progress and often also diversification. At present not much can be said in general about such issues. So I will in this chapter concentrate on conclusions based on the assumption that the trait space at least locally looks Euclidian.

2.7. Adaptive dynamics III: evolutionarily singular strategies

Evolutionarily singular strategies x^* can be calculated by setting the fitness gradient equal to zero. Figure 2.9 shows their classification according to dynamical type for the case of scalar traits[7].

[7] The classification is constructed by inserting the ecological consistency conditions $s(x|x) = 0$, $s(x_i|x_1, x_2) = 0$, $i = 1, 2$, $s(y|x_1, x_2) = s(y|x_2, x_1)$ and $s(y|x^*, x^*) = s(y|x^*)$ in the first and second-order directional Taylor polynomials of $s(y|x)$ and $s(y|x_2, x_1)$ around the singular point, and analysing what sort of dynamics would result from such simplified invasion fitness functions (see [30, 31]). (In the mutant direction the necessary smoothness conditions can be made part of the modelling assumptions. However, in the resident direction the environment is a derived quantity that in the simplest case is calculated from the community dynamical equilibrium equations. At a point (x^*, x^*) the conditions for the inverse function theorem do not hold good and the map $(x_1, x_2) \mapsto E_{\mathrm{attr}}$ is not a diffeomorphism. However, directional derivatives, and hence directional Taylor polynomials can be shown to exist [38].)

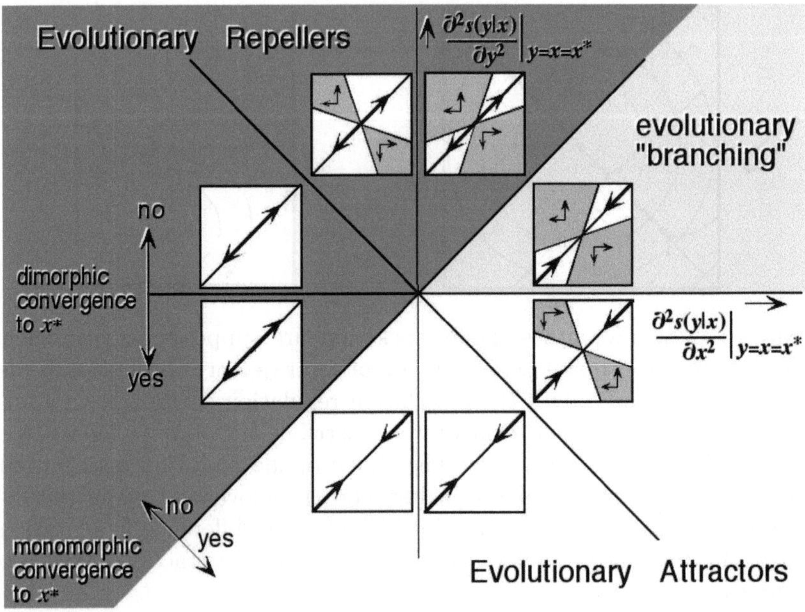

FIGURE 2.9. A classification of the ess-es for scalar traits. The cases in the lower half are all ESSes. The leftmost of these repels, the others attract. The latter ESSes are thus genuine evolutionary attractors. The branching points in the upper rightmost sector attract monomorphically but repel dimorphically.

Devising a good classification for higher-dimensional ess-es is an open problem. One of the reasons is that in higher dimensions the attractivity or non-attractivity of a singular point in general depends on the mutational covariance matrix, except in very special cases [46, 47, 48].

2.8. Adaptive speciation

The most interesting ess-es are branching points, where the eco-evolutionary process starts generating diversity. When approaching such points the evolutionary trajectory, although continually moving uphill, still gets itself into a fitness minimum. More precisely, it is overtaken by a fitness minimum. See Figure 2.10. The ecological cause can (by definition) be ascribed to so-called apparent competition, in the form of aversive direct interactions (such as fighting), competition for resources, having a common predator with a tendency to specialise on the most common types, etc. The following analogy may help intuiting the phenomenon. Somewhere gold has been found. As a result people converge to that special spot. However, after too many diggers have arrived, it becomes attractive to try one's luck at some distance.

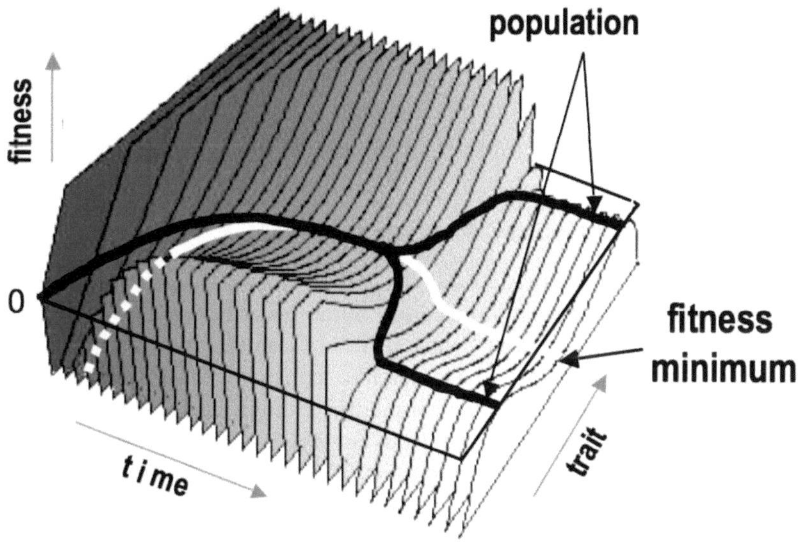

FIGURE 2.10. The development of the fitness landscape during a branching event.

The build up of diversity can take very different forms. In the clonal case the population just splits into two as depicted in Figure 2.1. In the Mendelian case the diversification starts with a broadening of the variation in the population. The fitness landscape locally has the shape of a parabola that increases away from x^*. This means that types more on the side have a higher fitness than those in the centre. It therefore pays not to get kids near the centre. The Mendelian mixer has the contrary tendency to produce intermediate kids out of dissimilar parents. Luckily, there are all sorts of mechanisms that may thwart this counterproductive mixing. The most interesting of these is the build up of some mechanism that lets the like extremes mate only among themselves, thus ensuring that the branches become separate genetic units. A very simple mechanism occurs in insects that diversify in their choice of host plants, with mating taking place on those hosts. More complicated mechanisms are mathematically explored in, e.g., [49, 50, 51, 52, 53].

My own conviction is that in cases where no automatic mating barrier puts itself in place a build up of other mechanisms engendering assortative mating is far from unexpected. Present day organisms are not simple particles, but are the product of some three and a half billion years of evolution. During that time their sensory and signalling apparatus has been evolutionarily honed for finding the most advantageous mates. As an example, spider mites, organisms one can barely see with the naked eye, were found (in an artificial arena) to mate assortatively on the basis of the food they had eaten previously [54]. Hence, I expect that there always will be an abundance of template mechanisms. Moreover, these mechanisms, once recruited to the task, will have a tendency to enhance each other in their effect. Therefore, I expect that the available generalised machinery often can rather easily be adapted so as to genetically separate the branches whenever evolution brings the population to a branching point. I should add, though, that most scientists working on the genetics of speciation do not appear to share this view.

3. Some meso-evolutionary predictions

To make meso-evolutionary predictions one has to look at recurring, close to model-independent, features of fitness landscapes. Some results derived by looking at the overall geometric features of well-behaved fitness landscapes are discussed below. I start with cases where there is no need to consider longer-term external environmental drivers, so that all environmental change is due to the community dynamics. (Any environmental fluctuations on a population dynamical time scale, like fluctuations in the day to day and year to year weather, are accounted for in the fitness function.)

3.1. The shortest time scale

1. When the tape of life is recorded in similar locations, say in the follow up of the colonisation of similar lakes after an ice age by the same fish species coming from the sea, the evolutionary trajectories will be similar, at least initially. One common pattern is that they move to the same ESS. The other common pattern is that they go to the same branching point. In both cases the trajectories in different localities move at similar speeds, and slow down in a similar manner when the ess is approached. If the latter is a branching point the trajectory lingers there for a while before the branches start to grow apart in a new phase of directional selection[8]. The very first divergence may be in many directions, but after that for a while two branches remain that initially diverge symmetrically, with similar speeds in different locations[9]. (I expect the lingering near the branching point to be more variable in duration since the exact speciation mechanism may depend on specific local opportunities, and therefore will be less repeatable than evolutionary features governed by the gross ecology.)

[8]In the clonal case in the small mutational step limit, the speed of diverging from a branching point is third order in the mutational step size, whereas the speed of directional movement is second order in the step size. However, this property is non-robust relative to relaxing the assumption of mutation limitation. For the clonal case this makes the above arguments about the consequences of the fitness landscape shapes less than robust. However, as with the quality of the approximation by the canonical equation, the adaptive dynamics results may without strict mutation limitation actually be saved by Mendelian inheritance as the need for the development of a mechanism for thwarting the Mendelian mixer may be expected to restore the time scale separation between directional selection and branching.

[9]The maximum number of branches that can be present initially equals one plus the dimension of the trait space. However, there are strong mathematical indications that soon only two branches remain that diverge in the direction of the dominant eigenvector of the second partial derivative of the fitness function in the invader direction (Stefan Geritz, pers. com.).

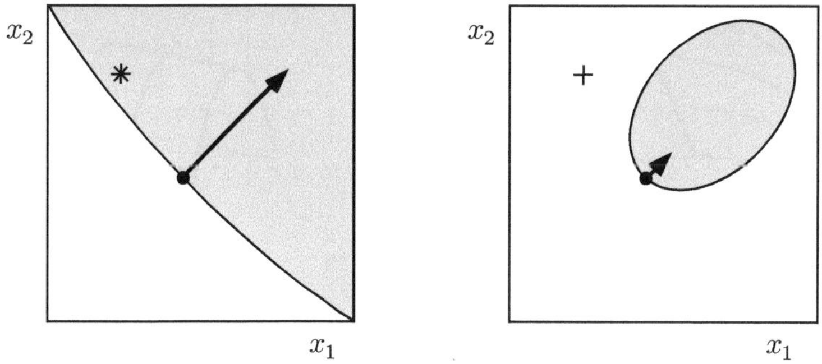

FIGURE 3.1. Left: Local fitness landscape shortly after the colonisation of a new territory, when a lot of slightly deleterious variation may get incorporated. In the example fitness increases with both x_1 and x_2, Yet, the mutant that is depicted will invade, even though the mutation decreases x_1, since under the circumstances this decrease is more than compensated by the increase in x_2. Right: Fitness landscape at a later stage of the adaptive process when such non-adaptive variation gets weeded out again.

2. Initially in the phases of directional selection a lot of non-adaptive variability will be incorporated, most of which will be weeded out at later stages. The reason is that initially the fitness landscape looks like a single steep hill, with moreover the resident far removed from the hilltop, so that the local fitness contours have low curvature, as in the left panel of Figure 3.1. As a result all sorts of mutants that are located at some distance from the resident in the almost neutral direction orthogonal to the fitness gradient can invade. Some of the trait changes caused by such mutations may actually be deleterious, but with that deleteriousness compensated by a pleiotropic advantageous change in some other traits. At a later stage, when the residents are closer to a top of the fitness landscape and the fitness contours have become correspondingly more curved, as in the right panel in Figure 3.1, the evolutionary path will become more and more constrained, and in the end all traits will end up at their most advantageous values for the environment realised at that time.

Real adaptive processes usually take place in a high-dimensional space with different speeds prevailing in different directions. Therefore, the weeding out will in general already start long before any final adaptive stops are reached. Figure 3.2 shows low-dimensional slices of fitness landscapes indicative of the relative balance of directional and stabilising selection in early and later stages of the adaptive process.

3.2. Intermediate time scales

3. When an empty habitat is colonised, initially speciation will occur frequently but the overall speciation rate may be expected to decreases rather quickly. The reason can be seen in Figure 2.1. Initially the fitness landscape tends to be a

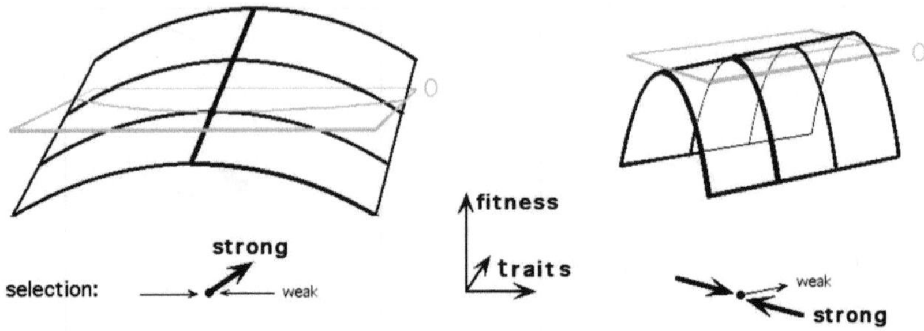

FIGURE 3.2. Left: Representative low-dimensional slice of the local fitness landscape shortly after the colonisation of a new territory, when a lot of slightly deleterious variation may get incorporated. Right: Representative low-dimensional slice of the local fitness landscape at a later stage of the adaptive process where such variation gets removed again.

single big hill, with submerged outer slopes and the resident population somewhere along the shoreline. With a increasing number of diversification events the landscape becomes anchored to zero at more and more points. Hence, at any branching point the steepness of the surrounding upward slopes (as expressed by the second derivative) becomes less and less[10]. With a decrease of the fitness slopes the speed at which branching points are reached and left becomes slower. In the Mendelian case the lower local curvature decreases the selection on mechanisms promoting assortative mating, which should slow down the speed of diversification even further (and actually may be expected to dominate that speed except in cases with full mutation limitation).

4. Speciation should be rare in environments that fluctuate on time scales between that of the faster scale of directional evolution and the slower scale of speciation. The reason is depicted in Figure 3.3. Speciation can only start from some very special points in trait space. (In the past this fact was even used as an argument for the improbability of adaptive speciation [55, 56]. Although it may seem ecologically unusual for a population to sit at precisely such a special point, one of the lessons from AD is that evolution may actually guide a population towards precisely such a point; see [57].) Slow fluctuations of the physical environment change the picture, in that the branching points will not stay in place but will move in response to the changes in the parameters of the community dynamics.

[10] At this moment I have but a heuristic argument for this statement. A rigorous proof should be based on the assumptions that (1) there is a uniform bound on the trait vectors beyond which the fitness landscape disappears below sea level, and (2) the derivatives of the fitness landscape, in particular the third one, are uniformly bounded. As the properties of the fitness landscape reflect only individual level mechanisms these are fair assumptions. (This in contrast to any assumptions about the dependence of the fitness function $s(Y|X)$ on the residents X as the latter dependency is through the community attractor, the properties of which need not depend everywhere smoothly on the resident traits.)

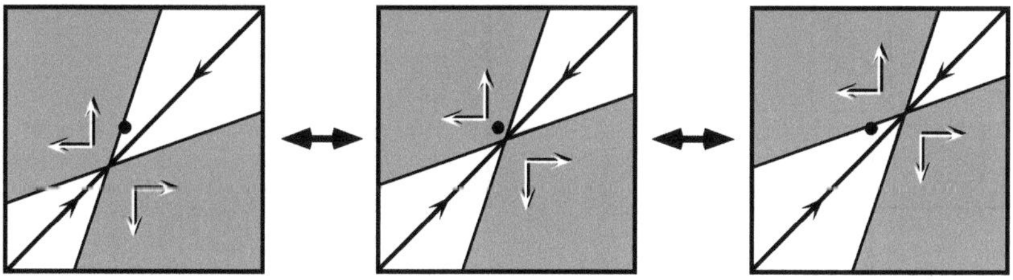

FIGURE 3.3. How a TEP may change as a result of slow fluctuations in the physical environment. As a result, if a species starts to branch (middle panel) the two incipient species may often in a short while find themselves outside the coexistence region (left and right panel).

This means that two incipient species will after some while find themselves no longer in the coexistence cone emanating from the branching point, as that cone has moved to another position, and one or the other of them will go extinct, aborting the diversification process. In simulations all this shows up as an abundance of diversification attempts which all turn into dead ends, resulting in an adaptive path with the shape of a closely pruned tree.

3.3. The longest time scales

5. In the nineteen-seventies paleontologists made a point about the common occurrence in the fossil record of so-called punctuated equilibria: short periods of rapid morphological change amidst longish periods of near morphological stasis [58, 59]. This phenomenon was ascribed by the inventors of the term to the occurrence of "morphological revolutions". With the paleontologists, I believe in the reality of punctuated equilibria, but I beg to differ with their explanation. I think that the time scales are too short and the morphological disparity that is involved too small for macro-evolutionary mechanisms really to play a role. Moreover, AD predicts the same phenomenon for mathematically well-established ecological reasons.

Given the customary speeds of directional evolution one may expect most slow changes in the fossil record to be due to the evolutionary tracking of slowly changing adaptive equilibria. These slow changes may, however, be punctuated by rare far shorter periods of fast directional evolution. The change in the ESSes can only be ascribed to slow overall environmental changes, e.g., changes in the climate or in the averaged action of the surrounding biosphere with its millions of simultaneously evolving species. For extended periods these changes will affect the ESSes only in a quantitative manner. However, mathematics tells that such relatively stable regimes will obtain only for so long as the parameter path does not cross any bifurcation points. Figure 3.4 depicts the two common bifurcations of ESSes. The corresponding punctuation events are of two types, "just so", due to the collision of an ESS and an evolutionary repellor, and coupled to speciation, when an ESS changes into a branching point, in good accordance with the findings of the paleontologists.

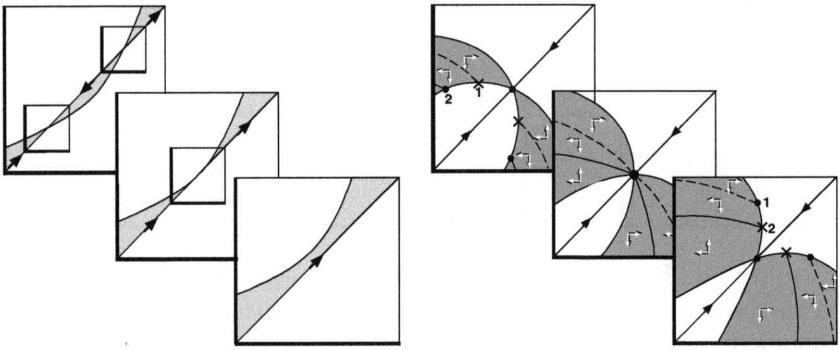

FIGURE 3.4. Robust AD bifurcations that will show up in the fossil record as "punctuated equilibria". Left: Three PIPs corresponding to a saddle node bifurcation in which an ESS is annihilated by an evolutionary repellor. After a longish period in which the adaptive trajectory tracks the slowly shifting ESS, this ESS abruptly disappears and the trajectory punctuates, i.e., embarks on a much more rapid path to some other attractor. Right: Three TEPs corresponding to a bifurcation of an ESS into a branching point. In this case the adaptive tracking of the ESS stops due to a change in character of the ess, resulting in a punctuation event that starts with speciation.

4. Macro-evolution

4.1. Evolutionary tinkering, tangled maps, and the need for intermediate abstractions

On a longer, macro-evolutionary, time scale, one has to take into consideration that the overall properties realised during evolution can in principle be realised by very different mechanisms. In general the first mechanism that does a sufficient job inherits the earth. Therefore, analysing which mechanisms should be most easy to realise has considerable predictive power. Moreover, evolution does not necessarily solve problems in the best possible manner. Evolution tinkers; it only optimises under very special circumstances, and then only very locally.

In the short term it may not matter which mechanism realises a certain desirable trait, but in the longer term different mechanisms lead to different mutational covariance matrices and hence to different evolutionary routes. In the language of dynamical systems: the real state space of the evolutionary process is not phenotype space, of which the trait spaces of AD are convenient abstractions, but genotype space. The mutational covariances reflect both the topology of genotype space, as generated by mutational distances, and the genotype to phenotype map generated by the developmental mechanics. This reflection may be not too inadequate locally in genotype space, and therefore locally in evolutionary time. However, for larger scale considerations different approaches are needed, both to delineate the domain of applicability of the simpler framework, and to step beyond its confines.

At the present state of our understanding the knowledge available about the detailed nitty gritty at the molecular level does not seem helpful yet for the questions I have in mind, however interesting it may be in other respects. The reason is the evolved complexity of the developmental process and the resulting tangledness of the genotype to phenotype map.

An indirect proof that the genotype to phenotype map is inexorably tangled is that assuming such tangledness appears the only way to resolve the discrepancy between the domination of adaptive processes as perceived by ecologists, functional morphologists, and the like, and the relatively satisfactory description that random models give of evolution at the DNA level. This resolution is moreover nicely compatible with the two main failings of the random model: (1) the different speeds of evolutionary change of different pieces of the genome which seem closely related to their functionality just one or a few translation steps away from the genome, but not further, (2) the far higher than Poisson variance in the number of substitutions, which most probably reflect the iteration of selective sweeps (see, e.g., [60]).

The previous arguments indicate the need for intermediate abstractions. In the remainder of this chapter I shall discuss some ideas to that effect. Most of these ideas were developed together with Frietson Galis, who is a functional morphologist by training, but over the last years has switched interest to Evo-Devo, where she concentrates precisely on the level that is of most interest in the present context. All the considerations below have to do with so-called internal selection processes (a definition follows below), and once again with the geometry of fitness landscapes. The third player in the game will be the consequences of high dimensionality both of phenotype, but in particular of genotype space. Some of these ideas go back to Sir Ronald Fisher, others are borrowed from Arno Wouters, Günter Wagner, Walter Fontana and Sergey Gavrilets.

4.2. Phenotypes (and genotypes)

Below I will generally leave unspecified whether the fitness landscapes under consideration are over a genotype or a phenotype space. The basic idea is that for most phenotype spaces of interest the genotype to phenotype map comes about as the concatenation of (a great number of) maps between other spaces (cf. Figure 2.2), coordinates of which can also be chosen as phenotypic coordinates. Below I will argue that the fitness landscape over phenotype space more often than not is very 'ridgy'. In general, most of the ridgyness over a phenotype space may be expected to automatically cascade back to any underlying spaces; only more ridgyness can be added further down, although in principle some steepness may be alleviated (think of how the chain rule acts). Ultimately this ridgyness then also cascades back to genotype space.

A related point is that I repeatedly fell back on some implicit smoothness assumptions, and that I will keep doing so. The reason why I think that this is justified is that I believe that the evolutionary changes that I am considering are mostly not so much changes in the coding regions of classical genes as well as in their regulation. Protein coding regions are in general preceded by a large number of relatively short regions where all sorts of regulatory material can dock. Changes in these docking regions, and changes in genes producing regulatory proteins lead to changes in the production rate of the gene product. Genes are more or less active in different parts of the body, at different times during development and under different micro-environmental conditions. So the lowest layer in the cascade of phenotype spaces is a space of gene expression levels as a function of these variables. This is the level from which we may start to think about the smallness of mutational steps. The influence of any specific regulatory site tends to be rather minor, and most changes in regulatory proteins will have only a minor effect on their affinity for the docking region. When I referred to the smoothness of a genotype to phenotype map in Subsection 2.5 Footnote 6, I referred to the map from this high-dimensional vector of gene expression levels to the phenotypic coordinates under consideration.

From the preceding paragraph it has probably already become clear that also the phenotypes of this story may not always be what one may naïvely expect. In keeping with the general ecological view that formed the basis for the ecological definition of fitness, phenotypes should in principle be interpreted as so-called reaction norms (another term is conditional strategies), i.e., maps from micro-environmental conditions to phenotypes in the naïve sense, i.e., characteristics of individuals. These reaction norms supposedly come as families with the phenotypic traits as identifying parameters. Only in the simplest cases these reaction norms are degenerate, taking only a single value, which we then may use as the phenotypic trait, so that the general and naïve sense phenotypes coincide.

4.3. Internal selection

Functional and constructional morphologists usually talk in terms of whether certain mechanisms work properly or not, and discuss evolution as a sequence of mechanisms all of which should work properly, and which only slightly change in every single transformational step. Translated into the language of fitness landscapes, this means that only the properly working mechanisms give fitnesses in the ecologically relevant range, while the improperly working ones give very low fitnesses for all relevant environmental conditions (cf. [61]). This leads to a picture of narrow, slightly sloping, ridges in a very high-dimensional fitness landscape. The slopes on top of the ridges are the domain of ecology, their overall location is largely ecology independent (cf. Figure 4.1).

As a simple example you may think of human leg length. Few ecologists will ever consider the length of the right and left leg as separate traits. The reason is that these lengths are kept equal by a very strong selection pressure, which keeps in place a developmental system that produces legs of precisely equal length, notwithstanding the fact that during development there is no direct coupling between the processes operating in the two leg primordia. Hence in a trait space

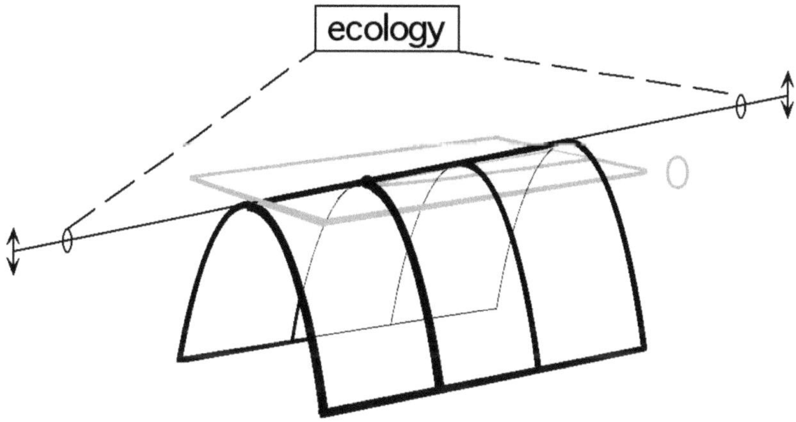

FIGURE 4.1. The interplay between internal and ecological selection. Internal selection sets the large-scale pattern of a deep ocean from which ridges that rise to around sea level. Ecology modulates the elevation on the upper slopes of the ridges.

spanned by the lengths of the right and left leg ecologists concentrate on just the diagonal.

Please notice that whereas ridges in the familiar three-dimensional world are necessarily one-dimensional, the trait spaces dealt with in morphology are very high-dimensional so that the top of a ridge may be higher-dimensional, while away from the ridge the fitness decreases very steeply in a far larger number of orthogonal directions.

A picture similar to that of the functional morphologists emerges from the consideration of developmental processes. The long term conservation of developmental units, think of the phylotypic stage or of homology, can only be due to strong stabilising selection, caused by the fact that mutations causing large pattern changes generally have many side effects with dire consequences for fitness (see, e.g., [62, 63, 64, 65]). As a result, ecological selection generally acts only on quantitative changes in the shapes and sizes of homologous body parts. As the fitness differences that are the guardians of homology usually manifest themselves already very early in the lives of individuals, in mammals in the womb, people tend to speak here of internal selection. In this chapter I will keep this term, but will use it not in a mechanistic but in a structural sense, as a reference to features of the fitness landscape that, within the confines of a particular argument, can be considered as relatively unchanging, that is, that are roughly the same for all the environments that explicitly or implicitly figure in that argument.

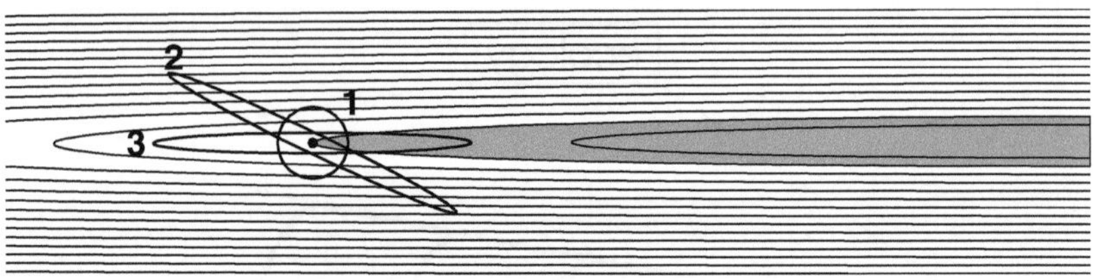

FIGURE 4.2. A steep, slightly sloping, fitness ridge together with three contours of potential unbiased mutation distributions. In distribution 1 all mutational directions are equally probable, distribution 3 is aligned with the ridge, while distribution 2 is similarly elongated, but misaligned. Clearly the corresponding evolutionary rates r_i are ordered as $r_3 > r_1 > r_2$. Effects like these become stronger with a larger number of off-ridge directions.

	Divergence time (min – max, est.)
Dipteran families	179 Ma – 330 Ma
Drosophila subgenera	60 Ma – 110 Ma
Mammalian orders	38 Ma – 70 Ma

TABLE 1. Minimum and maximum estimates of divergence times in Ma (From [66]).

4.4. A macro-evolutionary prediction

Any developmental system that leads to aligning the mutational steps with the direction of the ridge on which the resident is currently sitting, will evolve much faster than a system that is not aligned in such a manner (Figure 4.2). One way in which such a bias can come about is by having the development use cues that are related to the later function of the organ under consideration. In vertebrates, bones, muscles and nerve cells are modelled and/or grow in the embryo depending on their use. Unborn infants move for good reasons, and play is highly functional. As a consequence mammalian morphology evolves much faster than the morphology of, e.g., insects, in particular indirectly developing ones (see Table 1). Insects, on the other hand, do far better at the chemical end (think, e.g., of DDT resistance), due to their shorter generation times and far larger population sizes, and consequent higher availability of potentially useful mutations. In mammals teeth, which like butterfly adults have to develop mainly under ballistic developmental control, are so slow to evolve that they are used to define the higher levels in the Linnean classification.

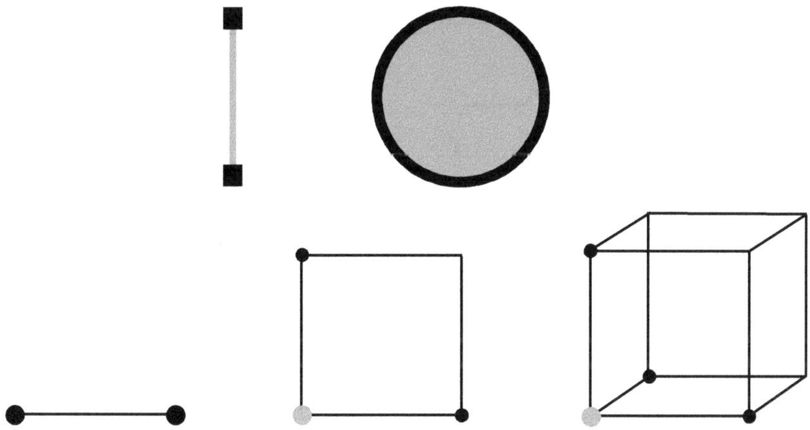

FIGURE 4.3. Top: Balls in 1 and 2 dimensions, with the points that are within a distance ε from the boundary coloured black. Clearly, the fraction of points that is within a ε-distance from the boundary is larger in 2 dimensions. With an increase in the number of dimensions this fraction increases further to go to 1 in the limit, independent of ε. Bottom: a similar construction on the Boolean cube in 1, 2 and 3 dimensions. Here as well, the fraction of points close to the boundary goes to 1 when the dimension goes up.

4.5. An Evo-Devo myth

The stabilising selection that underlies the long-term conservation of developmental units should have led evolutionarily to a considerable robustness of the developmental process. The selection is indirect, as a side effect of selection for robustness against environmental perturbations. It is, after all, disadvantageous after an unusually cold day to end up with a baby with two heads. The resulting robustness of the developmental system against perturbations will necessarily carry robustness against mutational perturbations in its wake. The inevitably resulting abundance of suspender and belt combinations and the overall tinkering nature of evolution have no doubt made their contribution to the tangledness of the genotype to phenotype map mentioned earlier. In the language of fitness landscapes, this robustness translates into the existence of extensive near neutral sets in genotype space: many mutational steps will have little effect due to the buffering by the developmental control system. The presence of extensive near neutral sets will also make for a mazelike character of the high fitness ridges in this space.

Contrary to naive expectation, however, the robustness of parts of the developmental process cannot be assumed by itself to conserve developmental units, or constrain their evolution. During the long periods of effective evolutionary stasis alluded to in Subsection 3.3, the population necessarily oozes as a diffuse cloud through the corresponding neutral set. In considering these sets one has to account

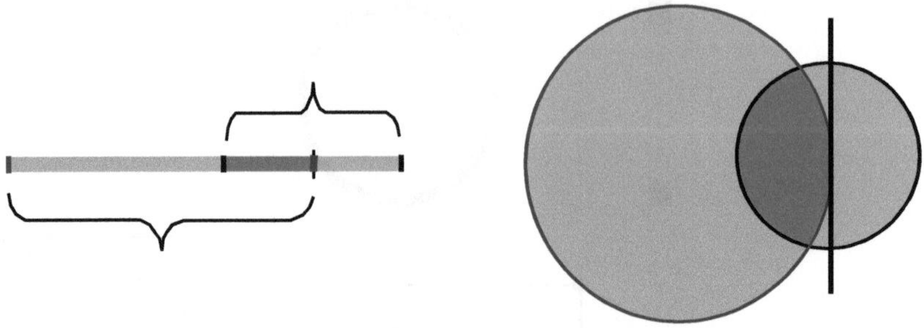

FIGURE 5.1. Left: Two balls in $\mathbb{R}^1$, with the centre of the smaller ball on the boundary of the larger ball. The ratio of the volume of their intersection to the volume of the smaller ball is $\frac{1}{2}$. Right: A similar configuration in $\mathbb{R}^2$. The volume of the intersection is now a smaller fraction of the volume of the smaller ball. For similar configurations in $\mathbb{R}^n$ this fraction quickly decreases to zero for larger n. Now think of the larger ball as the part above sea level of a fitness hill and of the smaller ball as a mutation distribution. Clearly the fraction of favourable mutants will go to zero with n.

for the effects of high dimensionality. Genotype space is very high-dimensional. For sets in high-dimensional spaces in general most points are very close to the boundary (see Figure 4.3). The diffusive oozing together with this geometric effect should make any longer term thwarting of phenotypic change next to impossible.

The upshot is that any long-term conservation can only be due to strong stabilising selection, or in other words, a fitness abyss that keeps the traits confined to narrow high fitness ridges. Hence homology. Evolution thus largely proceeds through the change in properties of homologous elements, the identity of which is conserved by stabilising internal selection.

5. Back to meso-evolution

5.1. Justifying adaptive dynamics

Fisher's old argument that the higher the dimensionality of the trait space the more difficult becomes the final convergence to an adaptive top (see Figure 5.1; see also [9, 67, 68, 69, 70]) seamlessly extends to the movement in a ridgy fitness landscape: the higher the number of orthogonal off-ridge directions, the more rare it is for a mutational step to end up above sea level. By a similar argument (see Figure 5.2) small mutational steps have a far higher propensity to end up above sea level than have large ones. Together these two arguments seem to underpin the requirements of AD that mutations in the ecologically relevant directions are scarce and the induced mutational steps generally small.

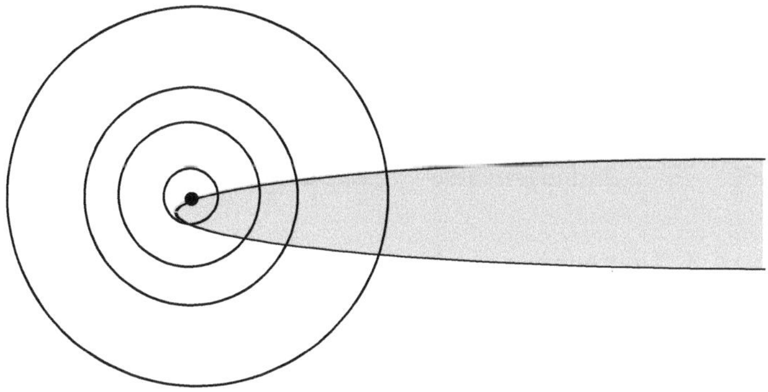

FIGURE 5.2. The shorelines of a fitness ridge and the contour lines of a mutation distribution. On land smaller mutational steps will be overrepresented relative to the larger ones.

Notwithstanding the appeal of the above arguments, they contain a biological flaw, the assumed rotational symmetry of the distribution of mutational steps. Real mutation distributions may be expected to show strong correlations between traits. Correlation structures can be represented in terms of principal components. I have not been able to find any good empirical data. However, the general experience with biological data is that almost always patterns are found like the ones shown in Figure 5.3. Figure 5.4 makes clear that the existence of mutational correlations will in general enhance rather than diminish the rareness and relative smallness of the mutational steps that end up above sea level, unless there is a very strong mechanistic link between the direction of the fitness ridges and the first principal axes of the mutation distribution, even stronger than the ones considered in Subsection 4.3[11].

The above conclusions seem to underpin nicely the assumptions of AD. Unfortunately, there are empirical observations that appear to contradict these conclusions. Populations brought into the lab always seem to harbour sufficient standing genetic variation to allow quick responses to selection, and so-called Quantitative Trait Loci are often found to underlie the variation in a trait. There are a number of reasons why I believe that these empirical observations may have less bearing on the issue than one might think. First, given the speed of evolution relative to the changes in the overall conditions of life, populations in the wild are probably most often hanging around some ESS. Moreover, real environments are inextricably noisy. Eco-evolutionary models with environmental noise almost invariably produce ESSes with very flat fitness maxima. This means that after a while a

[11]It would be of interest to try to derive the canonical equation of AD, or some similar equation, from such a fitness ridge perspective, instead of from the direct assumption that mutational steps are small. (Two limit procedures come to mind. In the first one the side slopes of the ridge are made steeper and steeper. In the second, probably more appropriate, one the dimension of the trait space and the number of orthogonal off-ridge directions are simultaneously increased.)

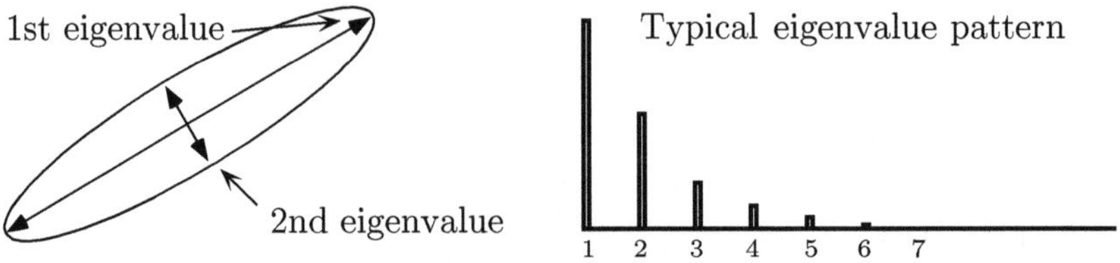

FIGURE 5.3. Contour line of a bivariate distribution, supposedly of mutational steps. The lengths of the two axes of the ellipse, called principal components, are proportional to the square root of the eigenvalues of the mutational covariance matrix. Right: Typical eigenvalue pattern found for large empirical covariance matrices.

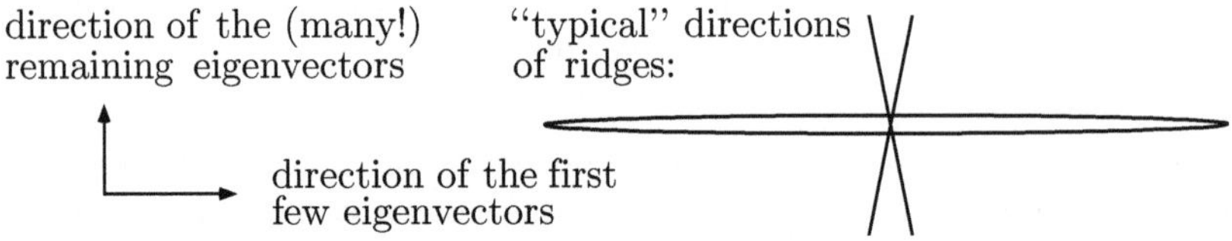

FIGURE 5.4. The mutation distribution will rarely be fully aligned with the fitness ridges. If one takes one's perspective from the mutation distribution and looks at the orientation of the ridges relative to the first few principal axes of this distribution, then, when the number of the dimensions of the trait space is very large and the ridge has a relatively low-dimensional top, the ridge will typically extend in a direction of relatively small mutational variation.

considerable amount of near neutral genetic variation will accumulate, which is exploited first when a population gets artificially selected on. The statics of AD corresponds to standard ESS theory. At ESSes the mutation limitation question is largely moot. Beyond the statics, AD's main interest is in the larger scale features of evolutionary trajectories after the colonisation of new territory or, even grander, a mass extinction. The scale of these features may be expected to require a further mutational supply of variation. Second, environmental fluctuations often lead to the accumulation of a lot of variability in traits directly involved in coping with those fluctuations. (Models incorporating environmental fluctuations often lead to adaptive branching in traits that modulate an individual's reaction to the environmental driver.) In the lab those fluctuations are removed, and selection is also otherwise relaxed but for the traits that are specifically selected for. This means that a lot more variation becomes available for the latter selection than would ever be available in the wild. The same holds for domesticated organisms. The selection of dogs for extreme sizes has had the side effect of producing lots of genetic diseases [71]. Third, AD style theory has shown that in the absence of assortative mating the initial increase of variability after the reaching of a branching point tends in the course of time to get redistributed over a smaller number of loci with increasing relative effect [72, 73]. The end effect will be QTLs, but these are produced through the cumulative effect of small genetic modifications.

5.2. Allopatric speciation

High-dimensional ridgyness also lies at the base of the usual ideas about so-called allopatric speciation, that is, the origination of reproductive incompatibility as a byproduct of long-term geographic separation. The underlying intuition is that separated populations independently wander around in the high fitness maze. After they have sufficiently diverged, any mixed offspring that occurs when they are secondarily confronted with each other ends up in the abyss. However, the models in this area have all been rigged to get the desired effect. Simple random genotype to phenotype maps followed by a more regular map to fitness almost never appear to do the job (Eke van Batenburg, Carolien de Kovel, pers. com.). Of course, real genotype to phenotype and hence to fitness maps are constrained by earlier evolution as well as by the technical requirements for producing well-functioning bodies. This leaves the interesting research question which combinations of ecologies and developmental maps are more and which are less prone to speciation, be it allopatric or sympatric.

Personally I believe that the ridgyness of the fitness landscape plays its role largely on a macro-evolutionary scale, while speciation typically is a meso-evolutionary phenomenon, and that therefore the developmental-biology-based intuition is unfounded. This does not mean that allopatric speciation is a non-phenomenon, as there are good alternatives to the developmentally-based fitness ridge scenario. One such scenario is based in wars of the sexes. For the sake of simplicity I will concentrate on the case of aquatic mass spawners, but similar scenarios can be dreamt up for species with more sophisticated mating systems. Ova and sperm find each other by means of chemical attractants. However, ova besieged by too many sperm are killed. Hence it pays an ovum to produce an attractant mix somewhat at the edge of the population distribution, to which not too many sperm are yet fully adapted. As a result, attractant mixes and receptor capabilities keep wandering in chemo-space. As the movement takes place in a very high-dimensional space, separated populations soon become incompatible. The prediction is that the genetic basis for reproductive incompatibilities usually will have to be sought in genes involved in such arms races, rather than in genes for building bodies.

Notwithstanding the somewhat *deus ex machina* character of its fundamental ingredient, for long the dogma has held sway that almost all speciation would be allopatric. Part of the strength of this dogma was due to its promulgation by Ernst Mayr, of modern synthesis fame. Mayr started as an ornithologist, working on bird faunas in the Pacific. I indeed agree, on a priori grounds, that those birds will in almost all cases have speciated allopatrically. The reason is that before on a more remote island a species has gone through the full process of splitting up ecologically as well as genetically, almost certainly an immigrant from another island will have arrived, fulfilling one of the missing ecological roles, and better than the novel local candidate. This new immigrant thus will undercut the ecological basis for the speciation process. On the other hand, birds have complicated sound and colour based mating signals, that may be subject to considerable drift, both as a result of the local circumstances influencing the effectiveness of different variants of the signals, and also as the result of processes with a dynamics similar to that of evolutionary arms races. So diversification in such patterns between different islands would not be unexpected. Moreover, bird taxonomists, in sinc with birds, are inclined to attach more importance to dissimilarities in song and coloration than to ecological similarities. However, the story for the snails or palms on those same islands may well be more in line with the adaptive speciation scenario.

An entirely different consideration is that paleontologically song and colour differences hold little interest as opposed to the morphological differences that go with ecological differentiation, in particular since the latter differences may form the fodder for further adaptive radiation. Hence, even when adaptive speciation scenarios do not hold true in the fine details, thinking in such scenarios may still help interpreting longer-term evolutionary patterns.

6. How to get on?

I have now come at the end of my ramblings. I have still one further message that I want to tout: the whole area of macro- and meso-evolutionary modelling is rife with open problems.

On the Evo-Devo side I have hit upon a few general geometrical arguments, in particular ones pertaining to the effect of high dimensionality. The geometrical features of those high-dimensional sets are sufficiently counterintuitive that I had to revise many of my preconceptions. I hope that my discourse has convinced you that such geometries should be the rule rather than the exception. Some open questions are: How can one best characterise high fitness mazes? Are there options for a reduced characterisation of the genotype to phenotype map within a maze? Is it possible to find abstract characterisations that help answering the question posed halfway the previous subsection?

The AD side is much further developed, which does not mean that there are no challenges left. These challenges are more about extending the mathematics than about defining the framework, although, of course, creatively doing the former necessarily asks for at least some smattering of the latter. Some open problems are extending the classification of singular points to higher-dimensional trait spaces and developing a full-fledged bifurcation theory for ESSes. Partial results abound, but we are far from having the full picture yet. More at the modelling end there are the question of developing a good background theory for dealing with constraints on the trait space, and the question what biological consistency conditions can be found that constrain the possible geometric structures occurring in AD on top of the ones mentioned in Footnote 7 (already many more are known than I have mentioned!).

I urge you, dear reader, to take up these challenges, and join in further developing this area.

References

[1] W. Ewens, *What changes has mathematics made to the Darwinian theory? In* F.A.C.C. Chalub and J.F. Rodrigues (eds.), *The Mathematics of Darwin's Legacy*, 7–26, Birkhäuser, Basel, 2011, *This issue.*

[2] R. Bürger, *Some mathematical models in evolutionary genetics. In* F.A.C.C. Chalub and J.F. Rodrigues (eds.), *The Mathematics of Darwin's Legacy*, 67–89, Birkhäuser, Basel, 2011, *This issue.*

[3] G.L. Jepsen, G.G. Simpson, and E. Mayr (eds.), *Genetics, Paleontology and Evolution.* Princeton University Press, USA, 1949.

[4] E. Mayr, *The Growth of Biological Thought: Diversity, Evolution, and Inheritance.* Belknap Press, USA, 1982.

[5] E. Mayr and W.B. Provine, *The Evolutionary Synthesis: Perspectives on the Unification of Biology.* Harvard University Press, USA, 1980.

[6] R. Amundson, *The Changing Role of the Embryo in Evolutionary Thought: Roots of Evo-Devo.* Cambridge. Cambridge University Press, UK, 2005.

[7] W.D. Hamilton, *Extraordinary sex ratios.* Science **156** (1967), 477.

[8] J. Maynard Smith and G.R. Price, *The logic of animal conflict.* Nature **246** (1973), 15–18.

[9] R.A. Fisher, *The Genetical Theory of Natural Selection.* Clarendon Press, Oxford, 1930.

[10] I. Eshel, M.W. Feldman, and A. Bergman, *Long-term evolution, short-term evolution, and population genetic theory.* J. Theor. Biol. **191** (1998), 391–396.

[11] I. Eshel, *Short-term and long-term evolution. In* U. Dieckmann and J.A.J. Metz (eds.), *Elements of Adaptive Dynamics*, Cambridge University Press, UK, *In press.*

[12] T.L. Vincent and J.S. Brown, *Evolutionary Game Theory, Natural Selection, and Darwinian Dynamics.* Cambridge University Press, UK, 2005.

[13] P.A. Abrams, *Modelling the adaptive dynamics of traits involved in inter- and intraspecific interactions: an assessment of three methods.* Ecol. Lett. **4** (2001), 166–175.

[14] R.A. Fisher, *On the dominance ratio.* Proc. Roy. Soc. Edin. **42** (1922), 321–341.

[15] J.A.J. Metz, *Fitness. In* S.E. Jörgensen and B.D. Fath (eds.), *Evolutionary Ecology*, volume 2 of *Encyclopedia of Ecology*, 1599–1612, Elsevier, UK, 2008.

[16] J.A.J. Metz and O. Diekmann (eds.), *The dynamics of physiologically structured populations*, volume 68 of *Lecture Notes in Biomathematics.* Springer-Verlag, Berlin, 1986.

[17] J.A.J. Metz and A.M. de Roos, *The role of physiologically structured population models within a general individual-based modeling perspective. In* D. DeAngelis and L. Gross (eds.), *Individual-based models and approaches in ecology: Concepts and Models*, 88–111, Chapman & Hall, USA, 1992.

[18] J.A.J. Metz, R.M. Nisbet, and S.A.H. Geritz, *How should we define "fitness" for general ecological scenarios?* Trends Ecol. Evol. **7** (1992), 198–202.

[19] R. Ferrière and M. Gatto, *Lyapunov exponents and the mathematics of invasion in oscillatory or chaotic populations.* Theor. Popul. Biol. **48** (1995), 126–171.

[20] P. Jagers, *Branching Processes with Biological Applications.* Wiley Series in Probability and Mathematical Statistics-Applied, London, UK, 1975.

[21] K.B. Athreya and S. Karlin, *Branching processes with random environments I – extinction probabilities.* Ann. Math. Stat. **42** (1971), 1499.

[22] K.B. Athreya and S. Karlin, *Branching processes with random environments II – limit theorems.* Ann. Math. Stat. **42** (1971), 1843.

[23] P. Haccou, P. Jagers, and V.A. Vatutin, *Branching Processes. Variation, Growth, and Extinction of Populations*, volume 5 of *Cambridge Studies in Adaptive Dynamics.* Cambridge University Press, UK, 2005.

[24] F.J.A. Jacobs and J.A.J. Metz, *On the concept of attractor for community-dynamical processes I: the case of unstructured populations.* J. Math. Biol. **47** (2003), 222–234.

[25] M. Gyllenberg, F.J.A. Jacobs, and J.A.J. Metz, *On the concept of attractor for community-dynamical processes II: the case of structured populations.* J. Math. Biol. **47** (2003), 235–248.

[26] O. Diekmann, M. Gyllenberg, and J.A.J. Metz, *Steady-state analysis of structured population models.* Theor. Popul. Biol. **63** (2003), 309–338.

[27] M. Gyllenberg, J.A.J. Metz, and R. Service, *When do optimisation arguments make evolutionary sense? In* F.A.C.C. Chalub and J.F. Rodrigues (eds.), *The Mathematics of Darwin's Legacy*, 233–268, Birkhäuser, Basel, 2011, *This issue.*

[28] I. Eshel, *Evolutionary and continuous stability.* J. Theor. Biol. **103** (1983), 99–111.

[29] J.A.J. Metz, S.D. Mylius, and O. Diekmann, *When does evolution optimize?* Evol. Ecol. Res. **10** (2008), 629–654.

[30] J.A.J. Metz, S.A.H. Geritz, G. Meszéna, F.J.A. Jacobs, and J.S. van Heerwaarden, *Adaptive dynamics, a geometrical study of the consequences of nearly faithful reproduction. In Stochastic and spatial structures of dynamical systems (Amsterdam,* 1995), Konink. Nederl. Akad. Wetensch. Verh. Afd. Natuurk. Eerste Reeks, 45, 183–231, North-Holland, Amsterdam, 1996.

[31] S.A.H. Geritz, É. Kisdi, G. Meszena, and J.A.J. Metz, *Evolutionarily singular strategies and the adaptive growth and branching of the evolutionary tree.* Evol. Ecol. **12** (1998), 35–57.

[32] U. Dieckmann and R. Law, *The dynamical theory of coevolution: A derivation from stochastic ecological processes.* J. Math. Biol. **34** (1996), 579–612.

[33] N. Champagnat, *A microscopic interpretation for adaptive dynamics trait substitution sequence models.* Stoch. Process. Their Appl. **116** (2006), 1127–1160.

[34] S.A.H. Geritz, M. Gyllenberg, F.J.A. Jacobs, and K. Parvinen, *Invasion dynamics and attractor inheritance.* J. Math. Biol. **44** (2002), 548–560.

[35] S.A.H. Geritz, *Resident-invader dynamics and the coexistence of similar strategies.* J. Math. Biol. **50** (2005), 67–82.

[36] F. Dercole and S. Rinaldi, *Analysis of Evolutionary Processes: the Adaptive Dynamics Approach and its Applications.* Princeton University Press, USA, 2008.

[37] N. Champagnat, *Convergence of adaptive dynamics n-morphic jump processes to the canonical equation and degenerate diffusion approximation.* Preprint of the University of Nanterre (Paris 10) No. 03/7. (2003).

[38] M. Durinx, J.A.J. Metz, and G. Meszena, *Adaptive dynamics for physiologically structured population models.* J. Math. Biol. **56** (2008), 673–742.

[39] S. Méléard and V.C. Tran, *Trait substitution sequence process and canonical equation for age-structured populations.* J. Math. Biol. **58** (2009), 881–921.

[40] J.A.J. Metz, *Invasion fitness, canonical equations, and global invasion criteria for Mendelian populations. In* U. Dieckmann and J.A.J. Metz (eds.), *Elements of Adaptive Dynamics*, Cambridge University Press, UK, *In press.*

[41] J.A.J. Metz and C.C. de Kovel, *The canonical equation of adaptive dynamics for diploid and haplo-diploid mendelian populations (in preperation).*

[42] T.J.M. Van Dooren, *Adaptive dynamics with Mendelian genetics. In* U. Dieckmann and J.A.J. Metz (eds.), *Elements of Adaptive Dynamics*, Cambridge University Press, UK, *In press.*

[43] G. Meszéna, M. Gyllenberg, F.J. Jacobs, and J.A.J. Metz, *Link between population dynamics and dynamics of Darwinian evolution.* Phys. Rev. Lett. **95** (2005), 078105.

[44] I. Salazar-Ciudad, *On the origins of morphological disparity and its diverse developmental bases.* Bioessays **28** (2006), 1112–1122.

[45] I. Salazar-Ciudad, *Developmental constraints vs. variational properties: How pattern formation can help to understand evolution and development.* J. Exp. Zool. Part B **306B** (2006), 107–125.

[46] O. Leimar, *Evolutionary change and Darwinian demons.* Selection **2** (2001), 65–72.

[47] O. Leimar, *The evolution of phenotypic polymorphism: Randomized strategies versus evolutionary branching.* Am. Nat. **165** (2005), 669–681.

[48] O. Leimar, *Multidimensional convergence stability.* Evol. Ecol. Res. **11** (2009), 191–208.

[49] U. Dieckmann and M. Doebeli, *On the origin of species by sympatric speciation.* Nature **400** (1999), 354–357.

[50] S.A.H. Geritz and É. Kisdi, *Adaptive dynamics in diploid, sexual populations and the evolution of reproductive isolation.* Proc. R. Soc. Lond. Ser. B-Biol. Sci. **267** (2000), 1671–1678.

[51] M. Doebeli and U. Dieckmann, *Speciation along environmental gradients.* Nature **421** (2003), 259–264.

[52] P.S. Pennings, M. Kopp, G. Meszena, U. Dieckmann, and J. Hermisson, *An analytically tractable model for competitive speciation.* Am. Nat. **171** (2008), E44–E71.

[53] J. Ripa, *When is sympatric speciation truly adaptive? An analysis of the joint evolution of resource utilization and assortative mating.* Evol. Ecol. **23** (2009), 31–52.

[54] M. Egas, M.W. Sabelis, F. Vala, and I. Lesna, *Adaptive speciation in agricultural pests. In* U. Dieckmann, M. Doebeli, J.A.J. Metz, and D. Tautz (eds.), *Adaptive Speciation*, 249–263, Cambridge University Press, UK, 2004.

[55] J. Maynard Smith, *Sympatric speciation.* Am. Nat. **100** (1966), 637.

[56] R.F. Hoekstra, R. Bijlsma, and A.J. Dolman, *Polymorphism from environmental heterogeneity – models are only robust if the heterozygote is close in fitness to the favored homozygote in each environment.* Genet. Res. **45** (1985), 299–314.

[57] É. Kisdi and S.A.H. Geritz, *Evolutionary branching and speciation: Insights from few-locus models. In* U. Dieckmann and J.A.J. Metz (eds.), *Elements of Adaptive Dynamics*, Cambridge University Press, UK, *In press.*

[58] N. Eldredge and S.J. Gould, *Punctuated equilibria: an alternative to phyletic gradualism. In* T. Schopf (ed.), *Models in Paleobiology*, 82–115, Freeman, Cooper and Company,, EUA, 1972.

[59] S.J. Gould and N. Eldredge, *Punctuated equilibria: the tempo and mode of evolution reconsidered.* Paleobiology **3** (1977), 115.151.

[60] J.H. Gillespie, *The Causes of Molecular Evolution.* Oxford University Press, UK, 1991.

[61] A.G. Wouters, *Explanation without a cause (verklaren zonder oorzaken te geven).* Questiones Infinitae, Publications of the Zeno Institute of Philosophy, XXIX, Utrecht (1999).

[62] F. Galis and J.A.J. Metz, *Testing the vulnerability of the phylotypic stage: On modularity and evolutionary conservation.* J. Exp. Zool. **291** (2001), 195–204.

[63] F. Galis, J.J.M. van Alphen, and J.A.J. Metz, *Why do we have five fingers? The evolutionary constraint on digit numbers.* Trends Ecol. Evol. **16** (2001), 637–646.

[64] F. Galis, T.J.M. van Dooren, and J.A.J. Metz, *Conservation of the segmented germband stage: modularity and robustness or pleiotropy and stabilizing selection?* TIG **18** (2002), 504–509.

[65] F. Galis, T.J.M. Van Dooren, J.D. Feuth, J.A.J. Metz, A. Witkam, S. Ruinard, M.J. Steigenga, and L.C.D. Wijnaendts, *Extreme selection in humans against homeotic transformations of cervical vertebrae.* Evolution **60** (2006), 2643–2654.

[66] F. Rodríguez-Trelles, R. Tarrio, and F.J. Ayala, *Molecular clocks: whence and whither? In* P.C.J. Donoghue and M.P. Smith (eds.), *Telling the Evolutionary Time: Molecular Clocks and the Fossil Record*, 5–26, CRC Press, Boca Raton, USA, 2004.

[67] S.H. Rice, *A geometric model for the evolution of development.* J. Theor. Biol. **143** (1990), 319–342.

[68] D. Waxman and J.J. Welch, *Fisher's microscope and Haldane's ellipse.* Am. Nat. **166** (2005), 447–457.

[69] D. Waxman, *Fisher's geometrical model of evolutionary adaptation – beyond spherical geometry.* J. Theor. Biol. **241** (2006), 887–895.

[70] D. Waxman, *Mean curvature versus normality: A comparison of two approximations of fisher's geometrical model.* Theor. Popul. Biol. **71** (2007), 30–36.

[71] F. Galis, I. Van Der Sluijs, T.J. Van Dooren, J.A. Metz, and M. Nussbaumer, *Do large dogs die young?* J. Exp. Zool. Part B **308B** (2007), 119–126.

[72] M. Kopp and J. Hermisson, *The evolution of genetic architecture under frequency-dependent disruptive selection.* Evolution **60** (2006), 1537–1550.

[73] G.S. Van Doorn and U. Dieckmann, *The long-term evolution of multilocus traits under frequency-dependent disruptive selection.* Evolution **60** (2006), 2226–2238.

J.A.J. (Hans) Metz
Institute of Biology and Mathematical Institute
Leiden, Netherlands

and

Evolution and Ecology Program
IIASA
Laxenburg, Austria
e-mail: j.a.j.metz@biology.leidenuniv.nl

When Do Optimisation Arguments Make Evolutionary Sense?

Mats Gyllenberg, J.A.J. (Hans) Metz and Robert Service

Abstract. The simplest behaviour one can hope for when studying a mathematical model of evolution by natural selection is when evolution always maximises the value of some function of the trait under consideration, thus providing an absolute measure of fitness for the model. We survey the role of such models, known as optimisation models in the literature, and give some general results concerning the question of when a model turns out to be an optimisation model. The results presented vary from more abstract results with a game-theoretical flavour to more detailed considerations of life history models. We also give a number of concrete examples and discuss the role of optimisation models in the wider framework of adaptive dynamics.

Mathematics Subject Classification (2000). Primary: 92D15; secondary: 91A30.
Keywords. Evolution by natural selection, order relation, rock-scissors-paper, adaptive dynamics, Evolutionarily Steady Strategy.

1. Introduction: philosophical preliminaries

The primary goal of science is understanding, i.e., creating orderly pictures of the world compatible with our observations. The test for, as well as the usefulness of such pictures is that they allow extrapolating from previous observations. This predictive capability is the basis for the hypothetico-deductive method (our main safeguard from wishful self-delusion) as well as for applications. Of course, regression analysis performs the same feats. The difference is that the predictive capabilities of good scientific theories extend far beyond the experimental situations from which they were inductively derived. Yet, contrary to the standard tenet, predictive capabilities are not the only proof of quality. For example, population genetics has made immense contributions to our understanding by showing that Mendel's particulate mechanism is compatible with the observations made on

This research benefited from the support of the Chair Modélisation Mathématique et biodiversité VEOLIA-Ecole Polytechnique-MNHN-F.X. The authors would like to thank Claus Rueffler for useful comments.

quantitative traits and that imperceptible selective pressures can have large effects over geologically very short time periods, thus overthrowing earlier naïve ideas to these effects. Yet, the reach of both pieces of theory in the cases where they can be tested in concrete instances effectively go but little beyond that of mere regression. What mattered is that the theory introduced a proper way of looking at the problems, and thereby made extrapolation possible. By now an impressive population genetical theory of short term adaptive evolution has been constructed, as surveyed in the Chapters by Warren Ewens [1] and Reinhard Bürger [2]. At the predictive end this theory does not reach much beyond the earlier feats. Still, these theoretical developments have greatly contributed to our understanding, largely since they allow analysing "what if" scenarios on a mechanistically founded and further essentially tautological basis, and thus enrich our naturally rather poor intuition.

The main reason for the limited predictive reach of population genetics theory is that genotypic fitnesses come in as unspecified phenomenological parameters, or, more recently, are specified through an assumed special genotype to phenotype map combined with a simple phenomenological ecological model. Only in the theory of neutral evolution there is a specification based on assumed strong first principles: all fitnesses are equal to one (or zero in the continuous time perspective to which we shall adhere below). The latter assumption leads to both that theory's mathematical strength and its limitations as a tool: although used in all sorts of inferences, the outcomes of those inferences hinge on this specific, often poorly tested, and moreover very non-generic, assumption.

To get further at the predictive side we need a more realistic handle on fitness. Mechanistically, fitness is determined by how the traits of phenotypes influence their population dynamical performance. Not only that, often those traits have our main interest in the first place. To link that primary focus on traits to population genetics we have to assess the map from genotype to phenotype and the ecological background for the population dynamical performance. In general, little can still be said about the former, notwithstanding the great molecular advances. Hence, most theoretical efforts concentrate on deriving predictions based only on the relation between traits and performance, sidestepping the need for population genetical modeling and the concomittant need for knowing the genetic basis of those traits. The main tool was initially, and to a large extent still is, optimisation models in which a presumed fitness proxy, like the strength of a bone or an energy intake rate is maximised as a function of a set of trait vectors delimited by constraints, justified by their users with a hand-waving reference to Fisher's fundamental theorem (cf. the Chapters by Ewens [1] and Bürger [2]).

Better thought out approaches are based on ESS theory (see the Chapter by Metz [3, Subsection 2.2]). ESS is an abbreviation of Evolutionarily Stable Strategy. However, as the definition of an ESS does not bring with it that an ESS is stable in the standard mathematical sense (see, e.g., [3, Figure 2.4]), we prefer to interpret the abbreviation as Evolutionarily Steady Strategy. ESSes are the traps of any evolutionary process driven by the invasion of new mutants. They are defined as such values of the trait vector, here called strategies, that no mutant playing an alternative strategy can invade in the environment produced by them. The important difference with the optimisation approach is that the ESS argument accounts for the fact that fitnesses not only depend on the traits of phenotypes but also on the environment in which those phenotypes live, and that this environment is not constant but codetermined by the phenotypes that are currently around (for if this were not the case populations would either grow to infinity or go extinct, except when the fitnesses are zero).

The ESS approach not only is better founded, it also has the advantage that we do not need an ever applicable fitness concept, which may exist in the simplified world of population genetical theory but does not do so so readily in the more messy ecologies of real life. Instead it suffices to have a fitness concept that characterises the potential for population growth of mutant phenotypes in an environment set by resident phenotypes, not yet influenced by the mutant.

In this Chapter we investigate how optimisation approaches fit with the ESS viewpoint. There are three reasons for embarking on such an effort. The first one is practical. We want to develop a feel for the reach of the optimisation results. In general, optimisation approaches appear to work rather well, notwithstanding their basically flawed methodology. This paradox will be resolved at the end of Subsection 2.3. The optimisation literature is not so much wrong as imprecise in that its reach is far less than suggested, and its results have been put only to correspondingly restricted tests. The next two reasons come into play when we want to do away with those strong implicit restrictions. The first one is again practical. Given the relative simplicity of optimisation procedures we want a handle on how to rig an eco-evolutionary model so that its ESSes can be calculated from an optimisation principle, as well as insight into the robustness of the results from such limited models. Finally, on the fundamental side there is the wish for insight on a meta-level. In precisely what manner do the various approaches in the literature fit together? In the textbooks one can find various hand-waving answers. We aim for precise ones.

2. Setting the stage: technical preliminaries

As the concepts and their technical implementation described below are not only needed as context for the discussion of evolutionary optimisation arguments but also are of considerable importance in their own right, we have strived for a self-contained exposition. We assume throughout that populations are large and relatively well mixed, i.e., any individual is directly or indirectly affected by a large number of other individuals each of which on average has but a small effect on the demographic behaviour of the focal individual. (For some further extensions, see [4, Section 'Aggregates'].) Moreover, for the ease of the exposition we assume till Subsection 2.4 that reproduction is clonal.

2.1. Fitness and fitness proxies

The effectiveness of ESS calculations is based on the close to universal existence of a scalar quantity, called fitness, characterising the speed at which a phenotype can invade in a given environment. (NB: This fitness concept is essentially different from the population genetical one; see the chapter by Metz [3, Subsection 2.2.].) Here we define environment as anything outside an individual that influences its population dynamical behaviour, which by definition consists of impinging on the environment, giving birth and dying (see, e.g., [4, 5, 6, 7]). We then can construct a Markovian representation of that behaviour in terms of a state space, transition probabilities that depend on the course of the environment and outputs that either deterministically depend on, or occur in a Poisson (or Poisson cluster) process with rates that depend on the individual's state and the condition of the environment. Given the course of the environment, individuals independently move through their state spaces, the population state is a measure over this space, and the expectation of this measure, which is again a measure, moves according to a positive linear evolutionary system. The theory of such systems then tells that generally the expected size of a population in an ergodic environment will in the long run on average grow or decline exponentially (for details, see [4, 8]). The per capita rate of this growth, to be denoted generally as ρ, or in accordance with standard custom in the special case of constant environments as r, is the sought after fitness.

Example. *Finite state individuals and clonal reproduction.* The following example lacks in biological realism but is the simplest one demonstrating the basic principles. A partial justification may be the hunch that any useful model has to be approximable (uniformly for all relevant environments) by a finite state model.

The expected growth of a population of independent finite state individuals in continuous time in a constant environment is given by

$$\frac{\mathrm{d}}{\mathrm{d}t}N = (\mathbf{R} + \mathbf{T})N,$$

N the vector of spatial densities of individuals in different states. (In our present context N refers to the mutants. To keep the notational burden low we suppress dependences on the trait vector of the mutants and resident environment.) The matrices $\mathbf{R}$ and $\mathbf{T}$ are built up from per capita rates. The off-diagonal components of $\mathbf{T}$ equal the transition rates between the corresponding states, the diagonal components equal minus the overall rates of transitions from the states minus the state dependent death rates. The components of $\mathbf{R}$ equal the average per capita birth rates in dependence on the state of the parent split according to the state of the offspring. The fitness of the type of individuals under consideration corresponds to the rightmost eigenvalue of $\mathbf{R} + \mathbf{T}$ (which is necessarily real and goes with a positive eigenvector since otherwise the trajectory would leave the positive cone).

The theory of branching processes moreover tells that when a population is started with a single individual it will, barring some technical conditions, eventually either go extinct or grow exponentially, with the probability of the latter being positive if and only if its fitness is so (see [9, 10, 11, 12]).

The quantity ρ necessarily is a function of two variables, the trait vector of the individuals Y and the environment E, to be written as $\rho(Y|E)$. (Depending on the context we may suppress one or both of these arguments.)

The existence of such a fitness is the basis for all deliberations below. However, given its existence it is often possible to replace ρ by some more easily determined quantity that leads to the same outcome for the calculations that have our interest. For example, in optimisation calculations ρ can be replaced with any quantity that is monotonically related to it, and in many other types of ESS calculation one may replace ρ with any sign-equivalent quantity. We shall refer to such quantities as fitness proxies. An example of an often used fitness proxy of the first type is the average rate of energy intake. It should be noted though, that being a fitness proxy is always predicated on additional assumptions. For instance, it may help a forager little to increase its average energy intake in an environment where doing so drastically increases its exposure to predation. An important fitness proxy of the second type, restricted to non-fluctuating environments, is the logarithm of the average life-time offspring number $\ln(R_0)$, where R_0 is defined as the dominant eigenvalue (or more generally spectral radius, or, still more generally, Perron root) of the next generation operator. (The advantage of using births as reference points in the life cycle is that usually the set of birth states is considerably smaller than the full complement of states necessary to describe how an individual passes through its life. However, if that happens to be more convenient, other points in the life cycle where the individual by necessity can be in but a few states may also be used as basis for the bookkeeping.) This next generation operator is constructed by calculating from a model for the behaviour of individuals how many offspring are born on average in different birth states dependent on the birth state of the parent (see, e.g., [13]).

Example. *Finite state individuals and clonal reproduction, continued.* Order the individual states so that the birth states come first. To step back and forth between a population state and a birth based approach we need a matrix $\mathbf{K}$ injecting the vector of birth rates into the space of changes in densities of all individuals, young and old alike:

$$\mathbf{K}^T = \begin{pmatrix} 1 & 0 & \cdots & 0 & 0 & \cdots & \cdots & 0 \\ 0 & \ddots & \ddots & \vdots & \vdots & & & \vdots \\ \vdots & \ddots & \ddots & 0 & \vdots & & & \vdots \\ 0 & \cdots & 0 & 1 & 0 & \cdots & \cdots & 0 \end{pmatrix}.$$

As can be seen from this formula, $\mathbf{K}^{\mathrm{T}}$ maps the space of population rates back onto the space of birth rates.

Arguing from first principles one can calculate the next generation matrix by first calculating the expected times that the Markov chain stays in each possible state before the individual dies as $\mathbf{T}^{-1}\mathbf{K}$, after which the next generation matrix can be expressed as $\mathbf{L} = \mathbf{K}^{\mathrm{T}}\mathbf{R}\mathbf{T}^{-1}\mathbf{K}$.

For the general case we observe that due to our assumption of environmental constancy the state of an individual can always be replaced by a proxy state consisting of age together with the state at birth, giving the production of offspring dependent on the state at birth as an operator-valued function $\mathbf{\Lambda}(a)$. Integrating out over age gives the next-generation operator $\mathbf{L}$.

Example. *Finite state individuals and clonal reproduction, continued.* In our finite state model the state of an individual moves according to a Markov chain with killing, and the probability that individuals born in certain birth states at age a are alive and reside in certain states is given by the matrix $\mathrm{e}^{a\mathbf{T}}\mathbf{K}$. Hence the average birth rate at age a split according to the birth state of the parent and that of the kids is $\mathbf{\Lambda}(a) = \mathbf{K}^{\mathrm{T}}\mathbf{R}\mathrm{e}^{a\mathbf{T}}\mathbf{K}$. This expression shows that given the mechanism as embodied in the matrices $\mathbf{R}$ and $\mathbf{T}$ it is possible to calculate the average birth rate of an individual from only its state at birth and age. Hence, age and birth state together are a proxy state for the goal of calculating the average birth rates of individuals.

The vector of population birth rates B satisfies a vectorial version of Lotka's integral equation from mathematical demography

$$B(t) = \int_0^{t-t_0} \mathbf{\Lambda}(a)B(t-a)\mathrm{d}a + \mathbf{K}^{\mathrm{T}}\mathbf{R}e^{(t-t_0)\mathbf{T}}N(t_0)$$

(if the initial datum is given in terms of the population composition at time t_0), which for $t_0 \to -\infty$ reduces to

$$B(t) = \int_0^{\infty} \mathbf{\Lambda}(a)B(t-a)\mathrm{d}a.$$

Substitution of an exponential trial solution $B(t) = \mathrm{e}^{rt}U$ gives that the invasion fitness r can be calculated from $\mathbf{\Lambda}$ by solving the characteristic equation:

$$\text{dominant eigenvalue of } \tilde{\mathbf{\Lambda}}(r) = 1, \quad r \in \mathbb{R}$$

(or equivalently $\det(\mathbf{I} - \tilde{\mathbf{\Lambda}}(r)) = 0$, $r \in \mathbb{R}$), with

$$\tilde{\mathbf{\Lambda}}(z) := \int_0^{\infty} \mathrm{e}^{-za}\mathbf{\Lambda}(a)\mathrm{d}a = \mathbf{K}^{\mathrm{T}}\mathbf{R}(z\mathbf{I}-\mathbf{T})^{-1}\mathbf{K}$$

for those z for which the integral converges, and that U equals the, positive, eigenvector with eigenvalue 1 of $\tilde{\mathbf{\Lambda}}(r)$. The general theory of renewal equations tells that for $t_0 \to -\infty$ indeed $B(t)$ will grow like $\mathrm{e}^{rt}U$. From the fact that also $B(t) = \mathbf{K}^{\mathrm{T}}\mathbf{R}N(t)$ it follows that the r found in this manner is equal to the dominant eigenvalue r of $\mathbf{R} + \mathbf{T}$.

To prove the sign equivalence of r and $\ln(R_0)$ note that

$$\mathbf{L} := \int_0^\infty \mathbf{\Lambda}(a)\mathrm{d}a = \tilde{\mathbf{\Lambda}}(0) \quad \text{and} \quad R_0 = \text{dominant eigenvalue of } \mathbf{L}.$$

Since all components of $\tilde{\mathbf{\Lambda}}(z)$ are positive and decrease with z, also its dominant eigenvalue decreases with z. Hence, r is positive when $\ln(R_0) > 0$ and is negative when $\ln(R_0) < 0$.

The following argument shows that R_0 rightfully can be interpreted as an average lifetime offspring number. The average lifetime numbers of offspring by individuals born in different states equals $\mathbf{1}^{\mathrm{T}}\mathbf{L}$, where $\mathbf{1}$ is a vector that has all its components equal to 1. The natural probability distribution to average these numbers over is the stationary distribution generated by the generation process itself, i.e., the right eigenvector U of $\mathbf{L}$ corresponding to R_0, normalised such that $\mathbf{1}^{\mathrm{T}}U = 1$. Doing so gives $\mathbf{1}^{\mathrm{T}}\mathbf{L}U = \mathbf{1}^{\mathrm{T}}R_0U = R_0\mathbf{1}^{\mathrm{T}}U = R_0$.

A further, partial, proxy for $\ln(R_0)$ in cases where the next generation operator is representable by a matrix $\mathbf{L}$ is $Q := -\det(\mathbf{I} - \mathbf{L})$. Q is sign equivalent to $\ln(R_0)$ where it counts most, that is, close to $\ln(R_0) = 0$. Moreover, for path connected trait spaces and an $\mathbf{L}$ that depends continuously on the traits, if $r(X_i) = 0$ for $i = 1, \ldots, k$ then $Q(X_i) = 0$ for the same X_i and if moreover $Q(Y) < 0$ for all Y different from those X_i then also $r(Y) < 0$ for those Y (see [14]).

2.2. Resident environments and invasion fitness

The only environments that matter in ESS calculations are environments generated by the attractors of some, so-called resident, community. In nature populations are necessarily bounded. If this bound were too small our population would go extinct in too short a time for it to reach an ESS. Hence we assume that the population is infinite in numbers although bounded in density, i.e., number of individuals per unit of area or volume. The community then follows a deterministic dynamics with as state space for each population a closed bounded subset of the cone of positive measures over the state space of the individuals, and as total state space the product of the state spaces of the comprising species, plus the state spaces of the dynamics of any inanimate resources. With an infinitesimal amount of noise the states of such communities will approach an "extinction preserving chain attractor" (see [15, 16]); with larger amounts of noise the community will in general end up in a stochastic attractor, that is, a stationary distribution of community states. We will throughout assume that the community attractor generates an ergodic environment (to all appearances exceptions to this assumption are rare).

Notation. The environment generated by a coalition of clones $C = \{X_1, \ldots, X_k\}$ will be written as

$$E_{\mathrm{attr}}(C)$$

with the convention that we write just X for $C = \{X\}$.

Convention 2.1. We take the use of the expression $E_{\mathrm{attr}}(C)$ as implying that for a community starting with all types $X \in C$ there exists an attractor with the densities of all those types nonzero.

For ease of exposition we moreover proceed on the

Assumption 2.2. $E_{\mathrm{attr}}(C)$ is unique.

The domain of the function E_{attr} is therefore the space of all realisable coalitions.

Assumption 2.2 is not necessary for any of the developments below. All statements can be extended to the general case with but minor modifications, which, however, would tally up to a considerable amount of verbal clutter.

A combination of the preceding arguments leads to the

Definition 2.3. The quantity

$$s_C(Y) := \rho\,(Y|E_{\mathrm{attr}}(C)) \tag{2.1}$$

is called the *invasion fitness* of a new type Y in a C-community.

An essential observation is

Proposition 2.4.

$$s_C(X) = 0 \quad \text{for all } X \in C. \tag{2.2}$$

Proof. The presence of X as a resident means that the density of X does not go to zero and neither can it go to infinity. Therefore its average per capita growth rate is zero. □

Hence for all $X \in C$, $s_C(X) = s_X(X) = 0$, and $s_C(Y) > s_C(X)$ whenever $s_C(Y) > 0$, and similarly with $>$ replaced by $<$ or $=$.

2.3. Calculating ESSes

The usual way of calculating ESSes is by devising some procedure to maximise $\rho(X|E)$ over all potential trait values for any feasible E, resulting in a function $X_{\mathrm{opt}}(E)$. As a next step one determines for each trait value the environment that it generates as a resident, $E_{\mathrm{attr}}(X)$. Finally one varies X to find an evolutionarily unbeatable value X^*, i.e., an X^* such that

$$X_{\mathrm{opt}}(E_{\mathrm{attr}}(X^*)) = X^*. \tag{2.3}$$

For more complicated trait spaces and ecologies solving such a combined optimisation problem and equation tends to be far from easy. Hence, determining a single optimisation principle that has to be satisfied by the ESS can be a great help.

After calculating an ESS one should preferrably ascertain that the set of trait values X_0 from which it is approximated with non-zero probability through a sequence $X_0, X_1, X_2, \ldots$ such that $s_{X_i}(X_{i+1}) > 0$, possibly interspersed with polymorphisms, is sufficiently large to warrant consideration of X^* as a potential evolutionary prediction. Although the last condition is not part and parcel of the ESS concept (it should have been!), only the attracting ESSes, customarily called

CSSes, are relevant as evolutionary predictions (cf. [17, 18, 19, 20]). (The acronym CSS is an abbreviation of the not overly informative phrase 'Continuously Stable Strategy'.) One further advantage of showing the existence of an optimisation principle is that this implies that the corresponding ESSes are globally attractive in the case of clonal and haploid organisms and may be expected to have a fair attainability in diploid Mendelian ones.

An alternative way of spotting ESSes, which also immediately gives insight in their evolutionary attractivity, is through the plotting of a so-called Pairwise Invasibility Plot (PIP), i.e., a plot of the sign of the invasion fitness of potential mutants with the potential resident trait values on the abscissa and the potential mutant trait values on the ordinate. See Figure 3.1 and the Chapter by Metz [3, Section 2, in particular Figure 2.4]. As we shall see in Subsection 3.2, such plots also provide us with an easy diagnostic for the presence or non-presence of an optimisation principle.

For higher-dimensional trait spaces it is not possible to work with PIPs, but the basic idea that underlies the determination of an ESS from a PIP still goes through. Any ESS satisfies

$$G(X^*) = 0 \quad \text{with} \quad G(X) := \left.\frac{\mathrm{d}s_X(Y)}{\mathrm{d}Y}\right|_{Y=X}, \tag{2.4}$$

together with the condition that

$$\mathbf{H}(X^*) := \left.\frac{\mathrm{d}^2 s_X(Y)}{\mathrm{d}Y^2}\right|_{Y=X=X^*} \quad \text{is negative definite.} \tag{2.5}$$

(In adaptive dynamics the vector G^{T} is known as the selection gradient and the matrix $\mathbf{H}$ as the selection Hessian.)

In addition to ESSes, which are by definition monomorphic, there may exist Evolutionarily Steady Coalitions (ESCs), i.e., combinations of phenotypes C such that $s_C(Y) < 0$ for all $Y \notin C$. Finding non-monomorphic ESCs through an extension of the optimisation route is tricky. However, the adaptive dynamics toolbox works almost unchanged. It lets us calculate candidate ESCs by intersecting the adaptive "isoclines" (see the Chapter by Metz [3, Subsection 2.4]) defined by setting the selection gradients equal to 0,

$$G_i(C^*) = 0 \quad \text{for } i = 1, \ldots, k, \quad \text{with} \quad G_i(C) = \left.\frac{\mathrm{d}s_C(Y)}{\mathrm{d}Y}\right|_{Y=X_i}, \tag{2.6}$$

followed by a check that a so-found singular point indeed corresponds to local fitness maxima

$$\mathbf{H}_i(C^*) := \left.\frac{\mathrm{d}^2 s_{C^*}(Y)}{\mathrm{d}Y^2}\right|_{Y=X_i^*} \quad \text{is negative definite,} \quad \text{for } i = 1, \ldots, k. \tag{2.7}$$

The local attractivity of any found ESS or ESC can be gauged by approximating the trait substitution process with the so-called canonical equation of adaptive dynamics (see [3, Subsection 2.5]).

Combining the previous considerations shows that a good recipe for numerically finding possibly attracting candidate ESSes is running the canonical equation for a reasonable sample of initial conditions and mutational covariance matrices. Necessary and sufficient conditions for a guaranteed local convergence, independent of the mutational covariance matrix can be found in [21, 22, 23].

Resolution of the paradox from Section 1: As a final issue we point to the fact that an ESS maximises the invasion fitness in the environment as set by that strategy. Hence, if we just measure the environment we may predict the evolutionarily steady trait values that go with that environment by maximising fitness in that environment. This is why the predictions from optimisation theory work so well. Optimisation theory may not predict the outcome of evolution, for that would entail also predicting the environment that goes with the ESS, but it often very satisfactorily predicts the strategies that may be present in that environment. However, such limited predictions are of little practical use when it comes to gauging the potential consequences of purposeful or inadvertent environmental manipulation like controlled fishing regimes or human induced global warming.

2.4. Genetics

Not many species that have our interest reproduce clonally. Luckily many results from the clonal theory go through almost unaltered under Mendelian inheritance. For the community dynamics one has to distinguish individuals according to their genotypes, and incorporate their mating opportunities with different genotypes into the description of the environment (cf. [24]; this in the case of casual matings, with more extended pair formation it becomes necessary to extend the state space of individuals to keep track of their marriage status). Alleles reproduce clonally and as such have invasion fitnesses. So in principle we can calculate ESSes based on the population dynamics of hypothetical mutant alleles affecting the phenotype.

To link with the usually encountered arguments that tend to be implicitly based on the assumption of clonal reproduction, we define for Mendelian diploids a mock fitnesss of phenotypes by introducing a parallel clonal model with individuals passing through their lives like their Mendelian counterparts and having a reproduction equal to the average of the contributions through the micro- and macro-gametic routes (for humans semen and ova) of those counterparts. The definition of R_0 can be similarly extended. As an example we give the recipe for the calculation of R_0 when, except possibly for a sex difference, there is but a single birth state.

Example. *Mendelian diploids with everybody born equal.* In the case of diploid hermaphrodites with but a single birth state, R_0 equals half the sum of the average numbers of offspring fathered or mothered. The factor $1/2$ comes from the wish to define R_0 such that the outcome from naive evolutionary calculations based on this "offspring number" for individuals matches the outcome from more detailed genetically based calculations.

When the sexes are separate, the sex difference comes on top of the physiological structure, spatial position, etc. In diploids, if everybody is born equal but for their sex, the corresponding next-generation operator is

$$\mathbf{L} = \tfrac{1}{2}\begin{pmatrix} \ell_{\mathrm{ff}} & \ell_{\mathrm{fm}} \\ \ell_{\mathrm{mf}} & \ell_{\mathrm{mm}} \end{pmatrix}.$$

with ℓ_{ff} the lifetime number of daughters of a female, ℓ_{fm} the lifetime number of daughters of a male, ℓ_{mf} the lifetime number of sons of a female, and ℓ_{mm} the lifetime number of sons of a male, all for the mutant, as they happen to occur in the environmental and genetic background provided by the resident population.

The simplest case is when the sex determination is independent of the trait in which the mutant differs from the resident as then we can write $\ell_{\mathrm{ff}} = p_{\mathrm{f}}f$, $\ell_{\mathrm{mf}} = p_{\mathrm{m}}f$, $\ell_{\mathrm{fm}} = p_{\mathrm{f}}m$, $\ell_{\mathrm{mm}} = p_{\mathrm{m}}m$, with m and f the numbers of offspring fathered and mothered over a lifetime, and p_{m} and p_{f} the probability of being born a male or a female. Therefore $\mathbf{L}$ has rank one and

$$R_0 = \tfrac{1}{2}\left(p_{\mathrm{f}}f + p_{\mathrm{m}}m\right).$$

This result could also have been obtained more directly by observing that everybody is born stochastically equal, having the same probabilities of being born male or female. We then get R_0 by just averaging over the possibilities.

As a curiosity we mention that when the trait in which the mutant differs has an influence on the sex determination we can still end up with the same formula by defining p_{m} and p_{f} to be the asymptotic probabilities of being born a male or a female, i.e., by choosing for p_{m} and p_{f} the components of the right eigenvector U of $\mathbf{L}$, and defining m and f again as the number of offspring fathered or mothered over a lifetime, i.e., $f = \ell_{\mathrm{ff}} + \ell_{\mathrm{mf}}$, $m = \ell_{\mathrm{fm}} + \ell_{\mathrm{mm}}$. Then, by using $R_0 = \mathbf{1}^{\mathrm{T}}\mathbf{L}U$, exactly the same formula for R_0 is obtained. Only the similarity of the expressions is pleasing: to calculate p_{m} and p_{f} we first have to calculate R_0.

With the above definitions various fitness-based deductions for the clonal case go through for Mendelian inheritance. In particular, for genetically homogeneous populations the fitness of a resident equals zero (since genetically homogeneous populations breed true and resident populations by definition do not in the long run grow or decline). Moreover, the invasion of a new mutant in a homogeneous population is correctly predicted, as that mutant initially only occurs in heterozygotes that breed true by backcrossing with the homogeneous resident.

The situation for ESCs is more complicated as there may be so-called genetic constraints. So it may happen, for example, that an invading mutant heterozygote also has a positive fitness in the environmental background provided by its own homozygote. Luckily, in the so-called Ideal Free (IF) case, as in the clonal case, all phenotypes comprising an ESC have fitness zero, at least when there is only a single birth state and the ESC engenders a community dynamical equilibrium. This IF case is defined by the requirement that there are no genetic constraints whatsoever, that is, mutants can occur that produce any feasible type as heterozygotes in the genetic backgrounds supplied by the resident population. Unfortunately at the present state of knowledge about genotype to phenotype maps there is no way of predicting for what traits persistent genetic constraints can indeed be ruled out.

Remark. Whether in general the conditions $s_C(X) = 0$ for all $X \in C$ and $s_C(Y) < 0$ for all $Y \notin C$ imply that C is uninvadable by any non-neutral mutant is still unresolved. For constant environments and individuals that can be born in only one state the proof is easy as there the invasion R_0 of a mutant allele can be written as a positively weighted sum of the R_0 of the phenotypes of all the different heterozygotes in which it may occur [24, 25, 26]. Hence, if the fitness of all $X \in C$ equals zero and no alternative phenotype has positive fitness, any non-neutral mutant has negative invasion fitness. It would be nice to have the issue resolved more generally.

Haploids basically follow the clonal rules. See [14] for haplo-diploids like hymenopterans in which the females are diploid and the males haploid (the supplementary material to [27] lists the many known haplo-diploid taxa). Polyploids as well as more complicated life cycles with both haploid and diploid phases as seen in mosses, ferns and various sorts of algae still remain to be studied.

3. Results

3.1. General considerations

Assumption 3.1. For any ESCs

$$\rho\left(X|E_{\mathrm{attr}}(C)\right) = 0 \quad \text{for all } X \in C, \tag{3.1}$$

that is, the ESCs that we consider are of clonal or haploid organisms, or in the case of Mendelian diploids they are ESSes, or more generally ESCs in which the fitness of all resident phenotypes is zero as is the case under the IF assumption at least in cases where there is only one state at birth as well as population dynamical equilibrium.

Notation. $\mathcal{X}$ will denote the set of potential trait vectors, $\mathcal{C}$ the set of all coalitions $C \in \mathrm{domain}(E_{\mathrm{attr}})$ satisfying equation (3.1), and $\mathcal{E} := E_{\mathrm{attr}}(\mathcal{C})$.

Convention 3.2. When we speak of all C this is meant to refer only to $C \in \mathcal{C}$, and when we speak of all E this is meant to refer only to $E \in \mathcal{E}$.

that is, we focus only on feasible E, i.e., E in the range of E_{attr}, and as far as these E are concerned we restrict the discussion to clonal or haploid organisms, or in the case of Mendelian diploids to monomorphisms while for polymorphisms we restrict ourselves to environments that go with ESCs satisfying Assumption 3.1.

Definition 3.3. We call a function $\phi : \mathcal{X} \to \mathbb{R}$ an optimisation principle when under any constraint the outcome of evolution can be determined by maximising ϕ.

We shall below abbreviate this as ϕ being maximised by evolution. The proviso "for any constraint" in Definition 3.3 mirrors the usual practice of combining an optimisation principle, derived from the population dynamics, with a discussion of the dependence of the evolutionary outcome on the possible constraints.

Remark. The above definition of an optimisation principle may seem unduly restrictive. Why not just ask for an optimisation principle to have its maxima coinciding with any ESSes? The point is that mathematically speaking the latter sort of optimisation principles always exist but are totally uninformative. Just calculate the ESSes for a model and take any function that has its maxima at those points.

Optimisation principles correspond more or less to the textbook intuition for the meaning of fitness, which generally fails to account for the fact that the fitnesses of all possible types are bound to change with any change in the character of the residents.

By letting the constraint set consist of just two possible trait values it follows that

Lemma 3.4. *If ϕ is an optimisation principle then*

$$\phi(Y) < \phi(X) \Leftrightarrow s_X(Y) < 0 \quad \text{and} \quad \phi(Y) = \phi(X) \Leftrightarrow s_X(Y) = 0.$$

As a consequence, in the clonal and haploid cases the existence of an optimisation principle ϕ allows one to rule out mutual exclusion ($s_X(Y) < 0$ and $s_Y(X) < 0$) and protected polymorphisms ($s_X(Y) > 0$ and $s_Y(X) > 0$) since both lead to the contradictory conclusion $\phi(Y) < \phi(X) < \phi(Y)$.

So far we only considered ϕ as a function on $\mathcal{X}$. For many models with an optimisation principle (to wit, all those models that we have encountered) it is possible to extend ϕ to the whole of $\mathcal{C}$. Such models are characterised by the fact that for $C = \{X_1, \ldots, X_n\}$ one has $\phi(X_1) = \cdots = \phi(X_n)$. Therefore we can set $\phi(C)$ equal to the common value of the $\phi(X_i)$. This extension is consistent in that one still has $\phi(Y) > \phi(C) \Leftrightarrow s_C(Y) > 0$ for any $Y \in \mathcal{X}$ and $C \in \mathcal{C}$. A consequence is that in the clonal and haploid cases invasion implies substitution: if $s_X(Y) > 0$ then we know that X and Y cannot form a stable coalition C, since this would lead to the contradiction that $\phi(X) = \phi(C) = \phi(Y)$ while $\phi(X) < \phi(Y)$. Thus the outcome of a successful invasion can only be that the resident dies out and is replaced by the invading type. Therefore each successful mutant X increases $\phi(X)$, and hence any ESS attracts.

In Mendelian diploids the argument given for convergence to an ESS does not work, as invasion needs not imply substitution. Population dynamically stable coalitions may arise when a heterozygote mutant invades that also enjoys positive fitness in the environment set by a population of the corresponding homozygotes. This is usually excluded for smooth genotype to phenotype maps when only small mutational steps are allowed as then the genotype to phenotype map is locally additive and invasion fitness is necessarily smooth in the invader trait, and will usually be smooth in the resident traits away from population dynamical bifurcation points. Moreover, in the case of potentially larger mutational steps, under the IF assumption a further mutant may appear that realises the phenotype of the heterozygote in any genetic background. In that case, at least when the environment is constant and the organisms have but a single birth state, such an evolutionary stalemate will after a while be broken up again. It may be expected that this results in an eventual convergence to the ESS. The strongest expression of this conviction can be found in [28]. However, whether this is indeed the case without exception is not fully clear yet. See further [19, 29, 30, 31, 32]; note though that all these authors restrict themselves to non-fluctuating environments and single birth states.

3.2. Highest abstraction level: traits only

Let us now turn to the most general context in which a discussion of optimisation is meaningful. Thus we forget about the maps $E_{\text{attr}} : \mathcal{X} \to \mathcal{E}$ and $\rho : \mathcal{X} \times \mathcal{E} \to \mathbb{R}$ and simply suppose that $(X, Y) \mapsto s_X(Y)$ is a given invasion fitness function $\mathcal{X} \times \mathcal{X} \to \mathbb{R}$ on which we impose, a priori, only the condition $s_X(X) = 0$ for all $X \in \mathcal{X}$. The question then arises what conditions are necessary and sufficient for the existence of a function $\phi : \mathcal{X} \to \mathbb{R}$ such that the conclusion

$$\phi(Y) < \phi(X) \Leftrightarrow s_X(Y) < 0 \quad \text{and} \quad \phi(Y) = \phi(X) \Leftrightarrow s_X(Y) = 0.$$

of Lemma 3.4 holds. For the purposes of this section we forget about any interpretation in terms of evolutionary outcomes and call any such function ϕ an optimisation principle for s. The question of characterising those s which admit an optimisation principle is addressed in [33] and [34] the results of which we now briefly discuss.

Clearly we are only interested in the sign of the invasion fitness function $s_X(Y)$. In the case of one-dimensional traits it is customary to represent the sign of the invasion fitness function by means of a pairwise invasibility plot (PIP) as in Figure 3.1. If x and y (we denote here, as in the rest of the paper, real numbers by lowercase letters and general vectors by uppercase letters) are plotted on the customary axes, then points (x, y) where $s_x(y) > 0$ are coloured grey, points where $s_x(y) < 0$ white and the neutral boundaries $s_x(y) = 0$ black. Since a PIP contains all information about the sign of the invasion fitness function s, any conditions we impose on s should only depend on the corresponding PIP.

The first observation, already mentioned implicitly in Subsection 3.1, is the following:

Lemma 3.5 (Sign-antisymmetry). *If there exists an optimisation principle for s then*

$$s_X(Y) > 0 \Leftrightarrow s_Y(X) < 0.$$

In terms of PIPs this is the property of skew-symmetry: by flipping the diagram with respect to the line $x = y$ the positive grey regions are mapped to negative white ones and vice versa, while the neutral black lines are mapped onto themselves.

A second observation that is also a straightforward consequence of the definitions is:

Lemma 3.6 (Sign-transitivity). *If there exists an optimisation principle for s then*

$$s_X(Y) \geq 0 \text{ and } s_Y(Z) \geq 0 \quad \Rightarrow \quad s_X(Z) \geq 0.$$

The content of Lemma 3.6 can be reformulated using the familiar game of rock-scissors-paper.

Definition 3.7. An ordered triple $(X, Y, Z) \in \mathcal{X}^3$ is called a *rock-scissors-paper-cycle* if

$$s_X(Y) > 0, \quad s_Y(Z) > 0, \quad \text{and} \quad s_Z(X) > 0. \tag{3.2}$$

The triple (X, Y, Z) is called a *weak rock-scissors-paper-cycle* if it satisfies (3.2) with two of the three $>$ signs replaced by $\geq$ signs.

Under the additional assumption of sign-antisymmetry, sign-transitivity for s is exactly the statement that there are no weak rock-scissors-paper-cycles. The advantage of this viewpoint is that it emphasizes that one can prove the nonexistence of an optimisation principle by exhibiting just a single (weak) rock-scissors-paper-cycle.

We note that the simple proofs of Lemmas 3.5 and 3.6 are based on the properties of antisymmetry and transitivity enjoyed by the order relation of the real numbers. Also note that, for the purposes of the discussion on the existence of optimisation principles, we are only interested in the sign of the invasion fitness function. These observations motivate us, following [34], to consider directly the binary relation $\preceq_s$ on $\mathcal{X}$ defined as follows:

Definition 3.8. Given an invasion fitness function $s : \mathcal{X} \to \mathcal{X}$ define the weak invadability relation $\preceq_s$ by $X \preceq_s Y \Leftrightarrow s_X(Y) \geq 0$.

We note that our assumption $s_X(X) = 0$ implies that this relation is always reflexive, i.e., $X \preceq_s X$. The content of Lemma 3.6 is then that if s admits an optimisation principle then $\preceq_s$ must be a transitive relation. A binary relation $\sim$ is called total, if for any X, Y in the relevant domains at least one of the alternatives $X \sim Y$ or $Y \sim X$ holds. The relation $\preceq_s$ defined above is total if and only if there is no case of mutual exclusion $s_X(Y) < 0$, $s_Y(X) < 0$. A total and transitive binary relation is called a total preorder.

On an abstract level we now see that an optimisation principle can be considered a representation of the relation $\preceq_s$ by means of an order preserving map $\phi : \mathcal{X} \to \mathbb{R}$ so that $X \preceq_s X \Leftrightarrow \phi(X) \leq \phi(Y)$.

One of the theorems that led to the birth of order theory was Georg Cantor's theorem that a countable dense (between any two points lies a third) linear order without a largest or smallest element is isomorphic to the rational numbers as an ordered set. By extending and adapting the proof of this fact, it can be shown that the necessary conditions we have listed, along with the rather nonrestrictive requirements that the trait space be a separable metric space and the invasion fitness is at least separately continuous, are also sufficient for the existence of an optimisation principle [34]. More specifically, one has:

Theorem 3.9. *Let $\mathcal{X}$ be a trait space which is a separable metric space. Let $s : \mathcal{X} \times \mathcal{X} \to \mathbb{R}$ be a given invasion fitness function which is continuous in each variable separately. Then, if there exists no optimisation principle for s, at least one of the following alternatives holds:*

(i) *There is a pair of traits X, Y satisfying mutual exclusion $s_X(Y) < 0$, $s_Y(X) \leq 0$, or*
(ii) *there is a weak rock-scissors-paper-cycle.*

Note the minor technicality that for mutual exclusion we must allow one "$\leq$" sign just as we were forced to introduce weak rock-scissors-paper-cycles. If instead we just ruled out $s_X(Y) < 0$, $s_Y(X) < 0$ then one could give the counterexample consisting of $\mathcal{X} = \{X, Y\}$ and $s_X(Y) = -1$, $s_Y(X) = 0$. Also note that protected coexistence $s_X(Y) > 0$, $s_Y(X) > 0$ is technically covered by (iii), since if X, Y enjoy protected coexistence, then (X, X, Y) is a weak rock-scissors-paper-cycle.

We note now a second criterion for PIPs which follows from the order theoretic viewpoint. For any trait X one may consider the set $\mathrm{Inv}(X)$ of trait types Y satisfying the weak invadability condition $Y \preceq_s X$. One can show easily that $\preceq_s$ is a preorder if and only if the sets $\mathrm{Inv}(X)$ (some of which may coincide) are totally ordered by inclusion. Given X and Y, if neither $\mathrm{Inv}(X) \subseteq \mathrm{Inv}(Y)$ nor vice versa, there must be elements X', Y' such that $X' \in \mathrm{Inv}(X)$, $X' \notin \mathrm{Inv}(Y)$, $Y' \notin \mathrm{Inv}(X)$ and $Y' \in \mathrm{Inv}(Y)$. In terms of PIPs this means that some four points (x, x'), (x, y'), (y, x'), (y, y') form the corners of an axis parallel rectangle such that one pair of opposite corners lies in the white region of the PIP while the other pair lies in the grey/black region; see Figure 3.1, Panel (g). Hence, a PIP accords with the existence of an optimisation principle if and only if it is skew symmetric and there exist no such 'opposite corner' configurations.

Theorem 3.9 has a useful corollary. We begin with a preliminary one-dimensional formulation:

Corollary 3.10. *Suppose the trait space is an interval I on the real line and suppose $s : I \times I \to \mathbb{R}$ is separately continuous in the resident and invader traits. Then there is an optimisation principle for s if and only if the following two conditions hold for all $x, y \in I$:*

(i) $\mathrm{sign}(s_x(y)) = -\mathrm{sign}(s_y(x))$,
(ii) *if* $s_x(y) = 0$ *then* $\mathrm{sign}(s_x(z)) = \mathrm{sign}(s_y(z))$ *for all* z.

Proof. The necessity of the two conditions given is clear: the first condition is just a restatement of the conclusion of Lemma 3.5 while the failure of the second implies the presence of a weak rock-scissors-paper-cycle.

To prove the sufficiency of the stated conditions, we show that the presence of a weak rock-scissors-paper triple leads to a contradiction under the assumptions

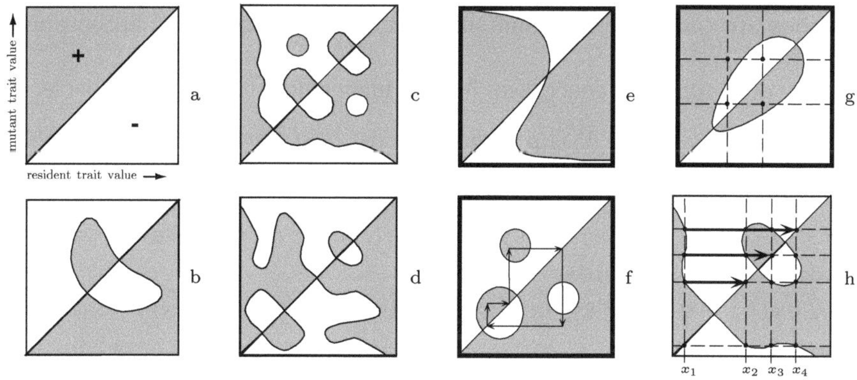

FIGURE 3.1. How the presence of an optimisation principle can be detected from PIPs. Panels (a) to (d) and (h) are examples of PIPs for models with an optimisation principle. Panel (e) is not skew symmetric and therefore there is no optimisation principle. In Panel (f) the presence of a rock-scissors paper trait triple is indicated by arrows. Panel (g) shows the 'opposite corners' obstruction to optimisation: an axis parallel rectangle with one pair of opposite corners in the positive region of the PIP while the other pair lies in the negative region. (You are encouraged to find yourself an opposite corners configuration in Panel (f) and a rock-scissors-paper triple in Panel (g).) Finally Panel (h) shows how the transitivity condition directly manifests itself in the PIP: above any resident trait value x_1 there is a certain alternation of plus and minus regions. If we read of the trait values that are selectively neutral relative to x_1 then these trait values should have exactly the same pattern of plus and minus regions above them. (Check for yourself that this condition is not fulfilled in Panels (e) to (g).)

(i) and (ii). First note that the presence of "=" in the weak rock-scissors-paper inequalities immediately implies a contradiction to (i)–(ii).

We are left to consider the following case: $s_x(y) > 0, s_y(z) > 0, s_z(x) > 0$. For sake of definiteness we assume that on the real line one has $x < y < z$, the other cases being similar. Let $x' = \sup\{t < y : s_x(t) = 0\}$. By continuity $s_x(x') = 0$. Now using (i) and (ii) one sees that $s_{x'}(z) < 0$. Thus by Rolle's theorem one has $s_{z'}(z) = 0$ for some $x' < z' < y$. Similar reasoning shows that $s_x(x'') = 0$ for some $x' < z' < x'' < y$, which contradicts the definition of x'. □

In terms of PIPs the conditions mentioned in Corollary 3.10 mean that whenever (x, y) lies on the black neutral line $s_x(y) = 0$, and one draws vertical lines passing through the points x and y on the horizontal axis, the patterns of white and grey on these lines are identical; see Figure 3.1, Panel (h). Of course, due to skew symmetry, one observes the same pattern in corresponding horizontal lines. Together these conditions are thus necessary and sufficient for the existence of an optimisation principle.

The next corollary generalises 3.10 to higher-dimensional trait spaces.

Corollary 3.11. *Suppose $\mathcal{X}$ is a path connected trait space which is separable and metrisable and the invasion fitness function s is separately continuous in the resident and invader traits. Then the following two conditions for all $X, Y \in \mathcal{X}$ are together necessary and sufficient for the existence of an optimisation principle:*

(i) $\text{sign}(s_X(Y)) = -\text{sign}(s_Y(X))$,
(ii) *if* $s_X(Y) = 0$ *then* $\text{sign}(s_X(Z)) = \text{sign}(s_Y(Z))$ *for all* $Z \in \mathcal{X}$.

Proof. It is enough to show that under condition (i) the presence of a weak rock-scissors-paper-cycle leads to a violation of condition (ii). Thus suppose condition (i) holds and that there is a weak rock-scissors-paper-cycle in $\mathcal{X}$. By path connectedness one can find a path $\gamma : I \to \mathcal{X}$ parametrised by an interval $I \subset \mathbb{R}$ such that the cycle is contained in the image γI. Applying Corollary 3.10 to the trait space I equipped with the invasion fitness function $(x, y) \mapsto s_{\gamma(x)}(\gamma(y))$ we find $x, y, z \in I$ such that $s_{\gamma(x)}(\gamma(y)) = 0$, but $\text{sign}(s_{\gamma(x)}(\gamma(z))) \neq \text{sign}(s_{\gamma(y)}(\gamma(z)))$, so (ii) fails for $X = \gamma(x)$ and $Y = \gamma(y)$. □

The following derived criterion will prove useful in Section 4, Application 4.3.

Corollary 3.12. *If there is an optimisation principle and $s_X(Y) = 0$, then the sets $\{Z|s_X(Z) = 0\}$ and $\{Z|s_Y(Z) = 0\}$ are equal.*

As a final point we note that when the trait space is multidimensional, the presence of an optimisation principle implies that all of the information necessary for deducing evolutionary outcomes is captured by a one-dimensional quantity. Even for a one-dimensional trait, s has to satisfy very strict requirements. From a mathematical viewpoint such a situation is highly nongeneric. Yet, the practical fact is that optimisation models are frequently encountered in the literature, in textbooks in particular. The cause is no doubt our penchant for making simplifications to ease the math, but one should not be misled into believing that optimisation is the default scenario in the real world.

3.3. Medium abstraction level: traits and environments

Delving deeper into the determining factors for having an optimisation principle we consider as first step in the mechanistic direction the separate functions $(Y, E) \mapsto \rho(Y|E)$ and $X \mapsto E_{\text{attr}}(X)$. This problem was first investigated for the clonal and haploid cases in [35]. [33] covers also the diploid case under the same assumptions as made in this chapter. Below we give a summary of the results. Proofs can be found in the references.

The first result in [33, 35] may not come entirely unexpected given the results from Subsection 3.2. An eco-evolutionary model has an optimisation principle if and only if "the trait values affect fitness effectively in a one-dimensional monotone manner". The term "effectively" here means that the specified properties only need to pertain to the range of fitness values closely surrounding the change from negative to positive. More precisely,

Proposition 3.13. *An eco-evolutionary model has an optimisation principle if and only if*

A. *there exists a function $\phi : \mathcal{X} \to \mathbb{R}$ and a function $g : \mathbb{R} \times \mathcal{E} \to \mathbb{R}$, increasing in its first argument, such that*

$$\operatorname{sign} \rho(X|E) = \operatorname{sign} g(\phi(X), E). \tag{3.3}$$

In that case ϕ, or any increasing function of ϕ, is an optimisation principle.

A perhaps more surprising result is

Proposition 3.14. *Condition* **A** *is equivalent to*

B. *there exists a function $\psi : \mathcal{E} \to \mathbb{R}$ and a function $h : \mathcal{X} \times \mathbb{R} \to \mathbb{R}$, increasing in its second argument, such that*

$$\operatorname{sign} \rho(X|E) = \operatorname{sign} h(X|\psi(E),) \tag{3.4}$$

which can be paraphrased as "the environment acts effectively in a one-dimensional monotone manner".

Propositions 3.13 and 3.14 show once again that optimisation principles, although frequently encountered in the literature, are exceptions rather than the rule.

Definition 3.15. We shall call a function $\psi : \mathcal{E} \to \mathbb{R}$ as in condition **B** a pessimisation principle.

Conditions **A** and **B** are related to each other by

Proposition 3.16. *If an optimisation principle, or equivalently a pessimisation principle, exists, it is possible to choose the functions ϕ and ψ such that*

$$\operatorname{sign} \rho(X|E) = \operatorname{sign} (\phi(X) + \psi(E)), \tag{3.5}$$

where ϕ and ψ are connected through the relation

$$\phi(X) = -\psi(E_{\text{attr}}(X)). \tag{3.6}$$

(Optimisation and pessimisation principles are only uniquely determined up to increasing transformations.) Hence, given a pessimisation principle ψ it is possible to construct a matching optimisation principle ϕ via the construction (3.6) and vice versa.

Corollary 3.17. *When a pessimisation principle ψ exists, evolution minimises $\psi(E_{\text{attr}}(X))$ under any constraint on X.*

Better still, $\psi(E_{\text{attr}}))$ decreases with each increase in its matching $\phi(X)$. Moreover, fitness increases with ψ where it counts, i.e., around zero. Hence the choice of the term "pessimisation principle". When a pessimisation principle exists, in the end the worst attainable world remains, together with the type(s) that can just cope with it. The following example may give a more concrete feel for the issue.

Example. *The textbook scenario for so-called r-selection.* Consider a structured population in continuous time regulated through an additional death rate d_{E} which is the same for all states of an individual and with all other demographic parameters independent of E. Then minus the mean death rate, $-\langle d_{\mathrm{E}}(E(t))\rangle_{\mathrm{time}}$, associated with an environment provides a pessimization principle (i.e., evolution maximises the mean death rate), with the asymptotic relative growth rate ρ_0 calculated on the assumption that $d_{\mathrm{E}} = 0$ as matching optimization principle. A special case is where the environment is constant except for occasional instantaneous decimating catastrophes, provided the latter kill totally indiscriminately (so that ρ_0 equals the intrinsic rate of population increase or Malthusian parameter r for that constant environment). But for the (essential, but generally unmentioned) indiscriminateness, this is the condition touted in the textbooks as supporting r-maximisation.

The practical importance of Proposition 3.16 is that, while condition **A** is close to trivial, the equivalent condition **B** and relation (3.6) often provide a useful tool for either deriving optimisation principles or proving the non-existence of such principles for large families of eco-evolutionary models.

The two optimisation principles most frequently touted in the evolutionary ecology literature are the intrinsic rate of population increase r and the lifetime offspring number R_0. The results discussed above can be used to characterise the ecological scenarios for which evolution "just maximises" r or R_0. Here "just maximising a function of X and E" should be interpreted as maximising that function by varying X for an unspecified choice of E (the latter as reflection of the absence of any mention of E in the usual statements in the non-epidemiological literature).

Remark. A convention of logic is that when a statement is not explicitly indicated as pertaining to a specific individual case, or subset of cases, it should be interpreted as pertaining to all possible cases. This convention is itself but a formalisation of the human habit of interpreting open statements like "raven are black" as meaning that all raven are black and not as some raven being black, or raven being black only under certain circumstances.

Under the presupposition that the community dynamics engenders constant environments so that the Malthusian parameter r and the lifetime offspring number R_0 are well defined, the following results hold good.

Proposition 3.18. *Evolution just maximises r if and only if*

C. *the combination of life histories and ecological embedding is such that r can be written as*

$$r(X|E) = g\left(r(X|E_0), E\right)$$

for some function g that increases in its first argument, and E_0 some fixed, but otherwise arbitrary, environment,

Proposition 3.19. *Evolution just maximises R_0 if and only if*

D. *the combination of life histories and ecological embedding is such that* $\ln(R_0)$ *can be written as*

$$\ln\left(R_0(X|E)\right) = g\left(\ln\left(R_0(X|E_0)\right), E\right)$$

for some function g that increases in its first argument, and E_0 some fixed, but otherwise arbitrary, environment.

In contrast to criteria **A** and **B**, criteria **C** and **D** are relatively easy to check in specific situations.

A fair fraction of textbook statements, if taken literally, applies only when condition **C** or **D** is fulfilled.

3.4. Lowest abstraction level: life histories

The next step in the mechanistic direction is to consider how invasion fitness ρ is built up from demographic parameters (in an evolutionary context usually referred to as life-history traits), and what conditions on those parameters correspond to the conditions that were found in Subsection 3.3.

Presently two manuscripts are floating around, by Roger Bowers [36] and by Claus Rueffler and co-workers [37], that relate life cycle structure to proporties of the associated invasion fitness function. Both manuscripts deal with finite state individuals in constant environments, i.e., community dynamical equilibria. The first one considers continuous time community models with all phenotypes within a species influencing the environment in the same manner. The dependence of the demographic parameters is considered to be essentially affine in the community composition, in a manner that depends on the, bivariate, traits. The second one considers discrete time population models with separable demographic parameters, that is, parameters that can each be decomposed into an inherited parameter times possibly a scalar function of E, referred to as regulatory function and denoted as R, without any further a priori assumptions on how E is determined. Some of the inherited demographic parameters are supposed to be under evolutionary control and then are called traits. The environment is supposed to be organised as the cartesian product of one-dimensional components, and the regulatory functions are supposed to be monotone in the order relation imposed by the positive cone. We only review the material that pertains to the existence or non-existence of optimisation principles.

We start with the plethora of sufficient conditions for the existence of an optimisation principle derived by Rueffler et al. [37]. Two immediate trivial cases are

Proposition 3.20. *If there exists a scalar function $\psi : \mathcal{E} \to \mathbb{R}$ such that all regulatory functions can be written as monotonically increasing functions of $\psi(E)$, then evolution maximises $-\psi(E_{\mathrm{attr}}(X))$.*

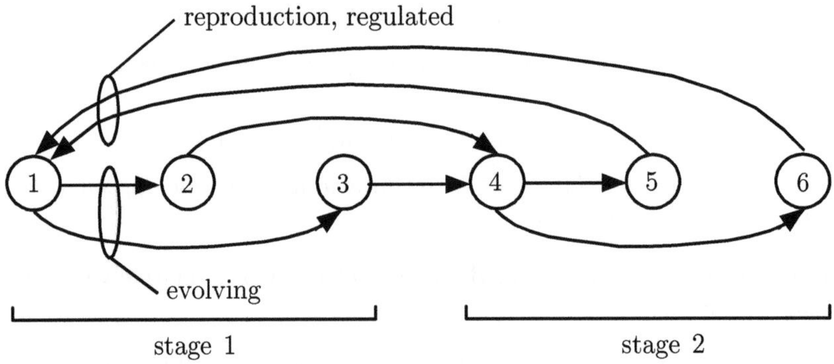

FIGURE 3.2. Life cycle that satisfies the conditions of Proposition 3.22 (and also condition 1 of Proposition 3.25).

Proposition 3.21. *If all demographic parameters come with the same R, then evolution minimises $R(E_{\mathrm{attr}}(X))$ and maximises the dominant eigenvalue of the population projection matrix for R set equal to some arbitrary fixed value.*

An example is when the only influence of the environment is through a death probability that is independent of the state or type of an individual on top of any state- or type-dependent ones. In that case evolution maximises that death probability and the value of r for any given fixed value of that probability.

In the next proposition we use the term *transition ratio* to cover both the state-transition-and-survival probabilities and the fertilities, and refer to both types of event together as *'transitions'*.

Proposition 3.22. *If*

(i) *the states can be partitioned into n disjunct classes G_i, $i \in \mathbb{N} \bmod n$, n even, with each class only connecting to the following class through a single state that may be an element of either class,*
with the transition ratios from states in odd numbered classes not being regulated and the transition ratios from states in the even numbered classes not evolving (the unregulated *resp.* non-evolving classes*), and*

(ii) *the transition ratios from the classes that can only be left from a single state satisfy the following restrictions*
for non-evolving classes: the transition ratios to the next class are all regulated in the same way, and
for unregulated classes: if the class does not connect to the next class only through single states in both classes, the transition ratios to the next class are non-evolving,

then there is an optimisation principle.

This optimisation principle can be calculated by treating the entrance stream into any of the classes as "births" for which we calculate R_0. This R_0 (i) is a fitness

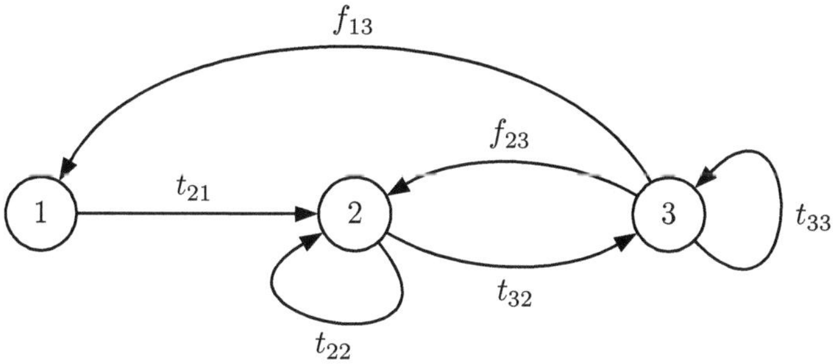

FIGURE 3.3. Life cycle of a perennial plant with a seedling (1), juvenile (2) and flowering (3) state. Q for this life cycle equals $-1 + (t_{22} + t_{33} + t_{21}t_{32}f_{13} + t_{32}f_{23}) - (t_{22}t_{33})$, with the bracketed terms corresponding to the first and second sums in (3.7). From [37].

proxy and (ii) can be written as a product of a function of the traits times a function of E. The result then follows from Proposition 3.19.

Remark. Without further restrictions on the models described in Proposition 3.22 the R_0 referred to above may be infinite. This happens as soon as one of the classes is disconnected from the rest of the life cycle and has a nonnegative growth rate. However, such a biological anomaly is implicitly excluded as this would lead to a contradiction with the assumption that the models allow a community dynamical equilibrium including the species under consideration.

Models of this type frequently occur in the literature. For examples see Figure 3.2 and Application 4.2.

The remainder of the results of Rueffler and coworkers are based on the fitness proxy Q described at the end of Subsection 2.1. The advantage of Q over r or R_0 is that Q is affine in each of the separate heritable demographic parameters (baseline state transition cum survival probabilities t_{ij} and fertilities times survival probabilities f_{ij}) and regulatory functions ($R_{\mathrm{t},ij}$ and $R_{\mathrm{f},ij}$, supposedly lying between zero and one), together collected in the demographic projection matrix $\mathbf{A}$. More in particular, it is possible to write

$$Q = -1 + \sum_{\mathcal{L}_{\mathbf{A}}} L - \sum_{\mathcal{L}_{\mathbf{A}}^{2*}} LM + \sum_{\mathcal{L}_{\mathbf{A}}^{3*}} LMN - \cdots , \tag{3.7}$$

with (i) L, M and N so-called loop transmissions, where a loop is a sequence of demographic parameters that lead from a given state to itself without passing more than once through some other state, with the loop transmission the product of the demographic parameters along the loop, (ii) $\mathcal{L}_{\mathbf{A}}$ the set of all loops associated with $\mathbf{A}$, and (iii) a $*$ hung on to an n-fold cartesian power of $\mathcal{L}_{\mathbf{A}}$ indicating that only

n-tuples are considered in which the sets of states occurring in those loops are mutually exclusive (cf. Figure 3.3). A loop is called evolving if it contains at least one trait, and regulated if it contains at least one demographic parameter affected by E.

Rueffler and co-workers use (3.7) together with

Proposition 3.23. *If it is possible to find functions* $g_1 : \mathcal{X} \to \mathbb{R}$, $g_2 : \mathcal{X} \to \mathbb{R}$, $e_1 : \mathcal{E} \to \mathbb{R}$ *and* $e_2 : \mathcal{E} \to \mathbb{R}$ *with* $\mathrm{sign}[g_2]$ *and* $\mathrm{sign}[e_1]$ *constant such that*

$$Q(X|E) = g_1(X)e_1(E) + g_2(X)e_2(E), \tag{3.8}$$

then $\mathrm{sign}[e_1(E)]g_1(X)/|g_2(X)|$ *is an optimisation principle.*

to delineate a zoo of special classes of sufficient conditions for the existence of an optimisation principle in terms of the life cycle graph. The taxonomy of this zoo is not simple. Moreover, it is not clear yet whether possibly any further special cases are still out in the wild. Hence we give only one relatively simple example and refer to [37] for the details.

Example. *Evolution of fertility patterns in perennial plants.* Assume that we are interested in the evolution in the two fertility parameters in the life-cycle graph in Figure 3.3. The biological interpretation is that f_{13} corresponds to fall reproduction, and f_{23} to spring reproduction so that the seeds can already germinate in the same year and appear as a juvenile at the next sampling time. Now assume that the environment influences only the winter seed survival, and by that f_{13}, and seedling survival t_{21}. Then we get, with a bar above a life history parameter indicating that it is constant and a tilde that it evolves,

$$Q(\tilde{f}_{13}, \tilde{f}_{12}|E) = [\bar{t}_{32}\tilde{f}_{23} + \bar{t}_{22} + \bar{t}_{33} - \bar{t}_{22}\bar{t}_{33} - 1] + [\bar{t}_{21}\bar{t}_{32}\tilde{f}_{13}][R_{\mathrm{t},21}(E)R_{\mathrm{f},13}(E)].$$

Hence evolution maximises $(\bar{t}_{21}\bar{t}_{32}\tilde{f}_{13})/(1 + \bar{t}_{22}\bar{t}_{33} - \bar{t}_{22} - \bar{t}_{33} - \bar{t}_{32}\tilde{f}_{23})$. (The term $(\bar{t}_{32}\tilde{f}_{23} + \bar{t}_{22} + \bar{t}_{33} - \bar{t}_{22}\bar{t}_{33} - 1)$ is negative since otherwise the population would not have an equilibrium.)

The problem of determining sufficient conditions for the existence of an optimisation principle can also be turned on its head in the form of deriving necessary conditions for the non-existence of an optimisation principle, or, in the terminology of [38], for selection to be frequency dependent. We start with a condition presented by Roger Bowers [36], although phrased in a different language.

Proposition 3.24. *Let the community be described by*

$$\frac{\mathrm{d}N}{\mathrm{d}t} = \mathbf{M}(X|N)N \tag{3.9}$$

with N *the vector of population densities of all species in the community differentiated according to the states of the individuals,* $\mathbf{M}$ *a corresponding block diagonal matrix, and the map* $N \mapsto \mathbf{M}(X|N)$ *affine. Then a necessary condition for evolution to be frequency dependent is that for each* X *the range of the map* $N \mapsto \mathbf{M}(X|N)$ *is at least two-dimensional.*

Claus Rueffler and co-workers strengthen this type of condition to

Proposition 3.25. *For selection to be frequency dependent it is necessary that E has more than one component and that at least two evolving and two regulated loops exist which occur in one of the following combinations:*

(i) *A pair of loops L, M exist that are both evolving and regulated such that*

$$Q(X|E) = L(X|E) + M(X|E) + \text{remainder}. \tag{3.10}$$

(ii) *Three loops L, M, N exist where L is both evolving and regulated, M is evolving and N is regulated such that*

$$Q(X|E) = L(X|E) + M(X) + N(E) + \text{remainder}. \tag{3.11}$$

(iii) *Four loops L, M, N and O exist where L and M are evolving and N and O are regulated and where L and N are unconnected such that*

$$Q(X|E) = L(X) + M(X) + N(E) + O(E) - L(X)N(E) + \text{remainder}, \tag{3.12}$$

where X, E and $(X|E)$ are added as arguments to loops to indicate whether they contain a trait, a regulated demographic parameter or both.

The goal is, of course, to delimit the existence versus non-existence of an optimisation principle in the form of necessary and sufficient conditions. However, Figure 3.2 depicts a life cycle that satisfies requirement 1 of Proposition 3.25 as well as the requirement of Proposition 3.22, showing that goal is not yet in sight.

4. Three applications

Application 4.1. *Evolution away from chaos*
Consider the population dynamical equations

$$n_i(t+1) = a_i \left(f\left(E(t)\right)\right)^{b_i} n_i(t) \quad \text{for } i = 0, \ldots, k, \tag{4.1}$$

with

$$E(t) = [c_0 n_0(t) + \cdots + c_k n_k(t)], \tag{4.2}$$

all a_i, b_i, and $c_i > 0$, and f decreasing from 1 to 0 for E increasing from 0 to ∞.

With the choice

$$f(E(t)) = [1 + E(t)]^{-1}, \tag{4.3}$$

and $k = 0$, this model becomes the model launched into fashion by, e.g., [39] as a touchstone for the appearance of chaotic fluctuations in single species population dynamics. Figure 4.1 shows the dependence of the dynamics on the parameter values.

The trait vector appearing in (4.1) and (4.2) is

$$X = (a, b, c). \tag{4.4}$$

The parameters a, $1/b$, and c can be interpreted in individual-based terms as respectively the per capita reproduction in a boom environment, the ability to

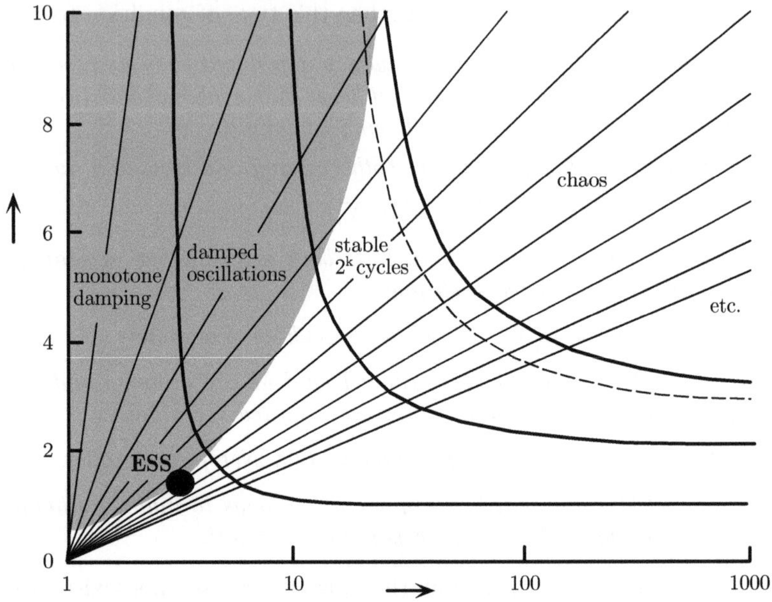

FIGURE 4.1. Types of dynamics of the monomorphic version of the model (4.1)–(4.3). The thin lines are contour lines of ϕ. A reasonable constraint set is indicated in grey together with the accompanying ESS.

cope with a bust environment and the per capita impingement on the common environment. From (4.1) we find

$$\rho(X|E) = L[a[f(E)]^b] = \ln[a] + b\psi(E), \tag{4.5}$$

with

$$\psi(E) = L[f(E)]$$

and L the log geometric mean operator

$$L(z) \quad := \quad \lim_{T\to\infty} T^{-1}\sum_{t=1}^{T} \ln(z(t)).$$

From $\rho(X|E_{\text{attr}}(X)) = 0$ we deduce that

$$\psi(E_{\text{attr}}(X)) = -b^{-1}\ln[a], \tag{4.6}$$

From Proposition 3.16 we conclude that evolution maximises

$$\phi(X) \quad := \quad \frac{\ln[a]}{b} \tag{4.7}$$

In accordance with Propositions 3.13 and 3.14 we can define the functions g and h occurring in the definitions of monotone one-dimensional action as

$$g(\phi(X), E) := \phi(X) + \psi(E) =: h(X, \psi(E)). \tag{4.8}$$

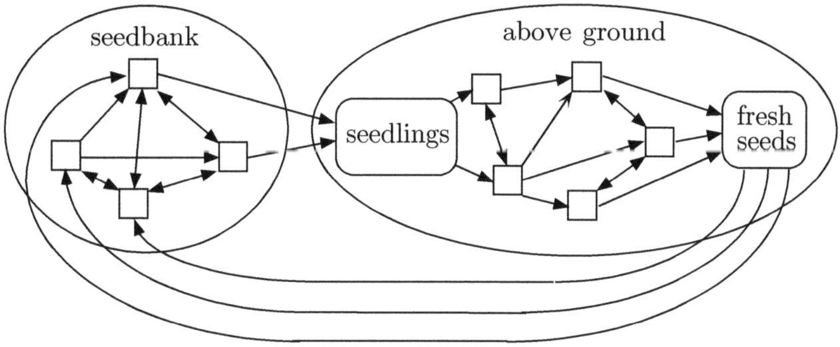

FIGURE 4.2. Life cycle of a perennial plant.

Note that

$$\rho(X,E) = b(\phi(X) + \psi(E)) \tag{4.9}$$

showing that g and h are only sign equivalent to ρ, but not equal to ρ. It can even be proved that for ρ given by (4.5) it is impossible to find pairs g and ϕ, or h and ψ, for which such an equality holds good. Hence, even in the domain of stable equilibria neither of the conditions **C** or **D** is fulfilled.

It may be expected that increasing a will in general go at the cost of increasing b. By combining the optimisation principle with a reasonable trade-off between a and b it is found that in general the ESS will lie in the region of stable equilibria (see Figure 4.1).

Application 4.2. *Evolution of germination strategies*
Figure 4.2 depicts the life history of a plant decomposed into two stages between which no information is transferred, as all seedlings are equal and so are all newly produced seeds. Within the two stages the seeds and plants are differentiated in, e.g., (age, depth in soil) – respectively (age, size above ground, size below ground)-classes. Sampling is done on a yearly basis, just before germination time. Hence the boxes "seedling" and "fresh seeds" do not correspond to states of the model as those conditions are but ephemeral on the considered time scale. Yet we put them in since they provide the unique connections between the stages that by that uniqueness prevent information being transferred from one stage to another. The environmental influences on an average individual plant can be decomposed into two components $(E_{\mathrm{a}}, E_{\mathrm{b}})$, with E_{b} representing the influences on the seeds (consisting of, e.g., seed predation pressures and fungal and bacterial attack rates at different depths; the 'b' refers to below ground), and with E_{a} representing influences on the plants (capturing all direct and indirect competitive influences within the community through shading, nutrient depletion and changing predation pressures; the 'a' refers to above ground). Seedlings can survive only in so-called safe sites, places that are temporarily without ground cover. Since the birth of a safe site necessarily coincides with the demise or a state change of one or more

plants in the community it may be assumed that E_a also determines the fraction of the area covered by safe sites. The quantity of evolutionary interest is the germination strategy. Especially deeper buried seeds have only a partial knowledge of whether they are in a safe site. Hence we can capture the germination strategy by the dependence of the germination probability on a seed's state and whether it is in a safe site. Therefore, for a full description of the eco-evolutionary model the following quantities and functions are needed, with a $\check{}$ indicating the mutant:

$N,\ \check{N}$: vector of densities of seeds in different states,
$P,\ \check{P}$: vector of densities of plants in different states,
$U(E_b)$: state distribution of a new seed just prior to germination time,
$\mathbf{S}(E_b)$: matrix of survival and state transition probabilities of seeds,
$h(E_a)$: density of safe sites available at germination time,
$k(E_a)$: fraction of total area covered by those safe sites,
$G,\ \check{G}$: state dependent probabilities that a seed in a safe site germinates,
$F,\ \check{F}$: state dependent probabilities that a seed outside a safe site germinates,
$\theta(E_a)$: average number of seedlings in a safe site that survive seedling competition divided by the average density of novel seedlings in safe sites,
$J(E_a)$: state distribution of young plants that have survived seedling competition,
$\mathbf{A}(E_a)$: matrix of survival and state transition probabilities of plants,
$Y(E_a)$: seed production by plants in different plant states.

Note that the probability distribution of the state of a newborn seed at the next germination time, encoded in the vector U, will in general be defective (i.e., has total mass smaller than one) due to seed mortality. Note also that in most concrete instances the probability distribution of plant states after seedling competition, encoded in the vector J, will probably be concentrated on but a single plant state: small juvenile. J by definition has full mass as the probabilities of seedling death are all accounted for in $\theta(E_a)$.

The resident population state satisfies the following recurrences

$$\begin{aligned} N' &= \mathbf{S}(E_b)(\mathbf{I} - k(E_a)\mathbf{diag}(G) - (1-k(E_a))\mathbf{diag}(F))N \\ &\quad + U(E_b)Y^{\mathrm{T}}(E_a)P, \\ P' &= \mathbf{A}(E_a)P + J(E_a)h(E_a)\theta(E_a)G^{\mathrm{T}}N, \end{aligned} \tag{4.10}$$

In words, the seeds of next year consist of this year's seeds that neither germinate nor succumb plus the surviving new seeds from this year. The plants of next year consist of this year's surviving plants plus the new plants, the density of which is calculated as the density of safe sites times this year's average numbers of survivors of seedling competition in a site. By the same token, the mutant population state satisfies

$$
\begin{aligned}
\check{N}' &= \mathbf{S}(E_\mathrm{b})(\mathbf{I} - k(E_\mathrm{a})\mathbf{diag}(\check{G}) - (1 - k(E_\mathrm{a}))\mathbf{diag}(\check{F}))\check{N} \\
&\qquad + U(E_\mathrm{b})Y^\mathrm{T}(E_\mathrm{a})\check{P}, \\
\check{P}' &= \mathbf{A}(E_\mathrm{a})\check{P} + J(E_\mathrm{a})h(E_\mathrm{a})\theta(E_\mathrm{a})\check{G}\mathrm{T}\check{N}.
\end{aligned}
\tag{4.11}
$$

These equations should be combined with equations for the remainder of the community to determine $(E_\mathrm{a}, E_\mathrm{b})$.

For the calculations below we assume that the resident population dynamics converges to an equilibrium.

As it turns out, the present model is still a bit too general to allow the ES germination strategy to be determined from an optimisation principle. However, it is only by considering more general models that it is possible to delineate the crucial assumptions needed for the results from the previous section to apply.

At the resident equilibrium the average lifetime offspring number of a resident equals 1. The calculation of this average lifetime offspring number can be broken down into a number of steps. First we calculate the average number of full seasonal cycles (measured between end-of-seedling-competition time points) that a survivor from the seedling stage lives through during its lifetime, split up according to the state the plant was in at the end-of-seedling-competition moments. From the general Markov chain results in [40] it follows that these numbers are given by the vector $(\mathbf{I} - \mathbf{A}(E_{\mathrm{a},G,F}))^{-1}J(E_{\mathrm{a},G,F})$, with $(E_\mathrm{a}, E_\mathrm{b})_{G,F}$ the equilibrium environment, to be determined from the full community dynamical equations for the resident strategy (G, F). Hence, the average number of seeds that a plant that just germinated in a safe site will produce over its lifetime is $Y^\mathrm{T}(E_{\mathrm{a},G,F})(\mathbf{I} - \mathbf{A}(E_{\mathrm{a},G,F}))^{-1}J(E_{\mathrm{a},G,F})\theta(E_{\mathrm{a},G,F})$. Similarly, the average number of germination moments that a seed experiences while in various seed states equals

$$
(\mathbf{I} - \mathbf{S}(E_{\mathrm{b},G,F})(\mathbf{I} - k(E_{\mathrm{a},G,F})\mathbf{diag}(G) - (1 - k(E_{\mathrm{a},G,F}))\mathbf{diag}(F)))^{-1}U(E_{\mathrm{b},G,F}).
$$

Therefore, the probability of a seed germinating in a safe site instead of dying or germinating elsewhere equals $G^\mathrm{T}(\mathbf{I} - \mathbf{S}(E_{\mathrm{b},G,F})(\mathbf{I} - k(E_{\mathrm{a},G,F})\mathbf{diag}(G) - (1 - k(E_{\mathrm{a},G,F}))\mathbf{diag}(F)))^{-1}U(E_{\mathrm{b},G,F})$. Multiplying these two numbers gives an expression for $R_0(G, F|(E_a, E_b)_{G,F})$ that we have to set equal to 1 as part of the process of calculating the resident equilibrium. As it turns out, there is no need to calculate this equilibrium in full. All that is needed later is an expression for $\theta(E_{a,G,F})$ as a function of the other resident parameters which can be determined from the equation $R_0 = 1$. The calculation of $R_0(\check{G}, \check{F}|(E_a, E_b)_{G,F})$ proceeds in a similar manner. After substituting the earlier found expression for $\theta(E_{a,G,F})$ and cancelling terms in the numerator and denominator we get

$$
\begin{aligned}
&R_0(\check{G}, \check{F}|(E_\mathrm{a}, E_\mathrm{b})_{G,F}) \\
&= \frac{\check{G}^\mathrm{T}\left(\mathbf{I} - \mathbf{S}(E_{\mathrm{b},G,F})\left(\mathbf{I} - k\left(E_{\mathrm{a},G,F}\right)\mathbf{diag}(\check{G}) - \left(1 - k\left(E_{\mathrm{a},G,F}\right)\right)\mathbf{diag}(\check{F})\right)\right)^{-1}U\left(E_{\mathrm{b},G,F}\right)}{G^\mathrm{T}\left(\mathbf{I} - \mathbf{S}\left(E_{\mathrm{b},G,F}\right)\left(\mathbf{I} - k\left(E_{\mathrm{a},G,F}\right)\mathbf{diag}(G) - \left(1 - k\left(E_{\mathrm{a},G,F}\right)\right)\mathbf{diag}(F)\right)\right)^{-1}U\left(E_{\mathrm{b},G,F}\right)}.
\end{aligned}
\tag{4.12}
$$

From this expression it can be seen that ESSes can be determined by optimising (G, F) in

$$\tilde{\phi}(G, F; k, \mathbf{S}, U) := G^{\mathrm{T}} \left(\mathbf{I} - \mathbf{S}\left(\mathbf{I} - k\mathbf{diag}(G) - (1-k)\mathbf{diag}(F)\right)\right)^{-1} U \tag{4.13}$$

in dependence on $(k, \mathbf{S}, U)$, and solving the community dynamical equilibrium equations together with

$$(G, F) = (G, F)_{\mathrm{opt}}(k(E_{\mathrm{a},G,F}), \mathbf{S}(E_{\mathrm{b},G,F}), U(E_{\mathrm{b},G,F})). \tag{4.14}$$

If and only if G and F do not influence the equilibrium values of the seed state transition and survival probabilities and the fraction of the area covered by safe sites, i.e., $\mathbf{S}(E_{\mathrm{b},G,F}) = \bar{\mathbf{S}}$, $U(E_{\mathrm{b},G,F}) = \bar{U}$ and $k(E_{\mathrm{a},G,F}) = \bar{k}$, the function $\phi : (G, F) \mapsto \tilde{\phi}(G, F; \bar{k}, \bar{\mathbf{S}}, \bar{U})$ is an evolutionary optimisation principle, in accordance with Proposition 3.22.

We are still in the midst of exploring this model. Some first results and more details can be found in [41].

Application 4.3. *Virulence evolution*
For a long time, it was close to dogma in epidemiological theorising (e.g., [42, 43]) that the main basis for the study of virulence evolution should be sought in the maximisation of R_0, defined in epidemiology as the number of secondary infections engendered by a primary infection *in an otherwise infection-free population.* To this end, R_0 is considered as a function of the disease's demographic parameters, which in turn are envisaged as functions of some underlying trait vector that is supposed to be under evolutionary control. Here we consider, following [44], how this dogma fares in the light of Propositions 3.13 and 3.14 (see also [45]).

The epidemiological models that we consider below have been chosen for the simplicity of the calculations they engender. In particular, their community dynamics possess unique internal point attractors.

We start by giving a full population dynamical description of the ecological context, before reverting to considerations focusing on infected individuals. It is the individual-based dynamics of the latter that provides the basis for the classification of the environmental feed-back loop relative to its consequences for the ESSs of disease traits. The details of the population dynamics surrounding infected individuals is relevant only in so far as it acts as an environment affecting the population dynamical behavior of the infected individuals.

To characterise the instantaneous environmental conditions to which infected individuals may be exposed, we follow standard custom by letting S denote the density of susceptible individuals and I the density of infected individuals. After specifying the dynamics of this instantaneous environment, the corresponding evolutionary environments can be calculated from the attractors of this dynamics. Infections occur according to the law of mass action, with a fixed rate constant β. Infected individuals do not recover but die at a per capita rate α, acting on top of the per capita death rate experienced by susceptible and infected individuals alike. In the absence of the disease, $I = 0$, the population grows in a density-dependent manner, with per capita birth rate $b_0 - h_{\mathrm{b}}(S, 0)$ and per capita death rate $d_0 + h_{\text{d}}(S,0)$, with $b_0 > d_0 > 0$. The functions h_{b} and h_{d} both increase in S and I, with $h_{\text{b}}(0,0) = h_{\text{d}}(0,0) = 0$. The full population dynamical equations are then given by

$$\frac{\mathrm{d}S}{\mathrm{d}t} = [b(S,I) - d(S,I) - \beta I]S, \quad \frac{\mathrm{d}I}{\mathrm{d}t} = [\beta S - \alpha - d(S,I)]I, \tag{4.15}$$

with

$$b(S,I) = b_0 - h_{\text{b}}(S,I), \quad d(S,I) = d_0 + h_{\text{d}}(S,I). \tag{4.16}$$

(The implicit assumption that infected individuals are not allowed to reproduce greatly simplifies the proofs of the attractivity of the equilibria.)

The parameters α and β are assumed to be under evolutionary control by the disease (evolution in host-controlled traits is not considered here). In agreement with the standard custom, we assume α and β to be connected by a constraint: β cannot become too high and α simultaneously not too low, which can be expressed as $g(\alpha, \beta) \leq m$ with g increasing in β and decreasing in α. As evolution acts to increase β and decrease α, it will quickly run into this constraint. From there on, evolution will effectively be restricted to the curve $g(\alpha, \beta) = m$, alternatively parameterised as $\beta = \beta(\alpha)$, or as $(\alpha(x), \beta(x))$ for some scalar physiological trait x.

Within the general class of models (4.15), we consider four special cases,

$$\begin{array}{lll} \text{(i)} & h_{\text{b}}(S,I) = \kappa(S+I), & h_{\text{d}}(S,I) = 0, \\ \text{(ii)} & h_{\text{b}}(S,I) = 0, & h_{\text{d}}(S,I) = \kappa S, \\ \text{(iii)} & h_{\text{b}}(S,I) = 0, & h_{\text{d}}(S,I) = \kappa S^2, \\ \text{(iv)} & h_{\text{b}}(S,I) = 0, & h_{\text{d}}(S,I) = \kappa(S+I). \end{array} \tag{4.17}$$

These model families have been rigged so that for model (i) and (ii) the environmental feedback for the disease is one-dimensional monotone. According to Propositions 3.13 and 3.14, these models thus support an optimisation principle. For model (i) the optimisation principle is equivalent (i.e., monotonically related) to R_0, while for model (ii) this is not the case. For model (iii) the environmental feedback acts one-dimensionally but not monotonically, and for model (iv) it acts two-dimensionally.

It should be understood that we chose the specific examples in Equation (4.17) primarily for didactic purposes. For their individual-based underpinning one may think of population regulation through fighting. For models (i) and (iv) fighting may be initiated by all individuals, whereas for models (ii) and (iii) infected individuals are assumed to suffer from fights without being able to initiate such fights themselves. Model (iii) is based on the assumption of aggression increasing linearly with aggressor density. Fighting, of course, may here be replaced by any other form of interference competition.

In this application we will distinguish the customary evolutionary R_0, i.e., the infection-time production of new disease cases by a mutant disease case introduced into a resident disease in equilibrium with its host, from the customary epidemiological R_0 by denoting the former as R. We start by expressing R as a general function of the mutant traits $\check{X} = (\check{\alpha}, \check{\beta})$ and of the variables (S, I) parameterising the potential environmental conditions,

$$\begin{array}{ll} \text{(i)} & R(\check{\alpha}, \check{\beta}|S, I) = \frac{\check{\beta}S}{\check{\alpha}+d_0}\,, \\ \text{(ii)} & R(\check{\alpha}, \check{\beta}|S, I) = \frac{\check{\beta}S}{\check{\alpha}+d_0+\kappa S}\,, \\ \text{(iii)} & R(\check{\alpha}, \check{\beta}|S, I) = \frac{\check{\beta}S}{\check{\alpha}+d_0+\kappa S^2}\,, \\ \text{(iv)} & R(\check{\alpha}, \check{\beta}|S, I) = \frac{\check{\beta}S}{\check{\alpha}+d_0+\kappa(S+I)}\,. \end{array} \tag{4.18}$$

It is only later that we will confine attention to the realisable environments, given by the equilibrium values $(\hat{S}(\alpha,\beta), \hat{I}(\alpha,\beta))$ produced by the possible residents $X = (\alpha, \beta)$.

For model (i), R increases with S. So the optimisation principle can be constructed directly from (3.6). Minimising $\hat{S}$, which can easily be seen from (4.15)–(4.17) to yield $\hat{S} = (\alpha + d_0)/\beta$, should thus be equivalent to maximising $\phi(\alpha, \beta) = -\hat{S} = -(\alpha + d_0)/\beta$. To calculate R_0 for this model, we observe that $R_0(\alpha, \beta) = R(\alpha, \beta|S_0, 0) = \beta S_0/(\alpha + d_0)$, with S_0 denoting the equilibrium value for S in the absence of the disease. It is not difficult to see that R_0 and the ϕ resulting from our general construction are indeed monotonically related, independent of the value of S_0.

For model (ii), R is again monotone in S. From $E_{\text{attr}}(X) = \hat{S} = (\alpha+d_0)/(\beta-\kappa)$, we find the matched optimisation principle $\phi = -(\alpha + d_0)/(\beta - \kappa)$. However, maximising ϕ is not equivalent to maximising $R_0 = \beta S_0/(\alpha + d_0 + \kappa S_0) = \beta(b_0 - d_0)/[\kappa(\alpha + b_0)]$, where we used $S_0 = (b_0 - d_0)/\kappa$ as for model (i). To see this non-equivalence, it suffices to observe that the contour lines, defined by $R_0(\alpha, \beta) = R_0(\alpha_0, \beta_0)$ and $\phi(\alpha, \beta) = \phi(\alpha_0, \beta_0)$ for given (α_0, β_0), differ, as can be seen from the lack of coincidence in their derivatives at (α_0, β_0), calculated via an implicit differentiation of the defining relations: $\mathrm{d}\alpha/\mathrm{d}\beta = (\kappa\alpha_0 + b_0)/(\beta_0\kappa)$ for R_0, which differs from $\mathrm{d}\alpha/\mathrm{d}\beta = (\alpha_0 + d_0)/(\beta_0 - \kappa)$ for ϕ.

The fact that invasion fitness in model (iii) is non-monotone in any scalar summary of the condition of the environment, and that the evolutionary environment in model (iv) is essentially two-dimensional, can already be guessed from (4.18). However, for a proof we have to deal with the fact that, for instance, in model (iii) R should be non-monotone relative to whatever summary variable even when its domain is restricted to the realisable values of S and in addition to an infinitesimal neighborhood of those combinations of $(\check{\alpha}, \check{\beta})$ and $\hat{S}(\alpha, \beta)$ for which $R(\check{\alpha}, \check{\beta}|\hat{S}(\alpha, \beta)) = 1$. The necessary technicalities can be found in [44, Appendix A]. For the present exposition it suffices to note that in cases (iii) and (iv) the directions $\mathrm{d}\alpha/\mathrm{d}\beta$ in (α_0, β_0) of the contour lines $\{(\alpha, \beta)|R(\alpha, \beta|S(\alpha_0, \beta_0), I(\alpha_0, \beta_0)) = 1\}$ and $\{(\alpha, \beta)|R(\alpha_0, \beta_0)|S(\alpha, \beta), I(\alpha, \beta)) = 1\}$, which can be determined by implicit differentiation, are generically different. Hence, by Corollary 3.12 neither case allows an optimisation principle.

5. Discussion

The title question was interpreted by us as: find necessary and sufficient conditions on eco-evolutionary models such that the ESSes for these models for all possible constraints on the trait space $\mathcal{X}$ can be calculated by optimising some function $\phi : \mathcal{X} \to \mathbb{R}$. At the highest level of abstraction this question was answered by naming two conditions that should be satisfied by the invasion fitness function $(X, Y) \mapsto s_X(Y)$: (i) it should be sign-antisymmetric, that is, $s_X(Y) > 0 \Leftrightarrow s_Y(X) < 0$, and (ii) there should be no weak rock-scissors-paper configurations, that is, triples (X, Y, Z) such that $s_X(Y) \geq 0$, $s_Y(Z) \geq 0$, and $s_Z(X) > 0$. At a lower level of abstraction this was found to be equivalent to both the trait and the environment acting effectively in a one-dimensional monotone manner. On a still lower level, that of life history parameters, the picture becomes more diffuse. As even for reasonably delimited classes of models no necessary and sufficient conditions are available yet, all we could do is give a brief summary of the insights that are available at the current time. This clearly is a highly interesting area for further research.

Of course, there exist still lower levels of abstraction, like the organisation of foraging or the construction of bones. Although quite a lot of optimisation modelling is done here in concrete applications, it for the time being appears inopportune to extend our approach to these levels, primarily since those applications customarily leave open the full eco-evolutionary context that would be the ultimate justification of the presumed optimisation principle. The best that one can say is that these applications should probably be considered as attempts at predicting not so much evolutionary outcomes as well as predicting properties of individuals from the environment in which they are observed to live, on the supposition that the combination of realised trait value and environment is currently sitting at an ESS, and that the optimised quantity is a fair fitness proxy.

References

[1] W. Ewens, *What changes has mathematics made to the Darwinian theory? In* F.A.C.C. Chalub and J.F. Rodrigues (eds.), *The Mathematics of Darwin's Legacy*, 7–26, Birkhäuser, Basel, 2011, *This issue.*

[2] R. Bürger, *Some mathematical models in evolutionary genetics. In* F.A.C.C. Chalub and J.F. Rodrigues (eds.), *The Mathematics of Darwin's Legacy*, 67–89, Birkhäuser, Basel, 2011, *This issue.*

[3] J.A.J. Metz, *Thoughts on the geometry of meso-evolution: Collecting mathematical elements for a post-modern synthesis. In* F.A.C.C. Chalub and J.F. Rodrigues (eds.), *The Mathematics of Darwin's Legacy*, 193–231, Birkhäuser, Basel, 2011, *This issue.*

[4] J.A.J. Metz, *Fitness. In* S.E. Jörgensen and B.D. Fath (eds.), *Evolutionary Ecology*, volume 2 of *Encyclopedia of Ecology*, 1599–1612, Elsevier, UK, 2008.

[5] J.A.J. Metz and O. Diekmann (eds.), *The dynamics of physiologically structured populations*, volume 68 of *Lecture Notes in Biomathematics.* Springer-Verlag, Berlin, 1986.

[6] J.A.J. Metz and A.M. de Roos, *The role of physiologically structured population models within a general individual-based modeling perspective. In* D. DeAngelis and L. Gross (eds.), *Individual-based models and approaches in ecology: Concepts and Models*, 88–111, Chapman & Hall, USA, 1992.

[7] J.A.J. Metz, R.M. Nisbet, and S.A.H. Geritz, *How should we define "fitness" for general ecological scenarios?* Trends Ecol. Evol. **7** (1992), 198–202.

[8] R. Ferrière and M. Gatto, *Lyapunov exponents and the mathematics of invasion in oscillatory or chaotic populations.* Theor. Popul. Biol. **48** (1995), 126–171.

[9] P. Jagers, *Branching Processes with Biological Applications.* Wiley Series in Probability and Mathematical Statistics-Applied, London, UK, 1975.

[10] K.B. Athreya and S. Karlin, *Branching processes with random environments I – extinction probabilities.* Ann. Math. Stat. **42** (1971), 1499.

[11] K.B. Athreya and S. Karlin, *Branching processes with random environments II – limit theorems.* Ann. Math. Stat. **42** (1971), 1843.

[12] P. Haccou, P. Jagers, and V.A. Vatutin, *Branching Processes. Variation, Growth, and Extinction of Populations*, volume 5 of *Cambridge Studies in Adaptive Dynamics.* Cambridge University Press, UK, 2005.

[13] O. Diekmann, J.A.P. Heesterbeek, and J.A.J. Metz, *On the definition and the computation of the basic reproduction ratio r_0 in models for infectious-diseases in heterogeneous populations.* J. Math. Biol. **28** (1990), 365–382.

[14] J.A.J. Metz and O. Leimar, *A simple fitness proxy for structured populations with continuous traits, with case studies on the evolution of haplo-diploids and genetic dimorphisms* (*in press*), J. Biol. Dyn.

[15] F.J.A. Jacobs and J.A.J. Metz, *On the concept of attractor for community-dynamical processes I: the case of unstructured populations.* J. Math. Biol. **47** (2003), 222–234.

[16] M. Gyllenberg, F.J.A. Jacobs, and J.A.J. Metz, *On the concept of attractor for community-dynamical processes II: the case of structured populations.* J. Math. Biol. **47** (2003), 235–248.

[17] I. Eshel, *Evolutionary and continuous stability.* J. Theor. Biol. **103** (1983), 99–111.

[18] P.D. Taylor, *Evolutionary stability in one-parameter models under weak selection.* Theor. Popul. Biol. **36** (1989), 125–143.

[19] I. Eshel, *On the changing concept of evolutionary population stability as a reflection of a changing point of view in the quantitative theory of evolution.* J. Math. Biol. **34** (1996), 485–510.

[20] S.A.H. Geritz, É. Kisdi, G. Meszena, and J.A.J. Metz, *Evolutionarily singular strategies and the adaptive growth and branching of the evolutionary tree.* Evol. Ecol. **12** (1998), 35–57.

[21] O. Leimar, *Evolutionary change and Darwinian demons.* Selection **2** (2001), 65–72.

[22] O. Leimar, *The evolution of phenotypic polymorphism: Randomized strategies versus evolutionary branching.* Am. Nat. **165** (2005), 669–681.

[23] O. Leimar, *Multidimensional convergence stability.* Evol. Ecol. Res. **11** (2009), 191–208.

[24] O. Diekmann, M. Gyllenberg, and J.A.J. Metz, *Steady-state analysis of structured population models.* Theor. Popul. Biol. **63** (2003), 309–338.

[25] I. Eshel and M.W. Feldman, *Initial increase of new mutants and some continuity properties of ESS in two locus systems.* Am. Nat. **124** (1984), 631–640.

[26] U. Liberman, *External stability and ESS: criteria for initial increase of new mutant allele.* J. Math. Biol. **26** (1988), 477–485.

[27] S.P. Otto and P. Jarne, *Evolution – haploids – hapless or happening?* Science **292** (2001), 2441–2443.

[28] P. Hammerstein and R. Selten, *Game theory and evolutionary biology. In* R. Auman and S. Hart (eds.), *Handbook of Game Theory With Economic Applications, Vol. II,* volume 11 of *Handbooks in Econom.*, 929–993, North-Holland, Amsterdam, 1994.

[29] P. Hammerstein, *Darwinian adaptation, population genetics and the streetcar theory of evolution.* J. Math. Biol. **34** (1996), 511–532.

[30] F.J. Weissing, *Genetic versus phenotypic models of selection: Can genetics be neglected in a long-term perspective?* J. Math. Biol. **34** (1996), 533–555.

[31] I. Eshel and M.W. Feldman, *Optimization and evolutionary stability under short-term and long-term selection. In* E. Sober and S. Orzack (eds.), *Adaptationism and Optimality,* 161–190, Cambridge University Press, Cambridge, UK, 2001.

[32] I. Eshel, *Short-term and long-term evolution. In* U. Dieckmann and J.A.J. Metz (eds.), *Elements of Adaptive Dynamics*, Cambridge University Press, UK, *In press.*

[33] J.A.J. Metz, S.D. Mylius, and O. Diekmann, *When does evolution optimize?* Evol. Ecol. Res. **10** (2008), 629–654.

[34] M. Gyllenberg and R. Service, *Necessary and sufficient conditions for the existence of an optimisation principle in evolution.* J. Math. Biol. **62** (2011), 259–369.

[35] J.A.J. Metz, S.D. Mylius, and O. Diekmann, *When does evolution optimise? on the relation between types of density dependence and evolutionarily stable life history parameters* (1996), IIASA Working Paper WP-96-04.

[36] R.G. Bowers, *On the determination of evolutionary outcomes directly from the population dynamics of the resident* (in press), J. Math. Biol.

[37] C. Rueffler, J.A.J. Metz, and T.J.M. Van Dooren, *What life cycle graphs can tell about the evolution of life histories* (in revision), J. Math. Biol.

[38] M. Heino, J.A.J. Metz, and V. Kaitala, *The enigma of frequency-dependent selection.* Trends Ecol. Evol. **13** (1998), 367–370.

[39] M.P. Hassell, J.H. Lawton, and R.M. May, *Patterns of dynamical behaviour in single-species populations.* J. Anim. Ecol. **45** (1976), 471–486.

[40] J.G. Kemeny and J.L. Snell, *Finite Markov Chains.* Van Nostrand, Princeton, NJ, 1960.

[41] J.A.J. Metz, P.G.L. Klinkhamer, and T.J. de Jong, *A different model to explain delayed germination.* Evol. Ecol. Res. **11** (2009), 177–190.

[42] R.M. Anderson and R.M. May, *Coevolution of hosts and parasites.* Parasitology **85** (1982), 411–426.

[43] R.M. Anderson and R.M. May, *Infectious Diseases of Humans: Dynamics and Control.* Oxford University Press, Oxford, UK, 1991.

[44] U. Dieckmann and J.A.J. Metz, *Surprising evolutionary predictions from enhanced ecological realism.* Theor. Popul. Biol. **69** (2006), 263–281.

[45] U. Dieckmann, *Adaptive dynamics of pathogens-host interaction. In* U. Dieckmann, J.A.J. Metz, M.W. Sabelis, and K. Sigmund (eds.), *Adaptive Dynamics of Infectious Diseases: In Pursuit of Virulence Management*, 39–59, Cambridge University Press, UK, 2002.

Mats Gyllenberg and Robert Service
Department of Mathematics and Statistics
University of Helsinki
Helsinki, Finland
e-mail: `mats.gyllenberg@helsinki.fi`
`robert.service@helsinki.fi`

J.A.J. (Hans) Metz
Institute of Biology and Mathematical Institute
Leiden, Netherlands

and

Evolution and Ecology Program, IIASA
Laxenburg, Austria
e-mail: `j.a.j.metz@biology.leidenuniv.nl`

Bibliography

Abrams, P.A. (2001), 'Modelling the adaptive dynamics of traits involved in inter- and intraspecific interactions: an assessment of three methods', *Ecol. Lett.* **4**(2), 166–175.

Akin, E. (1979), *The Geometry of Population Genetics*, Vol. 31 of *Lecture Notes in Biomathematics*, Springer-Verlag, Berlin.

Akin, E. (1982), 'Cycling in simple genetic systems', *J. Math. Biol.* **13**, 305–324.

Amundson, R. (2005), *The Changing Role of the Embryo in Evolutionary Thought: Roots of Evo-Devo*, Cambridge: Cambridge University Press, UK.

Anderson, R.M. & May, R.M. (1982), 'Coevolution of hosts and parasites', *Parasitology* **85**, 411–426.

Anderson, R.M. & May, R.M. (1991), *Infectious Diseases of Humans: Dynamics and Control*, Oxford University Press, Oxford, UK.

Arnold, S.J., Bürger, R., Hohenlohe, P.A., Ajie, B.C. & Jones, A.G. (2008), 'Understanding the evolution and stability of the *G*-matrix', *Evolution* **62**(10), 2451–2461.

Athreya, K.B. & Karlin, S. (1971*a*), 'Branching processes with random environments I – extinction probabilities', *Ann. Math. Stat.* **42**(5), 1499.

Athreya, K.B. & Karlin, S. (1971*b*), 'Branching processes with random environments II – limit theorems', *Ann. Math. Stat.* **42**(6), 1843.

Axelrod, R. & Hamilton, W.D. (1981), 'The evolution of cooperation', *Science* **211**(4489), 1390–1396.

Baake, E. & Wiehe, T. (1997), 'Bifurcations in haploid and diploid sequence space models', *J. Math. Biol.* **35**(3), 321–343.

Bacaër, N. (2009), *Histoires de Mathématiques et de Populations*, Editions Cassini, Paris.

Barabino, G.A., Platt, M.O. & Kaul, D.K. (2010), 'Sickle cell biomechanics', *Annu. Rev. Biomed. Eng.* **12**, 345–367.

Barles, G., Evans, L.C. & Souganidis, P.E. (1990), 'Wavefront propagation for reaction-diffusion systems of PDE', *Duke Math. J.* **61**(3), 835–858.

Barles, G., Mirrahimi, S. & Perthame, B. (2009), 'Concentration in Lotka-Volterra parabolic or integral equations: a general convergence result', *Methods Appl. Anal.* **16**, 321–340.

Barles, G. & Perthame, B. (2007), Concentrations and constrained Hamilton-Jacobi equations arising in adaptive dynamics, *in* D. Danielli, ed., 'Recent Developments in Nonlinear Partial Differential Equations', Vol. 439 of *Contemporary Mathematics Series*, pp. 57–68.

Barton, N.H. & Keightley, P.D. (2002), 'Understanding quantitative genetic variation', *Nat. Rev. Genet.* **3**(1), 11–21.

Barton, N.H. & Turelli, M. (1987), 'Adaptive landscapes, genetic-distance and the evolution of quantitative characters', *Genetical Research* **49**(2), 157–173.

Barton, N.H. & Turelli, M. (1991), 'Natural and sexual selection on many loci', *Genetics* **127**(1), 229–255.

Barton, N.H. & Turelli, M. (2004), 'Effects of genetic drift on variance components under a general model of epistasis', *Evolution* **58**(10), 2111–2132.

Beding, B. (2008), 'The stone-age whale hunters who kill with their bare hands', Daily Mail, 12th April.

Berestycki, H., Nadin, G., Perthame, B. & Ryzhik, L. (2009), 'The non-local Fisher-KPP equation: travelling waves and steady states', *Nonlinearity* **22**(12), 2813–2844.

Bernoulli, D. (1766), 'Essai d'une nouvelle analyse de la mortalité causée par la petite vérole', *Mem. Math. Phy. Acad. Roy. Sci. Paris, année 1760* .

Biebricher, C.K., Eigen, M. & William C. Gardiner, J. (1983), 'Kinetics of RNA replication', *Biochemistry* **22**, 2544–2559.

Biebricher, C.K., Eigen, M. & William C. Gardiner, J. (1984), 'Kinetics of RNA replication: Plus-minus asymmetry and double-strand formation', *Biochemistry* **23**, 3186–3194.

Biebricher, C.K., Eigen, M. & William C. Gardiner, J. (1985), 'Kinetics of RNA replication: Competition and selection among self-replicating RNA species', *Biochemistry* **24**, 6550–6560.

Billingsley, P. (1999), *Convergence of Probability Measures*, Wiley-Interscience, USA.

Binet, J. (1843), 'Mémoire sur l'intégration des équations linéaires aux différences finies, d'un ordre quelconque, à coefficients variables', *Comptes Rendue Hebdomadaires des Séances de l'Académie des Sciences (Paris)* **17**, 559–567.

Boehm, C. (1999), *Hierarchy in the Forest: The Evolution of Egalitarian Behavior*, Harvard University Press, USA.

Boesch, C. (2002), 'Cooperative hunting roles among tai chimpanzees', *Human Nature – an Interdisciplinary Biosocial Perspective* **13**(1), 27–46.

Bolker, B. & Pacala, S.W. (1997), 'Using moment equations to understand stochastically driven spatial pattern formation in ecological systems', *Theor. Popul. Biol.* **52**(3), 179–197.

Bolker, B.M. & Pacala, S.W. (1999), 'Spatial moment equations for plant competition: Understanding spatial strategies and the advantages of short dispersal', *Am. Nat.* **153**(6), 575–602.

Bovier, A. & Champagnat, N. (n.d.), 'Time scales in adaptive dynamics: directional selection, fast and slow branching', (*in preparation*).

Bowers, R.G. (*in press*), On the determination of evolutionary outcomes directly from the population dynamics of the resident. J. Math. Biol.

Bowles, S. (2003), *Microeconomics: Behavior, Institutions and Evolution*, Princeton University Press, USA.

Boyd, R. & Richerson, P.J. (1985), *Culture and the Evolutionary Process*, University of Chicago Press, USA.

Boyd, R. & Richerson, P.J. (1988), 'The evolution of reciprocity in sizable groups', *J. Theor. Biol.* **132**, 337–356.

Brakmann, S. & Johnsson, K., eds. (2002), *Directed Molecular Evolution of Proteins or How to Improve Enzymes for Biocatalysis*, Wiley-VCH, Weinheim, DE.

Brigatti, E., Schwammle, V. & Neto, M.A. (2008), 'Individual-based model with global competition interaction: fluctuation effects in pattern formation', *Phys. Rev. E* **77**(2, Part 1).

Bryant, J. (1994), Coordination theory, the stag hunt and macroeconomics, *in* J. Friedman, ed., 'Problems of Coordination in Economic Activity', Kluwer, Dordrecht, The Netherlands, pp. 207–225.

Bryden, J. & Jansen, V.A.A. (2010), 'The impact of clonal mixing on the evolution of social behaviour in aphids', *P. Roy. Soc. B: Biological Sciences* **277**, 1651–1657.

Bull, J.J., Ancel Myers, L. & Lachmann, M. (2005), 'Quasispecies made simple', *PLoS Comp. Biol.* **1**, 450–460.

Bull, J.J., Sanjuán, R. & Wilke, C.O. (2007), 'Theory for lethal mutagenesis for viruses', *J. Virology* **81**, 2930–2939.

Bulmer, M.G. (1980), *The Mathematical Theory of Quantitative Genetics*, Oxford: Clarendon Press.

Bürger, R. (1991), 'Moments, cumulants, and polygenic dynamics', *J. Math. Biol.* **30**(2), 199–213.

Bürger, R. (2000), *The Mathematical Theory of Selection, Recombination and Mutation*, Wiley Series in Mathematical and Computational Biology. Chichester: Wiley.

Bürger, R. (2005), 'A multilocus analysis of intraspecific competition and stabilizing selection on a quantitative trait', *J. Math. Biol.* **50**(4), 355–396.

Bürger, R. (2009*a*), 'Multilocus selection in subdivided populations I. Convergence properties for weak or strong migration', *J. Math. Biol.* **58**(6), 939–978.

Bürger, R. (2009*b*), 'Multilocus selection in subdivided populations II. Maintenance of polymorphism under weak or strong migration', *J. Math. Biol.* **58**(6), 979–997.

Bürger, R. (2010), 'Evolution and polymorphism under the multilocus Levene model without epistasis', *Theor. Pop. Biol.* **78**, 123–138.

Bürger, R. & Lande, R. (1994), 'On the distribution of the mean and variance of a quantitative trait under mutation selection drift balance', *Genetics* **138**(3), 901 912.

Calsina, A. & Cuadrado, S. (2004), 'Small mutation rate and evolutionarily stable strategies in infinite dimensional adaptive dynamics', *J. Math. Biol.* **48**(2), 135–159.

Calsina, A. & Perello, C. (1995), 'Equations for biological evolution', *Proc. R. Soc. Edinb. Sect. A-Math.* **125**(Part 5), 939–958.

Carrillo, J.A., Cuadrado, S. & Perthame, B. (2007), 'Adaptive dynamics via Hamilton-Jacobi approach and entropy methods for a juvenile-adult model', *Math. Biosci.* **205**(1), 137–161.

Castilloux, A.M. & Lessard, S. (1995), 'The fundamental theorem of natural selection in Ewens' sense (case of many loci)', *Theoret. Pop. Biol.* **48**, 733–742.

Champagnat, N. (2003), 'Convergence of adaptive dynamics n-morphic jump processes to the canonical equation and degenerate diffusion approximation', Preprint of the University of Nanterre (Paris 10) No. 03/7.

Champagnat, N. (2006), 'A microscopic interpretation for adaptive dynamics trait substitution sequence models', *Stoch. Process. Their Appl.* **116**(8), 1127–1160.

Champagnat, N., Ferrière, R. & Méléard, S. (2006), 'Unifying evolutionary dynamics: From individual stochastic processes to macroscopic models', *Theor. Popul. Biol.* **69**(3), 297–321.

Champagnat, N., Ferrière, R. & Méléard, S. (2008*a*), 'From individual stochastic processes to macroscopic models in adaptive evolution', *Stoch. Models* **24**(Suppl. 1), 2–44. 9th Symposium on Probability and Stochastic Process, Guanajuato, Mexico, Nov. 20–24, 2006.

Champagnat, N., Ferrière, R. & Méléard, S. (2008*b*), Individual-based probabilistic models of adaptive evolution and various scaling approximations, *in* R. C. Dalang, M. Dozzi & F. Russo, eds., 'Seminar On Stochastic Analysis, Random Fields And Applications V', Vol. 59 of *Progress In Probability*, pp. 75–113. 5th Seminar on Stochastic Analysis, Random Fields and Applications, Ascona, Switzerland, May 30–Jun 03, 2005.

Champagnat, N. & Méléard, S. (*to appear*), Polymorphic evolution sequence and evolutionary branching. Probab. Theory Relat. Fields.

Christiansen, F.B. & Loeschcke, V. (1980), 'Evolution and intraspecific exploitative Competition. 1. One-Locus theory for small additive gene effects', *Theor. Popul. Biol.* **18**(3), 297–313.

Claessen, D., Andersson, J., Persson, L. & de Roos, A.M. (2007), 'Delayed evolutionary branching in small populations', *Evol. Ecol. Res.* **9**(1), 51–69.

Cohen, J.E. (2004), 'Mathematics is biology's next microscope, only better; biology is mathematics' next physics, only better', *PLoS. Biol.* **2**(12), 2017–2023.

Crandall, M.G., Ishii, H. & Lions, P.L. (1992), 'User's guide to viscosity solutions of second order partial differential equations', *Bull. Amer. Math. Soc.* **27**(1), 1–67.

Creel, S. & Creel, N.M. (1995), 'Communal hunting and pack size in African wild dogs, Lycaon-pictus', *Anim. Behav.* **50**, 1325–1339.

Darwin, C. (1859), *On the Origin of Species by Means of Natural Selection or the Preservation of Favoured Races in the Struggle for Life*, John Murray, London.

Darwin, C. (1871), *The Descent of Man, and Selection in Relation to Sex*, John Murray, London.

Dawes, R.M. (1980), 'Social dilemmas. social dilemmas. social dilemmas', *Annu. Rev. of Psychol.* **31**, 169–193.

Dercole, F. & Rinaldi, S. (2008), *Analysis of Evolutionary Processes: the Adaptive Dynamics Approach and its Applications*, Princeton University Press, USA.

Desvillettes, L., Jabin, P.E., Mischler, S. & Raoul, G. (2008), 'On selection dynamics for continuous structured populations', *Commun. Math. Sci.* **6**(3), 729–747.

Dieckmann, U. (2002), Adaptive dynamics of pathogens-host interaction, *in* U. Dieckmann, J. A. J. Metz, M. W. Sabelis & K. Sigmund, eds., 'Adaptive Dynamics of Infectious Diseases: In Pursuit of Virulence Management', Cambridge University Press, UK, pp. 39–59.

Dieckmann, U. & Doebeli, M. (1999), 'On the origin of species by sympatric speciation', *Nature* **400**(6742), 354–357.

Dieckmann, U., Doebeli, M., Metz, J. A. J. & Tautz, D., eds. (2004), *Adaptive Speciation*, Cambridge University Press, UK.

Dieckmann, U. & Law, R. (1996), 'The dynamical theory of coevolution: A derivation from stochastic ecological processes', *J. Math. Biol.* **34**(5-6), 579–612.

Dieckmann, U. & Law, R. (2000), Relaxation projections and the method of moments, *in* 'The Geometry of Ecological Interactions: Symplifying Spatial Complexity', Cambridge University Press, UK, pp. 412–455.

Dieckmann, U. & Metz, J.A.J. (2006), 'Surprising evolutionary predictions from enhanced ecological realism', *Theor. Popul. Biol.* **69**(3), 263–281.

Diekmann, O. (2004), A beginner's guide to adaptive dynamics, *in* 'Mathematical Modelling of Population Dynamics', Vol. 63 of *Banach Center Publ.*, Polish Acad. Sci., Warsaw, pp. 47–86.

Diekmann, O., Gyllenberg, M. & Metz, J.A.J. (2003), 'Steady-state analysis of structured population models', *Theor. Popul. Biol.* **63**(4), 309–338.

Diekmann, O., Heesterbeek, J.A.P. & Metz, J.A.J. (1990), 'On the definition and the computation of the basic reproduction ratio r_0 in models for infectious-diseases in heterogeneous populations', *J. Math. Biol.* **28**(4), 365–382.

Diekmann, O., Jabin, P.E., Mischler, S. & Perthame, B. (2005), 'The dynamics of adaptation: An illuminating example and a Hamilton-Jacobi approach', *Theor. Popul. Biol.* **67**(4), 257–271.

Doebeli, M. & Dieckmann, U. (2003), 'Speciation along environmental gradients', *Nature* **421**(6920), 259–264.

Domingo, ed., E. (2005), 'Virus entry into error catastrophe as a new antiviral strategy', *Virus Research* **107**(2), 115–228.

Donoghue, P. C. J. & Smith, M. P., eds. (2003), *Telling the evolutionary time: molecular clocks and the fossil record.*, CRC Press, USA.

Drake, J.W. (1993), 'Rates of spontaneous mutation among RNA viruses', *Proc. Natl. Acad. Sci. USA* **90**, 4171–4175.

Drake, J.W., Charlesworth, B., Charlesworth, D. & Crow, J.F. (1998), 'Rates of spontaneous mutation', *Genetics* **148**, 1667–1686.

Dresher, M. (1961), *The Mathematics of Games of Strategy: Theory and Applications*, Prentice-Hall, Englewood Cliffs, NJ.

Dugatkin, L.A. (2007), 'Inclusive fitness theory from Darwin to Hamilton', *Genetics* **176**(3), 1375–1380.

Dunlap, R.A. (1997), *The Golden Ratio and Fibonacci Numbers*, World Sceintific, Singapore.

Durinx, M., Metz, J.A.J. & Meszena, G. (2008), 'Adaptive dynamics for physiologically structured population models', *J. Math. Biol.* **56**(5), 673–742.

Edwards, A.W. (1994), 'The fundamental theorem of natural selection', *Biological Reviews* **69**, 443–474.

Egas, M., Sabelis, M.W., Vala, F. & Lesna, I. (2004), Adaptive speciation in agricultural pests, *in* U. Dieckmann, M. Doebeli, J. A. J. Metz & D. Tautz, eds., 'Adaptive Speciation', Cambridge University Press, UK, pp. 249–263.

Eigen, M. (1971), 'Selforganization of matter and the evolution of biological macromolecules', *Naturwissenschaften* **58**, 465–523.

Eigen, M. & Schuster, P. (1978), 'The hypercycle. A principle of natural self-organization. Part B: The abstract hypercycle', *Naturwissenschaften* **65**, 7–41.

Eldredge, N. & Gould, S.J. (1972), Punctuated equilibria: an alternative to phyletic gradualism, *in* T. Schopf, ed., 'Models in Paleobiology', Freeman, Cooper and Company,, EUA, pp. 82–115.

Eshel, I. (1983), 'Evolutionary and continuous stability', *J. Theor. Biol.* **103**(1), 99–111.

Eshel, I. (1996), 'On the changing concept of evolutionary population stability as a reflection of a changing point of view in the quantitative theory of evolution', *J. Math. Biol.* **34**(5-6), 485–510.

Eshel, I. (*In press*), Short-term and long-term evolution, *in* U. Dieckmann & J. A. J. Metz, eds., 'Elements of Adaptive Dynamics', Cambridge University Press, UK.

Eshel, I. & Feldman, M.W. (1984), 'Initial increase of new mutants and some continuity properties of ESS in two locus systems', *Am. Nat.* **124**(5), 631–640.

Eshel, I. & Feldman, M.W. (2001), Optimization and evolutionary stability under short-term and long-term selection, *in* E. Sober & S. Orzack, eds., 'Adaptationism and Optimality', Cambridge University Press, Cambridge, UK, pp. 161–190.

Eshel, I., Feldman, M.W. & Bergman, A. (1998), 'Long-term evolution, short-term evolution, and population genetic theory', *J. Theor. Biol.* **191**(4), 391–396.

Ethier, S.N. & Kurtz, T.G. (1986), *Markov Processes, Characterization and Convergence*, John Wiley & Sons, USA.

Euler, L. (1760), 'Theoremata arithmetica nova methodo demonstrata', *Novi Comentarii Scientiarum Imperialis Petropolitanæ* **8**, 74–104. Reprinted in Euler's *Commentationes Arithmeticæ Collectæ*, Vol. 1, 274–286 and in Euler's *Opera Omnia*, Series 1, Vol. 2, 531–555.

Euler, L. (1767), 'Recherches générales sur la mortalité et la multiplication du genre humain', *Mem. Acad. R. Sci. Belles Lett. (Berlin)* **16**, 144–164.

Euler, L. (1988), *Introductio in Analysin Infinitorum,* 1748. *English Translation: John Blanton, Introduction to Analysis of the Infinite*, Vol. I and II, Springer-Verlag, Berlin.

Evans, L.C. & Souganidis, P.E. (1989), 'A PDE approach to geometric optics for certain semilinear parabolic equations', *Indiana Univ. Math. J.* **38**(1), 141–172.

Ewens, W.J. (1969), 'Mean fitness increases when fitnesses are additive', *Nature* **221**(5185), 1076.

Ewens, W.J. (1979), *Mathematical Population Genetics*, Vol. 9 of *Biomathematics Texts*, Springer-Verlag, Berlin.

Ewens, W.J. (1992), 'An optimizing principle of natural selection in evolutionary population genetics', *Theoret. Pop. Biol.* **42**, 333–346.

Ewens, W.J. (2004), *Mathematical Population Genetics. I: Theoretical Introduction. 2nd ed.*, Interdisciplinary Mathematics 27. New York, NY: Springer.

Ewens, W.J. (2011), 'What is the gene trying to do?', *Brit. J. Philos. Sci.* **62**(1), 155–176.

Fahlbeck, P. (1898, 1902), *Sveriges adel, statistisk undersökning öfver de å riddarhuset introducerade ätterna, I and II (The Swedish Nobility, a Statistical Investigation of the Families of the House of Nobility)*, Gleerups, Sweden.

Ferrière, R. & Gatto, M. (1995), 'Lyapunov exponents and the mathematics of invasion in oscillatory or chaotic populations', *Theor. Popul. Biol.* **48**(2), 126–171.

Fisher, R.A. (1918), 'The correlation between relatives on the supposition of Mendelian inheritance', *Trans. of the Roy. Soc. of Edinburgh* **52**, 399–433.

Fisher, R.A. (1922), 'On the dominance ratio', *Proc. Roy. Soc. Edin.* **42**, 321–341.

Fisher, R.A. (1930), *The Genetical Theory of Natural Selection*, Clarendon Press, Oxford.

Flood, M.M. (1952), 'Some experimental games, research memorandum RM-789', *RAND Corporation, Santa Monica, CA* .

Fontana, W. & Schuster, P. (1998), 'Continuity in evolution. On the nature of transitions', *Science* **280**, 1451–1455.

Fournier, N. & Méléard, S. (2004), 'A microscopic probabilistic description of a locally regulated population and macroscopic approximations', *Ann. Appl. Probab.* **14**(4), 1880–1919.

Freidlin, M.I. & Wentzel, A.D. (1984), *Random Perturbations of Dynamical Systems*, Springer-Verlag, Germany.

Galis, F. & Metz, J.A.J. (2001), 'Testing the vulnerability of the phylotypic stage: On modularity and evolutionary conservation', *J. Exp. Zool.* **291**(2), 195–204.

Galis, F., van Alphen, J.J.M. & Metz, J.A.J. (2001), 'Why do we have five fingers? The evolutionary constraint on digit numbers', *Trends Ecol. Evol.* **16**, 637–646.

Galis, F., Van Der Sluijs, I., Van Dooren, T.J., Metz, J.A. & Nussbaumer, M. (2007), 'Do large dogs die young?', *J. Exp. Zool. Part B* **308B**(2), 119–126.

Galis, F., Van Dooren, T.J.M., Feuth, J.D., Metz, J.A.J., Witkam, A., Ruinard, S., Steigenga, M.J. & Wijnaendts, L.C.D. (2006), 'Extreme selection in humans against homeotic transformations of cervical vertebrae', *Evolution* **60**(12), 2643–2654.

Galis, F., van Dooren, T.J.M. & Metz, J.A.J. (2002), 'Conservation of the segmented germband stage: modularity and robustness or pleiotropy and stabilizing selection?', *TIG* **18**, 504–509.

Galton, F. (1873), 'Problem 4001: On the extinction of surnames', *Educational Times* **26**, 17.

Galton, F. & Watson, H.W. (1874), 'On the probability of extinction of families', *EJ. Anthropol. Inst.* **4**, 138–144.

Gardner, A. & West, S.A. (2004), 'Spite and the scale of competition', *J. Evol. Biol.* **17**(6), 1195–1203.

Gauduchon, M. & Perthame, B. (2010), 'Survival thresholds and mortality rates in adaptive dynamics: conciliating deterministic and stochastic simulations', *Mathematical Medicine and Biology* **27**(3), 195–210.

Gavrilets, S. (2004), *Fitness landscapes and the origin of species*, Princeton Univ. Press, Princeton, NJ.

Genieys, S., Bessonov, N. & Volpert, V. (2009), 'Mathematical model of evolutionary branching', *Math. Comput. Model.* **49**(11-12), 2109–2115. Euro-Mediterranean Conference on Biomathematics, El Shorouk, Egypt, JUN, 2007.

Genieys, S., Volpert, V. & Auger, P. (2006), 'Adaptive dynamics: modelling darwin's divergence principle', *C. R. Biol.* **329**(11), 876–879.

Geritz, S.A.H. (2005), 'Resident-invader dynamics and the coexistence of similar strategies', *J. Math. Biol.* **50**(1), 67–82.

Geritz, S.A.H., Gyllenberg, M., Jacobs, F.J.A. & Parvinen, K. (2002), 'Invasion dynamics and attractor inheritance', *J. Math. Biol.* **44**(6), 548–560.

Geritz, S.A.H. & Kisdi, É. (2000), 'Adaptive dynamics in diploid, sexual populations and the evolution of reproductive isolation', *Proc. R. Soc. Lond. Ser. B-Biol. Sci.* **267**(1453), 1671–1678.

Geritz, S.A.H., Kisdi, É., Meszena, G. & Metz, J.A.J. (1998), 'Evolutionarily singular strategies and the adaptive growth and branching of the evolutionary tree', *Evol. Ecol.* **12**(1), 35–57.

Gillespie, J.H. (1991), *The Causes of Molecular Evolution*, Oxford University Press, UK.

Gimelfarb, A. (1989), 'Genotypic variation for a quantitative character maintained under stabilizing selection without mutations: Epistasis', *Genetics* **123**(1), 217–227.

Gimelfarb, A. & Willis, J.H. (1994), 'Linearity versus nonlinearity of offspring-parent regression: An experimental-study of *D*rosophila melanogaster', *Genetics* **138**(2), 343–352.

Gould, S.J. & Eldredge, N. (1977), 'Punctuated equilibria: the tempo and mode of evolution reconsidered', *Paleobiology* **3**, 115.151.

Gourley, S.A. (2000), 'Travelling front solutions of a nonlocal Fisher equation', *J. Math. Biol.* **41**(3), 272–284.

Grafen, A. (1985), 'A geometric view of relatedness', *Oxford Surveys in Evolutionary Biology* **2**, 28–89.

Grafen, A. & Archetti, M. (2008), 'Natural selection of altruism in inelastic viscous homogeneous populations', *J. Theor. Biol.* **252**(4), 694–710.

Graham, R.L., Knuth, D.E. & Patashnik, O. (1994), *Concrete Mathematics*, second edn., Addison-Wesley Publishing Co., Reading, MA.

Gyllenberg, M., Jacobs, F.J.A. & Metz, J.A.J. (2003), 'On the concept of attractor for community-dynamical processes II: the case of structured populations', *J. Math. Biol.* **47**(3), 235–248.

Gyllenberg, M. & Parvinen, K. (2001), 'Necessary and sufficient conditions for evolutionary suicide', *Bull. Math. Biol.* **63**(5), 981–993.

Gyllenberg, M. & Service, R. (2011), 'Necessary and sufficient conditions for the existence of an optimisation principle in evolution', *J. Math. Biol.* **62**, 259–369.

Haccou, P., Jagers, P. & Vatutin, V.A. (2005), *Branching Processes. Variation, Growth, and Extinction of Populations*, Vol. 5 of *Cambridge Studies in Adaptive Dynamics*, Cambridge University Press, UK.

Haldane, J.B.S. (1927), 'A mathematical theory of natural and artificial selection, Part V: Selection and mutation', *Proceed. of the Cambridge Phil. Soc.* **23**, 838–844.

Haldane, J.B.S. (1932), *The Causes of Evolution*, Longmans Green, London.

Haldane, J.B.S. (1955), 'Population genetics', *New Biol.* **18**, 34–51.

Hamilton, W.D. (1964), 'Genetical evolution of social behaviour I', *J. Theor. Biol.* **7**(1), 1.

Hamilton, W.D. (1967), 'Extraordinary sex ratios', *Science* **156**(3774), 477.

Hamilton, W.D. (1970), 'Selfish and spiteful behaviour in an evolutionary model', *Nature* **228**, 1218–1220.

Hamilton, W.D. (1971), Selection of selfish and altruistic behaviour in some extreme models, *in* J. F. Eisenberg & W. S. Dillon, eds., 'Man and Beast: Comparative Social Behaviour', Smithsonian Press, USA, pp. 57–91.

Hamilton, W.D. (1975), Biosocial anthropology, *in* 'Biosocial anthropology', Malaby Press, London, UK, pp. 133–155.

Hammerstein, P. (1996), 'Darwinian adaptation, population genetics and the streetcar theory of evolution', *J. Math. Biol.* **34**(5-6), 511–532.

Hammerstein, P. & Selten, R. (1994), Game theory and evolutionary biology, *in* R. Auman & S. Hart, eds., 'Handbook of Game Theory With Economic Applications, Vol. II', Vol. 11 of *Handbooks in Econom.*, North-Holland, Amsterdam, pp. 929–993.

Hammerstein, P.E. (2003), *Genetic and Cultural Evolution of Cooperation*, MIT press, Cambridge, MA., USA.

Hamming, R.W. (1986), *Coding and Information Theory*, 2nd edn., Prentice-Hall, Englewood Cliffs, NJ.

Hansen, T.F. & Wagner, G.P. (2001), 'Modeling genetic architecture: A multilinear theory of gene interaction', *Theor. Popul. Biol.* **59**(1), 61–86.

Hardin, G. (1968), 'The tragedy of the commons', *Science* **162**(5364), 1243–1248.

Hardy, G.H. (1908), 'Mendelian proportions in a mixed population', *Science* **28**, 49–50.

Hassell, M.P., Lawton, J.H. & May, R.M. (1976), 'Patterns of dynamical behaviour in single-species populations', *J. Anim. Ecol.* **45**(2), 471–486.

Hastings, A. (1981*a*), 'Simultaneous stability of $D = 0$ and $D \neq 0$ for multiplicative viabilities at two loci: an analytical study', *J. Theor. Biol.* **89**(1), 69–81.

Hastings, A. (1981*b*), 'Stable cycling in discrete-time genetic models', *Proc. Natl. Acad. Sci. USA* **78**(11), 7224–7225.

Hauert, C., Michor, F., Nowak, M.A. & Doebeli, M. (2006), 'Synergy and discounting of cooperation in social dilemmas', *J. Theo. Bio.* pp. 195–202.

Hauert, C., Traulsen, A., Brandt, H., Nowak, M.A. & Sigmund, K. (2007), 'Via freedom fo coercion: The emergence of costly punishment', *Science* **316**, 1905–1907.

Heino, M., Metz, J.A.J. & Kaitala, V. (1998), 'The enigma of frequency-dependent selection', *Trends Ecol. Evol.* **13**(9), 367–370.

Hermisson, J., Hansen, T.F. & Wagner, G.P. (2003), 'Epistasis in polygenic traits and the evolution of genetic architecture under stabilizing selection', *Amer. Nat.* **161**(5), 708–734.

Heyde, C.C. & Seneta, E. (1977), *I.J. Bienaymé, Statistical Theory Anticipated*, Springer-Verlag, New York.

Hirsch, W.M. & Smale, S. (1974), *Differential Equations, Dynamical Systems, and Linear Algebra*, Academic Press, New York.

Hoekstra, R.F., Bijlsma, R. & Dolman, A.J. (1985), 'Polymorphism from environmental heterogeneity – models are only robust if the heterozygote is close in fitness to the favored homozygote in each environment', *Genet. Res.* **45**(3), 299–314.

Hofbauer, J. (1985), 'The selection mutation equation', *J. Math. Biol.* **23**, 41–53.

Hofbauer, J. & Iooss, G. (1984), 'A Hopf-bifurcation theorem for difference-equations approximating a differential-equation', *Monatsh. Math.* **98**(2), 99–113.

Hofbauer, J. & Sigmund, K. (1988), *The Theory of Evolution and Dynamical Systems. Mathematical Aspects of Selection*, London Mathematical Society Student Texts, 7. Cambridge University Press, U,K.

Hofbauer, J. & Sigmund, K. (1990), 'Adaptive dynamics and evolutionary stability', *Appl. Math. Lett.* **3**(4), 75–79.

Hofbauer, J. & Sigmund, K. (1998), *Evolutionary Games and Population Dynamics*, Cambridge Univ. Press, Cambridge, UK.

Hofbauer, J. & Sigmund, K. (2003), 'Evolutionary game dynamics', *Bull. Amer. Math. Soc. (N.S.)* **40**(4), 479–519.

Israel, G. (1993), 'The emergence of biomathematics and the case of population dynamics, a revival of mechanical reductionism and darwinism', *Science in Context* **6**(2), 469–509.

Jabin, P.-E. & Raoul, G. (2009), 'On selection dynamics for competitive interactions', Preprint CMLA-ENS Cachan 17.

Jäckel, C., Kast, P. & Hilvert, D. (2008), 'Protein design by directed evolution', *Annu. Rev. Biophys.* **37**, 153–173.

Jacobs, F.J.A. & Metz, J.A.J. (2003), 'On the concept of attractor for community-dynamical processes I: the case of unstructured populations', *J. Math. Biol.* **47**(3), 222–234.

Jagers, P. (1975), *Branching Processes with Biological Applications*, Wiley Series in Probability and Mathematical Statistics-Applied, London, UK.

Jagers, P. & Klebaner, F. (2003), 'Random variation and concentration effects in PCR', *J. Theor. Biol.* **224**(3), 299–304.

Jagers, P., Klebaner, F.C. & Sagitov, S. (2007), 'On the path to extinction', *Proc. Natl. Acad. Sci. USA* **104**(15), 6107–6111.

Jansen, V.A.A. & Mulder, G.S.E.E. (1999), 'Evolving biodiversity', *Ecol. Lett.* **2**(6), 379–386.

Jansen, V.A.A. & Vitalis, R. (2007), 'The evolution of dispersal in a Levins' type metapopulation model', *Evolution* **61**, 2386–2397.

Janson, S. (1994), Large deviation inequalities for sums of indicator variables. Technical report, Dep. Mathematics, Uppsala University **34**.

Jepsen, G. L., Simpson, G. G. & Mayr, E., eds. (1949), *Genetics, Paleontology and Evolution*, Princeton University Press, USA.

Jervis, R. (1978), 'Cooperation under the security dilemma', *World Politics* **30**, 167–214.

Joffe, A. & Metivier, M. (1986), 'Weak-convergence of sequences of semimartingales with applications to multitype branching-processes', *Adv. Appl. Probab.* **18**(1), 20–65.

Jones, A.G., Arnold, S.J. & Bürger, R. (2003), 'Stability of the *G*-matrix in a population experiencing pleiotropic mutation, stabilizing selection, and genetic drift', *Evolution* **57**(8), 1747–1760.

Jones, A.G., Arnold, S.J. & Bürger, R. (2004), 'Evolution and stability of the *G*-matrix on a landscape with a moving optimum', *Evolution* **58**(8), 1639–1654.

Jones, B.L., Enns, R.H. & Rangnekar, S.S. (1976), 'On the theory of selection of coupled macromolecular systems', *Bull. Math. Biol.* **38**, 15–28.

Joyce, G.F. (2004), 'Directed evolution of nucleic acid enzymes', *Annu. Rev. Biochem.* **73**, 791–836.

Judson, H.F. (1979), *The Eighth Day of Creation. The Makers of the Revolution in Biology*, Jonathan Cape, London.

Karlin, S. & Feldman, M.W. (1978), 'Simultaneous stability of $D = 0$ and $D \neq 0$ for multiplicative viabilities at two loci', *Genetics* **90**(4), 813–825.

Karlin, S. & Taylor, H.M.A. (1975), *A First Course in Stochastic Processes*, 2nd edn., Academic, London, UK.

Kauffman, S.A. (1994), *The Origins of Order. Self-Organization and Selection in Evolution*, Oxford University Press, New York.

Kemeny, J.G. & Snell, J.L. (1960), *Finite Markov Chains*, Van Nostrand, Princeton, NJ.

Keynes, J.M. (1923), *A Tract on Monetary Reform*, MacMillan, London.

Kimura, M. (1983), *The Neutral Theory of Molecular Evolution*, Cambridge: University Press, UK.

Kingman, J.F. (1961*a*), 'On an inequality in partial averages', *Q. J. Math.* **12**(45), 78.

Kingman, J.F.C. (1961*b*), 'A mathematical problem in population genetics', *Proc. Camb. Phil. Soc.* **57**, 574–582.

Kingman, J.F.C. (1961*c*), 'A matrix inequality', *Quart. J. Math.* **12**, 78–80.

Kingman, J.F.C. (1982), 'The coalescent', *Stoch. Process. Appl.* **13**, 235–248.

Kirkpatrick, M., Johnson, T. & Barton, N. (2002), 'General models of multilocus evolution', *Genetics* **161**(4), 1727–1750.

Kisdi, É. (1999), 'Evolutionary branching under asymmetric competition', *J. Theor. Biol.* **197**(2), 149–162.

Kisdi, É. & Geritz, S.A.H. (*In press*), Evolutionary branching and speciation: Insights from few-locus models, *in* U. Dieckmann & J. A. J. Metz, eds., 'Elements of Adaptive Dynamics', Cambridge University Press, UK.

Klebaner, F.C. & Lazar, J. (1996), On the quasi-stationary distribution in randomly perturbed dynamical systems with a single attracting point, *in* R. Wilson, D. Pra Murthy & S. Osaki, eds., 'Proc. 2nd Australia-Japan Workshop on Stochastic Models in Engineering, Technology, and Management', The University of Queensland, Australia, pp. 348–354.

Klebaner, F.C., Sagitov, S., Vatutin, V.A., Haccou, P. & Jagers, P. (*to appear*), Stochasticity in the adaptive dynamics of evolution: the bare bones. Invited submission. J. Biol. Dynamics.

Klussmann, S., ed. (2006), *The Aptamer Handbook. Functional Oligonucleotides and Their Applications*, Wiley-VCh Verlag, Weinheim, DE.

Knuth, D.E. (1997), *The Art of Computer Programming. Vol. I: Fundamental Algorithms*, third edn., Addison-Wesley Publishing Co., Reading, MA.

Kollock, P. (1998), 'Social dilemmas: The anatomy of cooperation', *Annu. Rev. Sociol.* **24**, 183–214.

Kopp, M. & Hermisson, J. (2006), 'The evolution of genetic architecture under frequency-dependent disruptive selection', *Evolution* **60**(8), 1537–1550.

Kun, L.A. & Lyubich, Yu.I. (1980), 'Convergence to equilibrium in a polylocus polyallele population with additive selection', *Probl. Inform. Transmiss.* **16**, 152–161.

Lamarck, J.B. (1809), *Philosophie zoologique ou exposition des considérations relatives à l'histoire naturelle des animaux*, Dentu, Paris.

Lande, R. (1979), 'Quantitative genetic-analysis of multivariate evolution, applied to brain-body size allometry', *Evolution* **33**(1), 402–416.

Lande, R. & Arnold, S.J. (1983), 'The measurement of selection on correlated characters', *Evolution* **37**(6), 1210–1226.

Leimar, O. (2001), 'Evolutionary change and Darwinian demons', *Selection* **2**, 65–72.

Leimar, O. (2005), 'The evolution of phenotypic polymorphism: Randomized strategies versus evolutionary branching', *Am. Nat.* **165**(6), 669–681.

Leimar, O. (2009), 'Multidimensional convergence stability', *Evol. Ecol. Res.* **11**(2), 191–208.

Leimar, O. (*In press*), Multidimensional convergence stability and the canonical adaptive dynamics, *in* U. Dieckmann & J. A. J. Metz, eds., 'Elements of Adaptive Dynamics', Cambridge University Press, UK.

Leslie, P.H. (1945), 'On the use of matrices in certain population mathematics', *Biometrika* **33**(3), 183–212.

Lessard, S. (1997), 'Fisher's fundamental theorem of natural selection revisited', *Theor. Popul. Biol.* **52**(2), 119–136.

Li, C.C. (1967), 'Fundamental theorem of natural selection', *Nature* **214**(5087), 505.

Liberman, U. (1988), 'External stability and ESS: criteria for initial increase of new mutant allele', *J. Math. Biol.* **26**(4), 477–485.

Lorz, A., Mirrahimi, S. & Perthame, B. (n.d.), 'Dirac concentration in a multidimensional nonlocal parabolic equation', *In preperation.*

Lucas, K. & Roosen, P., eds. (2010), *Emergence, Analysis, and Evolution of Structures. Concepts and Strategies across Disciplines*, Understanding Complex Systems, Springer-Verlag, Berlin.

Lyubich, Yu.I. (1971), 'Basic concepts and theorems of evolutionary genetics of free populations', *Russ. Math. Surv.* **26**, 51–123.

Lyubich, Yu.I. (1992), *Mathematical Structures in Population Genetics*, Vol. 22 of *Biomathematics*, Springer-Verlag, Berlin.

Mackay, T.F.C. (2001), 'Quantitative trait loci in *Drosophila*', *Nat. Rev. Genet.* **2**(1), 11–20.

Macy, M.W. & Flache, A. (2002), 'Learning dynamics in social dilemmas', *Proc. Natl. Acad. Sci. USA* **99**(3), 7229–7236.

Malécot, G. (1948), *Les Mathématiques de l'Hérédité*, Masson, Paris.

Malthus, T.R. (1798), *An Essay on the Principle of Population, as it Affects the Future Improvement of Society, with Remarks on the Speculations of Mr Godwin, M. Condorcet and Other Writers*, John Murray, London.

Marrow, P., Law, R. & Cannings, C. (1992), 'The coevolution of predator prey interactions – ESSs and Red Queen dynamics', *Proc. R. Soc. Lond. Ser. B-Biol. Sci.* **250**(1328), 133–141.

Maynard Smith, J. (1964), 'Group selection and kin selection', *Nature* **201**(492), 1145.

Maynard Smith, J. (1966), 'Sympatric speciation', *Am. Nat.* **100**(916), 637.

Maynard Smith, J. (1974), 'Theory of games and evolution of animal conflicts', *J. Theor. Biol.* **47**(1), 209–221.

Maynard Smith, J. (1982), *Evolution and the Theory of Games*, Cambridge University Press, Cambridge, UK.

Maynard Smith, J. & Price, G.R. (1973), 'The logic of animal conflict', *Nature* **246**(5427), 15–18.

Maynard Smith, J. & Szathmáry, E. (1995), *The Major Transitions in Evolution*, Freeman, Oxford, UK.

Mayr, E. (1982), *The Growth of Biological Thought: Diversity, Evolution, and Inheritance.*, Belknap Press, USA.

Mayr, E. & Provine, W.B. (1980), *The Evolutionary Synthesis: Perspectives on the Unification of Biology*, Harvard University Press, USA.

Méléard, S. & Tran, V.C. (2009), 'Trait substitution sequence process and canonical equation for age-structured populations', *J. Math. Biol.* **58**(6), 881–921.

Mendel, G. (1866), 'Versuche über Pflanzenhybriden. (Experiments relating to plant hybridization)', *Verh. Naturforsch. Ver. Brünn* **4**, 3–17.

Meszéna, G., Gyllenberg, M., Jacobs, F.J. & Metz, J.A.J. (2005), 'Link between population dynamics and dynamics of Darwinian evolution', *Phys. Rev. Lett.* **95**(7), 078105.

Metz, J. A. J. & Diekmann, O., eds. (1986), *The dynamics of physiologically structured populations*, Vol. 68 of *Lecture Notes in Biomathematics*, Springer-Verlag, Berlin.

Metz, J.A.J. (2008), Fitness, *in* S. E. Jörgensen & B. D. Fath, eds., 'Evolutionary Ecology', Vol. 2 of *Encyclopedia of Ecology*, Elsevier, UK, pp. 1599–1612.

Metz, J.A.J. (*In press*), Invasion fitness, canonical equations, and global invasion criteria for Mendelian populations, *in* U. Dieckmann & J. A. J. Metz, eds., 'Elements of Adaptive Dynamics', Cambridge University Press, UK.

Metz, J.A.J. & de Kovel, C.C. (*in preperation*), 'The canonical equation of adaptive dynamics for diploid and haplo-diploid mendelian populations'.

Metz, J.A.J. & de Roos, A.M. (1992), The role of physiologically structured population models within a general individual-based modeling perspective, *in* D. DeAngelis & L. Gross, eds., 'Individual-based models and approaches in ecology: Concepts and Models', Chapman & Hall, USA, pp. 88–111.

Metz, J.A.J., Geritz, S.A.H., Meszéna, G., Jacobs, F.J.A. & van Heerwaarden, J.S. (1996), Adaptive dynamics, a geometrical study of the consequences of nearly faithful reproduction, *in* 'Stochastic and spatial structures of dynamical systems (Amsterdam, 1995)', Konink. Nederl. Akad. Wetensch. Verh. Afd. Natuurk. Eerste Reeks, 45, North-Holland, Amsterdam, pp. 183–231.

Metz, J.A.J., Klinkhamer, P.G.L. & de Jong, T.J. (2009), 'A different model to explain delayed germination', *Evol. Ecol. Res.* **11**(2), 177–190.

Metz, J.A.J. & Leimar, O. (*in press*), A simple fitness proxy for structured populations with continuous traits, with case studies on the evolution of haplo-diploids and genetic dimorphisms. J. Biol. Dyn.

Metz, J.A.J., Mylius, S.D. & Diekmann, O. (1996), When does evolution optimise? on the relation between types of density dependence and evolutionarily stable life history parameters. IIASA Working Paper WP-96-04.

Metz, J.A.J., Mylius, S.D. & Diekmann, O. (2008), 'When does evolution optimize?', *Evol. Ecol. Res.* **10**(5), 629–654.

Metz, J.A.J., Nisbet, R.M. & Geritz, S.A.H. (1992), 'How should we define "fitness" for general ecological scenarios?', *Trends Ecol. Evol.* **7**, 198–202.

Michod, R.E. & Hamilton, W.D. (1980), 'Coefficients of relatedness in sociobiology', *Nature* **288**(5792), 694–697.

Milinski, M., Semmann, D., Krambeck, H.J. & Marotzke, J. (2006), 'Stabilizing the Earth's climate is not a losing game: Supporting evidence from public goods experiments', *Proc. Natl. Acad. Sci. USA* **103**(11), 3994–3998.

Milinski, M., Sommerfeld, R.D., Krambeck, H.J., Reed, F.A. & Marotzke, J. (2008), 'The collective-risk social dilemma and the prevention of simulated dangerous climate change', *Proc. Natl. Acad. Sci. USA* **105**(7), 2291–2294.

Mulholland, H.P. & Smith, C.A.B. (1959), 'An inequality arising in genetical theory', *Am. Math. Mon.* **66**, 673–683.

Nagylaki, T. (1977), *Selection in One- and Two-Locus Systems*, Springer, Berlin.

Nagylaki, T. (1987), 'Evolution under fertility and viability selection', *Genetics* **115**, 367–375.

Nagylaki, T. (1989), 'Rate of evolution of a character without epistasis', *Proc. Natl. Acad. Sci. USA* **86**(6), 1910–1913.

Nagylaki, T. (1991), 'Error-bounds for the fundamental and secondary theorems of natural-selection', *Proc. Natl. Acad. Sci. USA* **88**(6), 2402–2406.

Nagylaki, T. (1992), *Theoretical Population Genetics*, Springer-Verlag, Berlin.

Nagylaki, T. (1993), 'The evolution of multilocus systems under weak selection', *Genetics* **134**(2), 627–647.

Nagylaki, T. (2009), 'Evolution under the multilocus Levene model without epistasis', *Theor. Popul. Biol.* **76**(3), 197–213.

Nagylaki, T. & Crow, J.F. (1974), 'Continuous selective models', *Theor. Popul. Biology* **5**, 257–283.

Nagylaki, T., Hofbauer, J. & Brunovský, P. (1999), 'Convergence of multilocus systems under weak epistasis or weak selection', *J. Math. Biol.* **38**(2), 103–133.

Nagylaki, T. & Lou, Y. (2007), 'Evolution under multiallelic migration-selection models', *Theor. Popul. Biol.* **72**(1), 21–40.

Nowak, M.A. (2006*a*), *Evolutionary Dynamics: Exploring the Equations of Life*, The Belknap Press of Harvard University Press, Cambridge, MA.

Nowak, M.A. (2006*b*), 'Five rules for the evolution of cooperation', *Science* **314**(5805), 1560–1563.

Nowak, M.A. & Sigmund, K. (2004), 'Evolutionary dynamics of biological games', *Science* **303**(5659), 793–799.

Ohtsuki, H., Hauert, C., Lieberman, E. & Nowak, M.A. (2006), 'A simple rule for the evolution of cooperation on graphs and social networks', *Nature* **441**(7092), 502–505.

Okasha, S. (2008), 'Fisher's fundamental theorem of natural selection – A philosophical analysis', *Brit. J. Phil. Sci.* **59**, 319–351.

Otto, S.P. (2009), 'The evolutionary enigma of sex', *Am. Nat.* **174**(Suppl. 1), S1–S14.

Otto, S.P. & Jarne, P. (2001), 'Evolution – haploids – hapless or happening?', *Science* **292**(5526), 2441–2443.

Pacheco, J.M., Santos, F.C. & Chalub, F.A.C.C. (2006), 'Stern-judging: A simple, successful norm which promotes cooperation under indirect reciprocity', *PLoS Computational Biology* **2**(12), e178.

Pacheco, J.M., Santos, F.C., Souza, M.O. & Skyrms, B. (2009), 'Evolutionary dynamics of collective action in n-person stag-hunt dilemmas', *P. Roy. Soc. B: Biological Sciences* **276**(1655).

Parvinen, K. (2005), 'Evolutionary suicide', *Acta Biotheor.* **53**(3), 241–264.

Pearson, K. & Lee, A. (1903), 'On the laws of inheritance in man. I. Inheritance of physical characters', *Biometrika* **2**, 357–462.

Pennings, P.S., Kopp, M., Meszena, G., Dieckmann, U. & Hermisson, J. (2008), 'An analytically tractable model for competitive speciation', *Am. Nat.* **171**(1), E44–E71.

Perthame, B. & Barles, G. (2008), 'Dirac concentrations in Lotka-Volterra parabolic PDEs', *Indiana Univ. Math. J.* **57**(7), 3275–3301.

Perthame, B. & Génieys, S. (2007), 'Concentration in the nonlocal Fisher equation: the Hamilton-Jacobi limit', *Math. Model. Nat. Phenom.* **2**(4), 135–151.

Phillipson, P.E. & Schuster, P. (2009), *Modeling by Nonlinear Differential Equations. Dissipative and Conservative Processes*, Vol. 69 of *World Scientific Series on Nonlinear Science A*, World Scientific, Singapore.

Price, G. (1972), 'Fisher's "Fundamental Theorem" made clear', *Ann. Hum. Genet.* **36**, 129–140.

Price, G.R. (1970), 'Selection and covariance', *Nature* **227**(5257), 520.

Proulx, S.R. & Day, T. (2002), 'What can invasion analyses tell us about evolution under stochasticity in finite populations?', *Selection* **2**, 2–15.

Queller, D.C. (1994), 'Genetic relatedness in viscous populations', *Evol. Ecol.* **8**, 70–73.

Queller, D.C. & Goodnight, K.F. (1989), 'Estimation of genetic relatedness using allozyme data', *Evolution* **43**, 258–275.

Raup, D. (1991), *Extinction: Bad Genes or Bad Luck?*, NW. W. Norton & Company, USA.

Reeve, J.P. (2000), 'Predicting long-term response to selection', *Genet. Res.* **75**(1), 83–94.

Rice, S.H. (1990), 'A geometric model for the evolution of development', *J. Theor. Biol.* **143**(3), 319–342.

Ripa, J. (2009), 'When is sympatric speciation truly adaptive? An analysis of the joint evolution of resource utilization and assortative mating', *Evol. Ecol.* **23**(1), 31–52.

Robertson, A. (1966), 'A mathematical model of the culling process in dairy cattle', *Anim. Prod.* **8**, 95–108.

Robertson, A. (1968), The spectrum of genetic variation, *in* R. Lewontin, ed., 'Population Biology and Evolutions', Syracuse University Press, EUA, pp. 5–16.

Rodríguez-Trelles, F., Tarrio, R. & Ayala, F.J. (2003), Molecular clocks: whence and whither?, *in* P. C. J. Donoghue & M. P. Smith, eds., 'Telling the evolutionary time: molecular clocks and the fossil record', CRC Press, USA, pp. 5–26.

Rodríguez-Trelles, F., Tarrio, R. & Ayala, F.J. (2004), Molecular clocks: whence and whither?, *in* P. C. J. Donoghue & M. P. Smith, eds., 'Telling the Evolutionary Time: Molecular Clocks and the Fossil Record', CRC Press, Boca Raton, USA, pp. 5–26.

Roughgarden, J.A. (1979), *Theory of Population Genetics and Evolutionary Ecology: an Introduction*, Macmillan, New York.

Rueffler, C., Metz, J.A.J. & Van Dooren, T.J.M. (*in revision*), What life cycle graphs can tell about the evolution of life histories. J. Math. Biol.

Saiki, R.K., amd Susanne Stoffel, D.H.G., Scharf, S.J., Higuchi, R., Horn, G.T., Mullis, K.B. & Erlich, H.A. (1988), 'Primer-directed enzymatic amplification of DNA with a thermostable DNA polymerase', *Science* **239**, 487–491.

Salazar-Ciudad, I. (2006*a*), 'Developmental constraints vs. variational properties: How pattern formation can help to understand evolution and development', *J. Exp. Zool. Part B* **306B**(2), 107–125.

Salazar-Ciudad, I. (2006*b*), 'On the origins of morphological disparity and its diverse developmental bases', *Bioessays* **28**(11), 1112–1122.

Santos, F.C. & Pacheco, J.M. (2005), 'Scale-free networks provide a unifying framework for the emergence of cooperation', *Phys. Rev. Lett.* **95**(9), 098104.

Santos, F.C., Pacheco, J.M. & Lenaerts, T. (2006), 'Evolutionary dynamics of social dilemmas in structured heterogeneous populations', *Proc. Natl. Acad. Sci. USA* **103**(9), 3490–3494.

Santos, F.C., Santos, M.D. & Pacheco, J.M. (2008), 'Social diversity promotes the emergence of cooperation in public goods games', *Nature* **454**(7201), 213–216.

Sasaki, A. & Ellner, S. (1995), 'The evolutionarily stable phenotype distribution in a random environment', *Evolution* **49**(2), 337–350.

Schelling, T.C. (1973), 'Hockey helmets, concealed weapons, and daylight saving: A study of binary choices with externalities', *J. Conflict Resolution* **17**(381), 381.

Scheuer, P.A.G. & Mandel, S.P.H. (1959), 'An inequality in population genetics', *Heredity* pp. 519–524.

Schlögl, F. (1972), 'Chemical reaction models for non-equilibrium phase transitions', *Z. Physik* **253**, 147–161.

Schneider, K.A. (2006), 'A multilocus-multiallele analysis of frequency-dependent selection induced by intraspecific competition', *J. Math. Biol.* **52**(4), 483–523.

Schneider, K.A. (2007), 'Long-term evolution of polygenic traits under frequency-dependent intraspecific competition', *Theor. Popul. Biol.* **71**(3), 342–366.

Schneider, K.A. (2008), 'Maximization principles for frequency-dependent selection I: the one-locus two-allele case', *Theor. Popul. Biol.* **74**(3), 251–262.

Schneider, K.A. (2010), 'Maximization principles for frequency-dependent selection II: the one-locus multiallele case', *J. Math. Biol.* **61**, 95–132.

Schnog, J.B., Duits, A.J., Muskiet, F.A.J., ten Cate, H., Rojer, R.A. & Brandjes, D.P.M. (2004), 'Sickle cell disease: A general overview', *The Netherlands Journal of Medicine* **62**, 364–374.

Schuster, P. (2006), 'Prediction of RNA secondary structures: From theory to models and real molecules', *Reports on Progress in Physics* **69**, 1419–1477.

Schuster, P. & Sigmund, K. (1983), 'Replicator dynamics', *J. Theor. Biol.* **100**, 533–538.

Schuster, P. & Sigmund, K. (1985), 'Dynamics of evolutionary optimization', *Ber. Bunsenges. Phys. Chem.* **89**, 668–682.

Schuster, P. & Swetina, J. (1988), 'Stationary mutant distribution and evolutionary optimization', *Bull. Math. Biol.* **50**, 635–660.

Schwämmle, V. & Brigatti, E. (2006), 'Speciational view of macroevolution: Are micro and macroevolution decoupled?', *Europhys. Lett.* **75**(2), 342–348.

Seneta, E. (1981), *Non-negative Matrices and Markov Chains*, second edn., Springer-Verlag, New York.

Shahshahani, S. (1979), 'A new mathematical framework for the study of linkage and selection', *Mem. Am. Math. Soc.* **211**, 34.

Sigler, L.E. (2002), *Fibonacci's Liber Abaci: A Translation into Modern English of Leonardo Pisano's Book of Calculation*, Springer-Verlag, New York. A translation of Leonardo Pisano's book of 1202 from Latin into modern English.

Singh, P. (1985), 'The so-called Fibonacci numbers in ancient and medieval India', *Historia Mathematica* **12**, 229–244.

Skyrms, B. (2001), 'The stag hunt', *Proceedings and Addresses of the American Philosophical Association* **75**(2), 31–41.

Skyrms, B. (2004), *The Stag Hunt and the Evolution of Social Structure*, Cambridge University Press, UK.

Souganidis, P.E. (1997), Front propagation: theory and applications, *in* 'Viscosity solutions and applications (Montecatini Terme, 1995)', Vol. 1660 of *Lecture Notes in Math.*, Springer, Berlin, pp. 186–242.

Souza, M.O., Pacheco, J.M. & Santos, F.C. (2009), 'Evolution of cooperation under N-person snowdrift games', *J. Theor. Biol.* **260**(4), 581–588.

Spiegelman, S. (1971), 'An approach to the experimental analysis of precellular evolution', *Quart. Rev. Biophys.* **4**, 213–253.

Stadler, B.R.M., Stadler, P.F., Wagner, G.P. & Fontana, W. (2001), 'The topology of the possible: Formal spaces underlying patterns of evolutionary change', *J. Theor. Biol.* **213**, 241–274.

Stander, P.E. (1992), 'Cooperative hunting in lions – the role of the individual', *Behavioral Ecology and Sociobiology* **29**(6), 445–454.

Steffensen, F.H. (1930), 'Om sandsyndligheden for at afkommet uddør', *Matematisk Tidskrift* **B**, 19–23.

Sugden, R. (1986), *The Economics of Rights, Co-operation and Welfare*, Basil Blackell, Oxford, UK.

Svirezhev, Y.M. & Passekov, V.P. (1990), *Fundamentals of Mathematical Evolutionary Genetics*, Kluwer, Dordrecht.

Svirezhev, Yu.M. (1972), Optimality principles in population genetics, *in* 'Studies in Theoretical Genetics', Novisibirsk: Inst. of Cytology and Genetics, pp. 86–102.

Takeuchi, N. & Hogeweg, P. (2007), 'Error-thresholds exist in fitness landscapes with lethal mutants', *BMC Evolutionary Biology* **7**, 15.

Taylor, P. (2009), 'Decompositions of Price's formula in an inhomogeneous population structure', *J. Evol. Biol.* **22**(1), 201–213.

Taylor, P.D. (1989), 'Evolutionary stability in one-parameter models under weak selection', *Theor. Popul. Biol.* **36**(2), 125–143.

Taylor, P.D. (1992*a*), 'Altruism in viscous populations – an inclusive fitness model', *Evol. Ecol.* **6**(4), 352–356.

Taylor, P.D. (1992*b*), 'Inclusive fitness in a homogeneous environment', *Proc. R. Soc. Lond. Ser. B-Biol. Sci.* **249**(1326), 299–302.

Taylor, P.D., Day, T. & Wild, G. (2007), 'Evolution of cooperation in a finite homogeneous graph', *Nature* **447**, 469–472.

Taylor, P.D. & Frank, S.A. (1996), 'How to make a kin selection model', *J. Theor. Biol.* **180**(1), 27–37.

Taylor, P.D., Lillicrap, T. & Cownden, D. (2010), Inclusive fitness analysis on mathematical groups. *Submitted.*

Tejero, H., Marín, A. & Moran, F. (2010), 'Effect of lethality on the extinction and on the error threshold of quasispecies', *J. Theor. Biol.* **262**, 733–741.

Thompson, C.J. & McBride, J.L. (1974), 'On Eigen's theory of the self-organization of matter and the evolution of biological macromolecules', *Math. Biosci.* **21**, 127–142.

Traulsen, A., Nowak, M.A. & Pacheco, J.M. (2006), 'Stochastic dynamics of invasion and fixation', *Phys. Rev. E: Stat. Nonlin. Soft. Matter. Phys.* **74**(1 Pt 1), 011909.

Traulsen, A., Nowak, M.A. & Pacheco, J.M. (2007), 'Stochastic payoff evaluation increases the temperature of selection', *J. Theor. Biol.* **244**(2), 349–356.

Traulsen, A., Pacheco, J.M. & Nowak, M.A. (2007), 'Pairwise comparison and selection temperature in evolutionary game dynamics', *J. Theor. Biol.* **246**(3), 522–529.

Turelli, M. (1988), 'Phenotypic evolution, constant covariances, and the maintenance of additive variance', *Evolution* **42**(6), 1342–1347.

Turelli, M. & Barton, N.H. (1990), 'Dynamics of polygenic characters under selection', *Theor. Popul. Biol.* **38**(1), 1–57.

Turelli, M. & Barton, N.H. (1994), 'Genetic and statistical analyses of strong selection on polygenic traits: What, me normal?', *Genetics* **138**(3), 913–941.

Turelli, M. & Barton, N.H. (2004), 'Polygenic variation maintained by balancing selection: Pleiotropy, sex-dependent allelic effects and $G \times E$ interactions', *Genetics* **166**(2), 1053–1079.

Van Dooren, T.J.M. (*In press*), Adaptive dynamics with Mendelian genetics., *in* U. Dieckmann & J. A. J. Metz, eds., 'Elements of Adaptive Dynamics', Cambridge University Press, UK.

Van Doorn, G.S. & Dieckmann, U. (2006), 'The long-term evolution of multilocus traits under frequency-dependent disruptive selection', *Evolution* **60**(11), 2226–2238.

van Strien, S. J. & Verduyn Lunel, S. M., eds. (1996), *Stochastic and spatial structures of dynamical systems. Proceedings of the meeting, Amsterdam, Netherlands, January* 1995, Verhandelingen, Afdeling Natuurkunde. Eerste Reeks. Koninklijke Nederlandse Akademie van Wetenschappen. North-Holland, Amsterdam: North-Holland.

Verhulst, P. (1845), 'Recherches mathématiques sur la loi d'accroisement de la population', *Nouv. Mém. de l'Académie Royale des Sci. et Belles-Lettres de Bruxelles* **18**, 1–41.

Verhulst, P. (1847), 'Deuxième mémoire sur la loi d'accroisement de la population', *Mém. de l'Académie Royale des Sci., des Lettres et de Beaux-Arts de Belgique* **20**, 1–32.

Verhulst, P.F. (1838), 'Notice sur la loi que la population pursuit dans son accroisement', *Corresp. Math. Phys.* **10**, 113 121.

Vincent, T.L. & Brown, J.S. (2005), *Evolutionary Game Theory, Natural Selection, and Darwinian Dynamics*, Cambridge University Press, UK.

von Kiedrowski, G., Wlotzka, B., Helbig, J., Matzen, M. & Jordan, S. (1991), 'Parabolic growth of a self-replicating hexanucleotide bearing a 3′-5′-phosphoamidate linkage', *Angew. Chem. Internat. Ed.* **30**, 423–426.

Wagner, G.P. & Krall, P. (1993), 'What is the difference between models of error thresholds and Muller's ratchet', *J. Math. Biol.* **32**, 33–44.

Wakeley, J. (2008), *Coalescent Theory: An Introduction*, Greenwood Village, Colorado, USA.

Wang, Z.C., Li, W.T. & Ruan, S.G. (2006), 'Travelling wave fronts in reaction-diffusion systems with spatio-temporal delays', *J. Differ. Equ.* **222**(1), 185–232.

Watson, J.D. & Crick, F.H.C. (1953), 'A structure for deoxyribose nucleic acid', *Nature* **171**, 737–738.

Waxman, D. (2006), 'Fisher's geometrical model of evolutionary adaptation – beyond spherical geometry', *J. Theor. Biol.* **241**(4), 887–895.

Waxman, D. (2007), 'Mean curvature versus normality: A comparison of two approximations of fisher's geometrical model', *Theor. Popul. Biol.* **71**(1), 30–36.

Waxman, D. & Welch, J.J. (2005), 'Fisher's microscope and Haldane's ellipse', *Am. Nat.* **166**(4), 447–457.

Weinberg, W. (1908), 'Über den Nachweis der Vererbung beim Menschen. (On the detection of heredity in man)', *Jahreshelfts. Ver. Vaterl. Naturf. Württemb.* **64**, 368–382.

Weissing, F.J. (1996), 'Genetic versus phenotypic models of selection: Can genetics be neglected in a long-term perspective?', *J. Math. Biol.* **34**(5-6), 533–555.

West, S.A., Pen, I. & Griffin, A.S. (2002), 'Conflict and cooperation – cooperation and competition between relatives', *Science* **296**(5565), 72–75.

Wiehe, T. (1997), 'Model dependency of error thresholds: The role of fitness functions and contrasts between the finite and infinite sites models', *Genet. Res. Camb.* **69**, 127–136.

Wilke, C.O., Wang, J.L., Ofria, C., Lenski, R.E. & Adami, C. (2001), 'Evolution of digital organisms at high mutation rates leads to survival of the flattest', *Nature* **412**, 331–333.

Wilson, D.S. (1975), 'Theory of group selection', *Proc. Natl. Acad. Sci. USA* **72**(1), 143–146.

Wilson, D.S., Pollock, G.B. & Dugatkin, L.A. (1992), 'Can altruism evolve in purely viscous populations?', *Evol. Ecol.* **6**(4), 331–341.

Wouters, A.G. (1999), 'Explanation without a cause (verklaren zonder oorzaken te geven)', Questiones Infinitae, Publications of the Zeno Institute of Philosophy, XXIX, Utrecht.

Wrenn, S.J. & Harbury, P.B. (2007), 'Chemical evolution as a tool for molecular discovery', *Annu. Rev. Biochem.* **76**, 331–349.

Wright, S. (1931), 'Evolution in mendelian populations', *Genetics* **16**, 97–159.

Wright, S. (1935), 'Evolution in populations in approximate equilibrium', *J. Genetics* pp. 257–266.

Wright, S. (1948), Genetics of populations, *in* 'Encyclopaedia Britannica, 14th ed. revised, vol 10', pp. 111–115.

Wynne-Edwards, V.C. (1962), *Animal Dispersion in Relation to Social Behaviour*, Oliver and Boyd, Edinburgh.

Yu, F. (2007), 'Stationary distributions of a model of sympatric speciation', *Ann. Appl. Probab.* **17**(3), 840–874.

Zeeman, M.L. (1993), 'Hopf bifurcations in competitive three-dimensional Lotka-Volterra systems', *Dynam. Stability Systems* **8**(3), 189–217.

Zhang, X.S., Wang, J.L. & Hill, W.G. (2004), 'Influence of dominance, leptokurtosis and pleiotropy of deleterious mutations on quantitative genetic variation at mutation-selection balance', *Genetics* **166**(1), 597–610.

Zhivotovsky, L.A. & Gavrilets, S. (1992), 'Quantitative variability and multilocus polymorphism under epistatic selection', *Theor. Popul. Biol.* **42**(3), 254–283.

Zwillinger, D. (1998), *Handbook of Differential Equations*, third edn., Academic Press, San Diego, CA.

Index

MIX
Papier aus verantwortungsvollen Quellen
Paper from responsible sources
FSC® C105338

If you have any concerns about our products,
you can contact us on
ProductSafety@springernature.com

In case Publisher is established outside the EU,
the EU authorized representative is:
Springer Nature Customer Service Center GmbH
Europaplatz 3, 69115 Heidelberg, Germany

Printed by Libri Plureos GmbH
in Hamburg, Germany